T0344371

Science, Alchemy and the Great Plague of London

Science, Alchemy and the Great Plague of London

Scott Shelley

Algora Publishing
New York

Library of Congress Cataloging-in-Publication Data —

Names: Shelley, William Scott, author.
Title: Science, alchemy, and the Great Plague of London / William Scott Shelley.
Description: New York: Algora Publishing, [2017] | Includes bibliographical references and index.
Identifiers: LCCN 2017042212 (print) | LCCN 2017042634 (ebook) | ISBN 9781628943146 (pdf) | ISBN 9781628943122 (soft cover: alk. paper) | ISBN 9781628943139 (hard cover: alk. paper)
Subjects: LCSH: Great Plague, London, England, 1664-1666. | Starkey, George, 1627-1665. | Plague—England—History—17th century. | Medicine, Medieval. | Diseases and history.
Classification: LCC RA644.P7 (ebook) | LCC RA644.P7 S54 2017 (print) | DDC 614.5/732—dc23
LC record available at https://lccn.loc.gov/2017042212

Printed in the United States

Acknowledgments

I would like to gratefully thank Dr. Mary Matossian and Dr. Dan Merkur for their suggestions and encouragement, and in obtaining the materials for this monograph the Wellcome Library, London; University of Washington Libraries; and Seattle Public Library, Interlibrary Loans.

Table of Contents

On Spiritual Alchemy and Chemistry 1

On The Great Plague of London: A New Diagnosis 7

Introduction 11

Chapter 1. Alchemy in the New World 17

Chapter 2. George Starkey and Eirenaeus Philalethes 23

Chapter 3. The Chymical Physicians Versus the Aristotelians 33

Chapter 4. Robert Boyle and George Starkey 59

Chapter 5. Isaac Newton, Robert Boyle, and George Starkey 85

Chapter 6. George Starkey's Philosophers' Stone 109

Chapter 7. Starkey, Boyle, and Newton's Divine Quest 117

Chapter 8. The Bubonic Plague Theory 121

Chapter 9. Ergot and *Fusarium* Poisoning 125

Chapter 10. The English Diet in the Seventeenth Century 131

Chapter 11. Environmental Conditions and the Plague 135

Chapter 12. The Great Plague: An Overview of the Epidemic 141

Chapter 13. The Great Plague in London 145

Chapter 14. The Medical Evidence: Diseases Preceding the Plague 155

Chapter 15. The Medical Evidence: Diseases Occurring with the Plague 161

Chapter 16. The Medical Evidence: Diseases Following the Plague 179

Chapter 17. The Medical Evidence: Symptoms of the Plague 193

Chapter 18. The Medical Evidence: Fever and the Plague 201

Chapter 19. The Death of George Starkey: A Victim of Plague 209

Chapter 20. George Starkey's Secret Remedy 217

Chapter 21. Epilogue 227

Chapter 22. Conclusion 229

Bibliography 235

On Spiritual Alchemy and Chemistry

Dan Merkur

The maxim that truth is the first casualty of war applies to the Scientific Revolution as to any other. In the sixteenth and seventeenth centuries, the cause of science was advanced within the practice of medicine by men who followed Paracelsus in combining surgery, an experimental attitude to the sickbed, and "chemical physick" as pharmaceutical drugs were then known. The practitioners of the "chemical philosophy" were physicians who attended the sick and themselves manufactured the chemical preparations that they administered. Their processes of drug manufacture were commonly termed both "alchemy" and "chemistry." The two terms were synonymous throughout the sixteenth century and only gradually began to be distinguished over the course of the seventeenth. It was only in the early eighteenth century that the term "alchemy" came to be restricted to the ostensible manufacture of gold, and the term "chemistry" was reserved for would-be scientific laboratory procedures. When, after 1700, the terms "alchemy" and "chemistry" diverged in meaning, the history of the chemical philosophy was rewritten in a manner that systematically deleted alchemy from the narratives. The result was a myth that is still being taught in grade schools down to the present time.

From the eighteenth century "Enlightenment" onward, practitioners who had in their own times been called both "alchemist" and "chemist" were classified either as alchemists or as chemists, but no longer as both. For example, Robert Boyle and Isaac Newton, whose contributions to the science of physics led to their celebration as archetypes of the scientist, were retrospectively portrayed as though they had had Enlightenment sensibilities. Historians of science—and textbooks in our school systems—ignored (and sometimes suppressed) the massive documentary evidence of alchemical interests in their writings, notebooks, and correspondence. The reputation of George Starkey was similarly revised, but in the opposite direction.

George Starkey had no less than taught chemistry to Boyle and, to a lesser extent, to Newton. He was a leading proponent of the Helmont school within alchemy, which had abandoned Paracelsus's ideas about (Platonic) "signs of nature" in favor of an empirical experimentalism. This shift from a theosophical chemistry to a chemistry that lacked a general theory was otherwise still embedded in the traditional secrecy, obscure literary style, and other persisting medievalisms of the chemists' trade. Starkey was a working physician who generally guarded his pharmacological recipes as secrets on which his livelihood depended. His work belonged to the phase in the emergence of scientific chemistry immediately prior to the innovation of academic interest in pure theory that Boyle and Newton were so famously to contribute. However, for eighteenth century historians of science, looking back on the mid-seventeenth century, Starkey had no cachet as a scientist, and they abandoned his memory.

The occult revival of the nineteenth century advanced the claim that alchemists had never been so foolish as to attempt the transmutation of lead into gold but had instead used the metallic, planetary, mythological, and sexual imagery of alchemy to express the process of the soul's transformation. In *A Suggestive Inquiry into the Hermetic Mysteries*, Mary Anne Atwood, an English writer, claimed that a practice of spiritual alchemy stemmed from Paracelsus, by way of the mystic Jacob Boehme and his English adherents, the theosophists John Pordage, Jane Lead, and others. In *Remarks on Alchemy and the Alchemists*, Ethan Allen Hitchcock discussed a somewhat different practice that, he suggested, represented the teachings of the Rosicrucians and Freemasons. Theories of spiritual alchemy have since multiplied among devotées of Western esotericism. In the early twentieth century, historians of chemistry began to document the chemical processes that were discussed in the Greek alchemical literature of late antiquity, the Muslim alchemists of the Middle Ages. It turned out that a technique for making a copper-lead alloy that could be used as gold in costume jewelry, or counterfeit coinage, had been transmitted down the centuries. A considerable variety of further chemical recipes also abounded in the literature.

Through the analytic psychologist Carl G. Jung, the occult hypothesis that alchemists had always had spiritual interests gained new life within academia, while simultaneously satisfying the findings of historians of chemistry. Jung supposed that the alchemists had used mental imagery techniques in order to access unconscious symbolism while they watched the chemical events within their laboratories, doing mystical riffs, as it were, on chemical melodies.

Alchemists notoriously wrote their texts in code. The difficulty of their cryptic writings is exemplified by the derivation of the word "gibberish" from the name of Jabir ibn Hayyan, the greatest alchemist of medieval Islam. Like the Pythagoreans and the Platonists, the alchemists of late antiquity had used allegories to convey secret teachings. The names of the planets, the names of gods, spirits, and legendary persons, flora and fauna, parts of the human body, man-made objects, human activities and mythological narratives, were all used as coded ways to speak of chemical substances and their processes. Jabir introduced the additional technique of "dispersion," writing a passage about a topic, stopping in the middle without warning or notice, and then later in the text resuming, again without warning or notice. In this way, readers expecting a continuous exposition would be diverted onto mistaken topics. Another literary technique of the alchemists, explicitly remarked by Maimonides, was self-contradiction. An author would both make a claim and also assert a second matter that was logically inconsistent with the first. The manifest incoherence of the text would stymie literal-minded readers but could be resolved into a coherent subtext by recognizing the metaphoric or symbolic character of the apparent contradiction. Boehme was perhaps the first to remark on a further technique, called by him cabalistic, that devised egregious wordplays involving far-fetched and sometimes multilinguistic not-quite homonyms.

Because reading an alchemical document involves solving a puzzle, as slowly and carefully as, for example, a crossword puzzle, the documents support as many and varied interpretations as, for example, do nocturnal dreams. The earlier academic students of the literature were content to skim the surfaces of the texts, gleaning what was reasonably clear while ignoring the remainder. Because different alchemists kept different matters secret, while speaking openly of others, what was clear in one writer sometimes clarified what was obscure in others; and academics have slowly progressed in building a lexicon to what amounts to a forgotten language. At the present time, we are still engaged in a fairly early stage of the work. The academic consensus uncritically perpetuates the occult-cum-Jungian hypothesis, while some few of us are advocating a new and more methodologically rigorous paradigm.

A handful of scholars, beginning perhaps with the occultist Arthur Edward Waite in 1926 but now extending to myself, in the history of religions, and William R. Newman and Lawrence M. Principe, historians of science who have specialized in the study of Starkey and Boyle, have argued that alchemical writings concerned chemical processes from their late antique beginning until the Renaissance. Spiritual alchemy originated, it would appear, in the early sixteenth century when Paracelsus blended his medical practice of alchemy with his interest in the Hermetic-Platonic tradition of Renaissance esotericism. Alchemical texts that seem to be exclusively spiritual, that is, that have no apparent interest in practical laboratory procedures, date no earlier than the late Elizabethan period.

Principe and Newman have argued that Boyle, Elias Ashmole and other English alchemists were concerned with one or more substances, termed the red stone, the angelic stone, and otherwise, whose effects were described as spiritual. Principe and Newman speak of a "supernatural school" within English alchemy of the seventeenth century. In *Gnosis*, I noted that this substance produced mystical experiences and was implicitly psychoactive; and in a later article I associated it with the motif of manna, the bread of the angels that descended miraculously from the heavens to feed the Israelites in the wilderness in the biblical story of the Exodus. The motif of manna or, more precisely, the hidden manna of Revelation 2:17, was associated with mystical experience in the twelfth century theology of the School of Laon, afterward spreading to Cistercian, Franciscan, and other writers. The biblical motif appears to have had a related significance in English alchemical literature from the Tudor period onward. Given that Elias Ashmole was a friend of Arthur Dee, Ashmole's tale that Arthur's father, Dr. John Dee, obtained his supply of the white powder from excavations in the Glastonbury churchyard speaks in an encoded alchemical way to the passage of the hidden manna from medieval theology to Renaissance alchemy.

William Scott Shelley has been pursuing a different trajectory through the alchemical literature. The term alkahest, or alcahest, appears to have originated with Paracelsus and passed from him to van Helmont and thence to Starkey. It designated an elixir rather than a stone; but because elixirs can be produced by dissolving stones, which is to say, solids, we are likely dealing with different code-names for the same or similar substances. Shelley admirably documents, and replicates for the reader, the difficulty and confusion that Boyle, Starkey, and Newton experienced as they worked their way through the alchemical literature, seeking general principles that would transform it into a science. He also documents unequivocally the sort of spiritual experiences that Starkey reported, his concern with psychoactive substances in the example of opium and the preparation of laudanum, his homeopathic theory of deriving a medicine from a pathogen (the principle that underlies modern vaccines), and his explicit discussion of how to prepare elixirs from grains and other vegetable matters. Already in the Elizabethan period it was commonly understood that grains such as darnel could be psychoactive; however, that the psychoactivity traced not to the grain but to the ergot fungus that infested it was apparently not appreciated at the time. Shelley's thesis that Starkey's alkahest was a preparation made from ergot is entirely persuasive.

Dan Merkur earned his doctorate in the history of religions at the University of Stockholm. He has taught at five universities and published eleven books and many articles. Both alchemy and psychedelic drugs have been research interests for over three decades. In addition to his academic work, he qualified as a clinician at the Toronto Institute for Contemporary Psychoanalysis. He is currently in private practice as a psychoanalyst in Toronto, Canada.

* * *

On The Great Plague of London: A New Diagnosis

Mary K. Matossian

In this book, Scott Shelley makes a new diagnosis of the so-called "Great Plague" of London, providing telling evidence that it was not a case of bubonic plague at all. He has studied thoroughly the symptoms and epidemiology of the diseases he identifies as the culprits.

In particular, he has exhaustively studied the weather in 1665 and compared it to the kinds of weather usually associated with ergot and Fusarium poisoning. Weather changes play an important role in these diseases because they control the amount of fungal colonization of cereals, both in the field and in storage.

In 1665 medical science in England was struggling to be born. Even educated people referred to epidemics indiscriminately as "plagues." They did not distinguish between infections and poisoning, much less the various kinds of infections and poisoning. When physicians found that they could not treat cases of "plague" successfully, all the people of an afflicted town who could afford to depart did so, hastily. Very few physicians took the trouble to describe the symptoms of the current "plague" in detail.

People had some awareness that to eat cereals suffering from mold colonization was harmful. They said that such cereals were contaminated by "blight," "rust" and "mildew." The poor ate such food only when there was no healthy alternative. Unfortunately, some kinds of Fusarium colonization were not easily detected by sight, smell, or taste. There were no microscopes available at this time. The scientific principles of food storage were still unknown.

In modern times historians of medicine have tended to ignore mold poisoning as a possible diagnosis of past epidemics. That is because mold poisoning has been rarely observed since World War II. The successful use of antibiotics against infections has led to the assumption that past epidemics were all caused by bacteria, protozoa, and viruses. Most physicians have ignored the possibility that the microfungi also can cause epidemics.

This was not the case with animal diseases. Those engaged in animal husbandry used to be rather careless about what they fed their animals. This often led to painful financial losses. When I was doing my research on mold poisoning epidemics in the 1980s, veterinarians were more receptive to my findings than physicians.

If mold poisoning was an important cause of epidemics in the past, why is it relatively rare in the present? There are several agents for this change. One is dietary change, particularly the substitution of potatoes for rye cereal in the diet of men and animals. In Western Europe and the British colonies in America, this began in the late eighteenth century. In the nineteenth century potato cultivation spread rapidly eastward in Europe and Russia. Potatoes, to be sure, were vulnerable to mold colonization. When that happened, the stink of rotten potatoes warned away would-be consumers. But healthy potatoes provided more calories and other nutrients per acre than cereals.

Scientific understanding of how microfungi grow and of the different kinds of fungi began to increase in the middle of the nineteenth century. In the 1930s in the United States, agronomists pioneered in the scientific study of food storage. Then the development of scientific agriculture spread rapidly among farmers. The governments of major states began to regulate the quality of foods entering the market. For all these reasons, epidemics of mold poisoning almost disappeared. They were no longer visible to most practicing physicians in search of a diagnosis.

So the history of medicine, which includes the history of human health, needs considerable rewriting. Scott Shelley's work is an important contribution to the subject. It points the way to more new insights.

Mary K. Matossian earned her B.A. Stanford, History, magna cum laude and Phi Beta Kappa; M.A. American University of Beirut, Lebanon, Near Eastern History; and Ph.D. Stanford, History, with Distinction. She is the author of four books, notably *Poisons of the Past: Molds, Epidemics, and History* (Yale University Press, 1989).

INTRODUCTION

> *There is a fiery Stone of Paradise,*
> *So call'd because of its Celestial hew,*
> *Named of Ancient years by Sages wise*
> Elixir, *made of Earth and Heaven new,*
> *Anatically mixt ; strange to relate,*
> *Sought for by many, but found out by few ;*
> *Above vicissitudes of Nature, and by fate*
> *Immortal, like a Body fixt to shew,*
> *Whose penetrative vertue proves a Spirit true.*[1]

> *And pass from darkness of Purgatory to light*
> *Of Paradise, in whiteness Elixir of great might.*[2]

Alchemy constitutes an obscure and commonly misunderstood endeavor that traverses the great scientific and intellectual traditions of Western Civilization; some of the most influential figures in history are included among those who participated in this occluded practice. Concealed within this secret discipline is perhaps the greatest mystery contained in the history of science—the identity of the numinous

[1] For the sake of accuracy, the material quoted is presented as it is found in the original source, unless indicated by [brackets]. Sir George Ripley; in Eirenaeus Philalethes (pseud.), "An Exposition upon Sir George Ripley's Preface," *Ripley Reviv'd: Or, An Exposition upon Sir George Ripley's Hermetico-Poetical Works*, London, Printed by *Tho. Ratcliff* and *Nat. Thompson*, for *William Cooper* at the *Pelican* in *Little-Britain*, 1678, pp. 88-89.

[2] Sir George Ripley; in Philalethes, "An Exposition upon the First Six Gates of Sir George Ripley's Compound of Alchymie," Ibid., p. 356.

Philosophers' Stone, the source of the coveted "Philosophers Elixer,"[3] also known as the "immortal liquor" called Alkahest.

In reference to this "Elixir of Life," a petition of King Henry VI for the grant of a license to practice alchemy contains the following:

> [I]n former times wise and famous Philosophers in their writings and books, under figures and coverings, have left on record and taught that from wine, from precious stones, from oils, from vegetables, from animals, from metals, and the cores of minerals, many glorious and notable medicines can be made ; and chiefly that a most precious medicine which some Philosophers have called the Mother and Empress of Medicines, others have named it the priceless glory, but others have called it the Quintessence, others the Philosophers' Stone and Elixir of Life ; of which potion the efficacy is so certain and wonderful, that by it all infirmities whatsoever are easily curable, human life is prolonged to its natural limit, and man wonderfully preserved in health and manly strength of body and mind, in vigour of limbs, clearness of memory, and perspicacity of talent to the same period [i.e., degree] ; All kinds of wounds too, which may be cured are healed without difficulty, and in addition it is the best and surest remedy against all kinds of poisons...[4]

The origins of alchemy and the knowledge of this mysterious remedy can be traced back to the secret recipes that were transmitted by oral tradition in the temples of the ancient Egyptians, part of which is extant in the botanical writings of Dioscorides and Pliny the Elder.[5] Alchemy was a term synonymous with chemistry and medicine to the most highly renowned "Chemical Physicians" of the Scientific Revolution, who claimed to have possession of this elusive medicine considered to be the panacea, the universal cure for all disease.

---

[3] Eirenaeus Philalethes (pseud.), *The Marrow Of Alchemy: Being an Experimental Treatise, Discovering the secret of the most hidden Mystery of the Philosophers Elixer*, London, Printed by *A. M.* for *Edw. Brewster* at the Signe of the Crane in *Pauls* Church-yard, 1654.

[4] Robert Steele, "Alchemy in England," *Antiquary*, 24 (1891), p. 102; Cf. D. Geoghegan, "A Licence of Henry VI to Practise Alchemy," *Ambix*, 6 (1957), pp. 10–17.

[5] "[T]he origins of alchemy can be traced back to the practice of the industrial arts among the ancient Egyptians in connection with metals and their alloys, glass, enamels, dyeing, medicine and perfumery. These arts were centred in the temples and the recipes and formulæ connected with them were at first transmitted by oral tradition as secret processes. About the beginning of the Christian era, many of these recipes, accompanied by magic formulæ, were written down in Greek on papyrus. A few of these papyri still survive, but the majority were apparently destroyed by an edict of Diocletian, about A.D. 290, when he ordered all Egyptian books of alchemy on gold and silver to be burnt...Part of the information contained in these documents has, however, come down to us through the writings of Dioscorides and Pliny, who lived in the first century of this era..." A. J. V. Underwood, "The Historical Development of Distilling Plant," *Transactions of the Institution of Chemical Engineers*, 13 (1935), p. 35.

The great scientific achievements in seventeenth century England occurred on an intellectual background dominated by alchemy, a practice intimately connected with the Scientific Revolution,[6] a period in history containing the monumental developments in the physical sciences that subsequently resulted in the tradition of alchemy virtually fading into obscurity and disrepute. The Scientific Revolution has traditionally been defined as the achievements in astronomy, mathematics, and physics of motion, and is viewed as the progress from Nicholas Copernicus (1473–1543) to Tycho Brahe (1546–1601) and Johannes Kepler (1571–1630), and from these figures to Galileo Galilei (1564–1642) and Isaac Newton (1642–1727).[7] Robert Boyle (1627–1691) has also traditionally been considered a founder of modern science; in his lifetime he developed the vacuum pump, discovered the gas law that bears his name, and was instrumental in founding the prestigious Royal Society of London. Boyle's younger colleague Isaac Newton conceived laws of motion and gravity,

[6] "[A]lchemy was not part of the Aristotelian system that the Scientific Revolution overturned. There had been alchemy during the Middle Ages, and inevitably it had expressed itself in Aristotelian terms, though it was never part of the Scholastic philosophy. The great age of alchemy was the late sixteenth and the seventeenth centuries. The majority of the classics in the art come from that period, and they expressed themselves in the extravagant language of postmedieval, Neoplatonic natural philosophies. For those who have not done so, I recommend that they read in the work of the alchemist who appears most frequently in Newton's alchemical manuscripts, Irenaeus Philalethes (the pseudonym of George Starkey...), and try to imagine how Saint Thomas would have reacted to it. However they interpret Newton's interest in alchemy, they are unlikely after that exercise to consider it as a lingering echo of the old order... The only satisfactory solution to the question of alchemy that I see is to accept it as part of the rejection of Aristotelian natural philosophy and to incorporate it into our account of the Scientific Revolution. I know very well that prevailing accounts of the Scientific Revolution do not do this." Richard S. Westfall, "The Scientific Revolution Reasserted," *Rethinking the Scientific Revolution*, M. J. Osler (ed.), Cambridge, Cambridge University Press, 2000, pp. 52-53.

[7] "The dominant tradition of twentieth-century scholarship developed around the idea of the Scientific Revolution...Recent developments in Newtonian scholarship—particularly the serious study of his theological and alchemical interests—have thus played a significant role in re-evaluations of that important historiographical concept, the Scientific Revolution. The traditional account of the Scientific Revolution was deeply influenced by nineteenth-century positivist assumptions that privileged mathematical physics. Consequently, astronomy, mathematics, and physics became the focus of discussions of the Scientific Revolution. According to this account, the Scientific Revolution began when Copernicus proposed a heliocentric system to replace the Ptolemaic astronomy that had held sway since the second century A.D. Because Copernicus and his followers advocated a realist interpretation of their heliocentric astronomy, the new system challenged Aristotelian cosmology, which was explicitly geocentric and geostatic. The new astronomy thus raised perplexing questions about physics: why do unattached objects not fly off the surface of a moving earth? What holds the planets in their orbits? The working out of the answers to these questions through the contributions of Kepler, Galileo, Descartes, and Newton is considered the heart of the Scientific Revolution." Margaret J. Osler, "The New Newtonian Scholarship and the Fate of the Scientific Revolution," *Newton and Newtonianism: New Studies*, J. E. Force and S. Hutton (eds.), Dordrecht, Kluwer Academic Publishers, 2004, pp. 1-2.

provided insight into optics, invented calculus, and in 1703 Newton became president of the Royal Society.

The pre-eminent and most highly influential alchemists of the Scientific Revolution were Paracelsus von Hohenheim (1493–1541), Michael Maier (1569–1622), Jean Baptiste van Helmont (1579–1644), and George Starkey (1628–1665). These esteemed figures were the most accomplished physicians of their time, and the numerous accounts of the remarkable success of their "Chymical" preparations exists as indisputable evidence of their extraordinary knowledge of medicine, which these physicians repeatedly attributed to their possession of the "immortal liquor." In order to conceal their knowledge these "Chemical Physicians" would often intentionally write in obscure and at times biblical language, employing ambiguous symbols, metaphors, and allegories to effectively place a hermetic seal on their secret activities.[8] These clandestine experiments of the alchemists culminated in the Scientific Revolution, initiated and facilitated by these Hippocratic physicians and Neoplatonic philosophers who successfully negotiated the confusing labyrinth of the "Black Art" and solved this Sphinx-like riddle.

In conjunction with these revolutionary accomplishments in the burgeoning physical sciences, medicine and chemistry, at that time called *physick*, also underwent an intellectual revolution, an extraordinary episode in history that constitutes the origins of modern medicine, chemistry, empirical psychology, and psychotherapy.[9]

[8] "The origins of alchemical symbolism are as ancient as they are obscure. However, it is generally believed that the first to use symbols to abbreviate the description of medical remedies were the Greeks...At the beginning of the fourth century, Greek alchemists, such as Zosimos and Stephanus of Alexandria, began to employ an enigmatic nomenclature and to mix their practical knowledge with speculative notions derived from Neo-Platonism, astrology, gnosticism, and magic...However, it was not until the sixteenth and seventeenth centuries that alchemical symbolism flourished in the variety of forms with which we are acquainted. The rediscovery of Hermeticism and Platonism and the intellectual battle against Aristotle are the main reasons for the philosophical and iconographical success enjoyed by alchemy during the Renaissance. Being a movement of cultural opposition, alchemy had to use a language completely different from that of the Aristotelian tradition; accordingly, alchemists used pictograms, symbols, emblems, etc. instead of arguing about the etymological origins of words and their precise definition...The mysticism and magic of the Egyptians, the reinterpretation of the Bible, and the allegories of the poets and the Platonists were indeed an integral and significant part of alchemical thought." Marco Beretta, *The Enlightenment of Matter: The Definition of Chemistry from Agricola to Lavoisier*, USA, Science History Publications, 1993, pp. 332, 335, 346.

[9] "I had long been aware that alchemy is not only the mother of chemistry, but is also the forerunner of our modern psychology of the unconscious. Thus Paracelsus appears as a pioneer not only of chemical medicine but of empirical psychology and psychotherapy." Carl Gustav Jung, "Alchemical Studies," *The Collected Works of C. G. Jung*, R. F. C. Hull (trans.), Princeton, Princeton University Press, 1967, vol. XIII, p. 189. For more on this subject and a balanced criticism of the Aristotelian model that Newman and Principe impose upon the anti-Aristotelian writings of

The principal figures of the Scientific Revolution in medicine and chemistry were Paracelsus, Jean Baptiste van Helmont, and the American physician George Starkey, the latter of whom, along with his alter ego Eirenaeus Philalethes, would serve as an authoritative guide to Robert Boyle and Isaac Newton in their quest to gain the forbidden knowledge hidden within alchemy.

George Starkey, see Hereward Tilton, *The Quest for the Phoenix: Spiritual Alchemy and Rosicrucianism in the Work of Count Michael Maier (1569–1622)*, Berlin, Walter de Gruyter, 2003, pp. 1-34.

Chapter 1. Alchemy in the New World

With the arrival of the Europeans in the New World, the tradition of alchemy spread to Colonial America where the search for the prized Philosophers' Stone was continued into the nineteenth century by New England's political and intellectual elite.[10] Among the influential Americans who practiced alchemy was John Winthrop, Jr., the founder and first governor of Connecticut, who was also a highly successful physician. Winthrop amassed an enormous collection of alchemical writings in his private library,[11] and the memory of his pursuits in alchemy survived in a folk-legend recorded by Ezra Stiles more than a century after Winthrop's death.[12]

Winthrop began his search for the Philosophers' Stone between 1625 and 1627 while studying law in London, and in 1631 he emigrated to New England and joined his father John Winthrop, Sr., the founder and first governor of the Massachusetts-Bay Colony, who also shared his son's interest in alchemy.[13] During his second tour of Europe in the 1640s, the well-traveled Winthrop visited the German pastor Johann

[10] Ronald Sterne Wilkinson, "New England's Last Alchemists," *Ambix*, 10 (1963), pp. 128–138; Patricia A. Watson, *The Angelical Conjunction: The Preacher-Physicians of Colonial New England*, Knoxville, University of Tennessee Press, 1991, pp. 97–121.

[11] Ronald Sterne Wilkinson, "The Alchemical Library of John Winthrop, Jr. (1606–1676) and his Descendants in Colonial America," *Ambix*, 11 (1963), pp. 33-51.

[12] Franklin Bowditch Dexter (ed.), *The Literary Diary of Ezra Stiles*, Volumes I-III, New York, Charles Scribner's Sons, 1901, vol. III, p. 266.

[13] Ronald Sterne Wilkinson, "'Hermes Christianus:' John Winthrop, Jr. and Chemical Medicine in Seventeenth Century New England," *Science Medicine and Society in the Renaissance: Essays to Honor W. Pagel*, Volumes I-II, A. G. Debus (ed.), New York, Science History Publications, 1972, vol. I, pp. 221-225; *Dictionary of American Biography*, D. Malone (ed.), New York, Charles Scribner's Sons, 1943, vol. XX, pp. 411-413.

Rist at Wedel, who attempted to dissuade Winthrop of his belief that common mercury was the first matter of the Philosophers' Stone, and recorded this conversation in his work entitled *The Most Noble Folly of the Whole World*:

> [Winthrop] was at first of the opinion that merely and only mercury was the true material of the philosophers' stone. I replied, yes, to be sure, it was mercury, yet not the common but the philosophical mercury. "I am also of that opinion," said the Englishman, "but one must so process and prepare the common mercury or quicksilver that a philosophical mercury develops from it." However, I soon showed him his grave error...This first opinion of the English gentleman, that common mercury is the true material of the much desired philosophers' stone, is still to this day accepted by many...[14]

After returning to England to secure a charter for his Connecticut colony, Winthrop was the first American given entry into the newly formed Royal Society of London; his admittance in 1662 made him an "Original Fellow."[15] Counted among Winthrop's many friends were Elias Ashmole, Henry Oldenburg, Christopher Wren, and Robert Boyle.[16] In New England, the alchemical associates of the younger Winthrop included Robert Child and George Starkey, and among his alchemical associates in Europe were Samuel Hartlib and the great Jean Baptiste van Helmont.[17] Winthrop was also a close alchemical associate of Reverend Gershom Bulkeley, who graduated Harvard College in 1655 and became a physician and the son-in-law of Harvard's second president Charles Chauncy, who also practiced chemical medicine and passed on this tradition to all six of his Harvard-educated sons.[18] With the assistance of his brother-in-law Bulkeley, Reverend Israel Chauncy became a founder and early

[14] Johann Rist, *Die alleredelste Tohrheit der gantzen Welt...*, Hamburg, In Verlegung Ioh. Naumanns, Buchh., 1664, pp. 238-240; in Harold Jantz, "America's First Cosmopolitan," *Proceedings of the Massachusetts Historical Society*, 84 (1972), pp. 6-7.

[15] Thomas Birch, *The History of the Royal Society of London*, London, Printed for A. Miller in the Strand, 1756, vol. I, p. 68; Winthrop Papers, Massachusetts Historical Society Library, vol. V, p. 30, vol. XI, p. 35; in Richard S. Dunn, *Puritans and Yankees: The Winthrop Dynasty of New England 1630–1717*, Princeton, Princeton University Press, 1962, p. 130.

[16] Cromwell Mortimer's dedication to John Winthrop, F. R. S., *Philosophical Transactions*, 40 (1741); Samuel Eliot Morison, *The Intellectual Life of Colonial New England*, New York, New York University Press, 1956, p. 250; Robert C. Black, *The Younger John Winthrop*, New York, Columbia University Press, 1966, p. 217.

[17] Wilkinson, "Hermes Christianus," pp. 222-223.

[18] John Langdon Sibley, *Biographical Sketches of Graduates of Harvard University*, Volumes I-III, Cambridge, Charles William Sever, 1873–1885, vol. I, pp. 302, 389-402, vol. II, p. 81; Thomas W. Jodziewicz, "A Stranger in the Land: Gershom Bulkeley of Connecticut," *Transactions of the American Philosophical Society*, 78 (1988), pp. 14–15.

administrator of Yale College; he was chosen the first president of Yale but declined the position.[19]

Leonard Hoar graduated Harvard in 1650 and received a "Doctor of Physick" from the University of Cambridge in 1671; he succeeded Reverend Chauncy to become Harvard's third president in 1672.[20] Reverend Hoar was a friend of Robert Boyle and Robert Morison, senior physician to Charles II and first Professor of Botany at the University of Oxford, home of the famous Oxford Botanic Garden, also known as the Physick Garden, alchemy commonly referred to as *physick* or *chymical physick*.[21] Shortly after becoming president of Harvard, Hoar wrote to Boyle of his plans to create a botanical garden similar to the one at Oxford, consisting of "A large well-sheltered garden and orchard for students addicted to planting; an ergasterium for mechanick fancies; and a laboratory chemical for those philosophers," and Hoar asked Boyle "to deign us any advice or device, by which we may become not only nominal, but real scholars."[22]

Samuel Danforth was another well-known American alchemist. He graduated Harvard in 1715 and was appointed to the Middlesex court of common pleas in 1741 where he served as judge and chief justice for thirty-four years.[23] Danforth's formal education at Harvard included the *Compendium Physicae* of Reverend Charles Morton, who became the vice-president of Harvard in 1697.[24] Morton's scientific textbook was adopted by the college in 1687 and circulated in manuscript among Harvard's undergraduates. In it there are several long paragraphs on "the finding of the Phylosophers stone," and Morton affirms that "Some have done it, such are cal'd the Adepti," which includes Lully, Paracelsus, van Helmont, "and others of whom we shall not farther Insist."[25]

[19] Yale's first graduate in 1702 was Nathaniel Chauncy, the son of Israel's brother Nathaniel. Sibley, *Graduates of Harvard*, vol. II, pp. 74-87; Franklin Bowditch Dexter (ed.), *Documentary History of Yale University: Under the Original Charter of the Collegiate School of Connecticut 1701–1745*, New Haven, Yale University Press, 1916, pp. 8–11, 20-23, 149; Samuel Eliot Morison, *Harvard College in the Seventeenth Century*, Volumes I-II, Cambridge, Harvard University Press, 1936, vol. II, pp. 546-548.

[20] Sibley, *Graduates of Harvard*, vol. I, pp. 228-252.

[21] I. Bernard Cohen, "The Beginning of Chemical Instruction in America: A Brief Account of the Teaching of Chemistry at Harvard Prior to 1800," *Chymia*, 3 (1950), pp. 18–19.

[22] Hoar to Boyle, 13 December 1672; in Morison, *Harvard College*, vol. II, pp. 644-646.

[23] Clifford K. Shipton, *Sibley's Harvard Graduates; Biographical Sketches of Those Who Attended Harvard College...*, Volumes IV-XIV, Cambridge, Harvard University Press [etc.], 1933–1968, vol. VI, pp. 80-86.

[24] Morison, *Harvard College*, vol. I, pp. 236-238, vol. II, p. 513.

[25] Charles Morton, *Compendium Physicae*, T. Hornberger (ed.), *Publications of the Colonial Society of Massachusetts*, 33 (1940), pp. xxiii, 121.

Danforth was a reputed Adept, and after his support of an unpopular Excise Bill in 1754, he was mocked in a political pamphlet as "Madam CHEMIA, (a very philosophical Lady) who some Years since (as is well known) discover'd that *precious Stone*, of which the Royal Society has been in quest a long Time, to no Purpose."[26] In 1773, Danforth wrote to his friend Benjamin Franklin in London and apparently offered Franklin a sample of his Philosophers' Stone. Franklin responded by thanking Danforth for his "kind Intentions of including me in the Benefits of that inestimable Stone, which, curing all Diseases (even old Age itself), will enable us to see the future glorious State of our America...I anticipate the jolly Conversation we and twenty more of our Friends may have 100 Years hence on this subject, over that well replenish'd Bowl at Cambridge Commencement."[27]

Also involved in the American alchemical tradition was the Reverend Ezra Stiles, who graduated Yale College in 1746, served as minister to the Second Congregational Church of Newport, Rhode Island, and in 1778 became president of Yale, where he remained until his death in 1795.[28] Reverend Stiles had a reputation as an alchemist and an Adept, which he strongly denied.[29] Following the death of Danforth in 1777, Stiles wrote in his *Diary* that his friend "was deeply studied in the Writings of the Adepts, believed the Philosophers Stone a Reality," although a few months earlier Stiles had noted that Danforth was merely a "conjectural & speculative" philosopher who had never attained the Stone.[30]

Reverend Stiles was also a close friend of Samuel West, a liberal theologian who graduated Harvard in 1754 and became a founder of the American Academy of Arts and Sciences.[31] During their frequent visits Stiles and West often discussed theology and alchemy, and according to Stiles's *Diary* entry in 1777, Dr. West "believed in the reality of the Philosopher's Stone," although the minister had failed to obtain it.[32] Reverend West became close friends with the patriots and

[26] Thomas Thumb (pseud.), *The Monster of Monsters...*, Boston, Printed by Zechariah Fowle, 1754, p. 15.
[27] Franklin to Danforth, London, 25 July 1773; in Albert Henry Smyth (ed.), *The Writings of Benjamin Franklin*, New York, Macmillan Company, 1907, vol. VI, p. 106.
[28] Franklin Bowditch Dexter, *Biographical Sketches of the Graduates of Yale College...*, New York, Henry Holt and Company, 1896, vol. II, pp. 92-97.
[29] *Diary of Stiles*, vol. II, pp. 173–174, vol. III, p. 345.
[30] Ibid., vol. II, pp. 174, 216.
[31] Shipton, *Sibley's Harvard Graduates*, vol. XIII, pp. 501-510.
[32] *Diary of Stiles*, vol. I, pp. 253, 261, vol. II, p. 174, vol. III, pp. 302, 330.

future President John Adams and Governor John Hancock while at Harvard, and was an outspoken defender of the American Revolution, and later helped to ratify the Federal Constitution in 1788, personally obtaining Hancock's crucial support for the document.[33] The minister was a guest of the future President John Quincy Adams that year, a man critical of weakness, who captivatingly listened to West expound philology during his visit, and described him as "very sociable and sensible...He keeps late hours and entertained me with conversation... till between twelve and one o'clock."[34]

Danforth's son, Dr. Samuel Danforth, graduated Harvard in 1758 and later served as president of the Massachusetts Medical Society from 1795 to 1798, and was probably the last New England alchemist when he died in 1827.[35] The younger Danforth followed Helmontian medicine and inherited his father's alchemical books and interest in chemical medicine, acquiring valuable equipment through his son Thomas who was sent to Europe to study medicine. This allowed Danforth to assemble one of the finest laboratories in Boston, but pressures from his medical career forced him to discontinue his pursuit of the "immortal liquor."[36] Aaron Dexter studied chemistry and medicine under the younger Danforth, and in 1783 was appointed the first Professor of Chemistry and Materia Medica in the newly organized Harvard Medical School; Dexter's first lecture was devoted to the history and origins of chemistry, which he began with the contributions of Paracelsus.[37]

Aeneas Munson graduated Yale in 1753, after which he studied divinity under the tutelage of Stiles as well as medicine, and was described by his pupil, Dr. Eli Ives, as a "pioneer in the science of Botany...unrivaled in his knowledge of indigenous materia medica," and a chemist who "made many chemical experiments."[38] Munson occupied the first chair of Materia Medica and Botany in the newly formed Yale Medical School and was a founder of the New Haven County Medical Society, chartered in 1792 as the Connecticut Medical

[33] Shipton, *Sibley's Harvard Graduates*, vol. XIII, pp. 501, 504-505, 514.

[34] "Diary of John Quincy Adams," *Proceedings of the Massachusetts Historical Society*, Second Series, 16 (1903), p. 438.

[35] Shipton, *Sibley's Harvard Graduates*, vol. XIV, pp. 250-254.

[36] James Thatcher, *American Medical Biography...*, Volumes I-II, Boston, Richardson & Lord and Cottons & Barnard, 1828, vol. II, pp. 233-238.

[37] Cohen, "Chemical Instruction in America," pp. 17, 36-41.

[38] Thatcher, *American Medical Biography*, vol. I, pp. 401-403; Henry Bronson, "Medical History and Biography," *Papers of the New Haven Colony Historical Society*, 2 (1877), pp. 268-269.

Society. He served as president of this institution from 1794 to 1801.[39] Stiles and Munson often conversed about Hermetic philosophy, and in 1792 Munson showed Stiles "a Piece of malleable whitish Metal" that he had produced, which Stiles called "fixt Mercury, the first I ever saw."[40] Both men were intimately familiar with other failed attempts at transmutation by their contemporaries, and in 1777 Stiles wrote in his *Diary*: "D[r] [Benjamin] Franklin told me there were several at Philad[elphi][a] &c. who were loosing their Time in chemical Experiments to no Effect."[41] It was Aeneas Munson's death in 1826 that marked the end to the nearly two-hundred-year-old pursuit of the Philosophers' Stone by America's political and intellectual leaders.[42]

[39] Dexter, *Graduates of Yale*, vol. II, pp. 311-313; Bronson, "Medical History," pp. 265-266.

[40] *Diary of Stiles*, vol. III, pp. 345, 471.

[41] Ibid., vol. II, p. 174, vol. III, p. 472.

[42] Although the practice of alchemy discontinued in America, interest in the tradition persisted; among the very few who published on the subject was the exemplary U.S. Army Major-General Ethan Allen Hitchcock, *Remarks upon alchymists...*, Carlisle, Penn., Printed at the Herald Office, 1855; Hitchcock, *Remarks upon Alchemy and the Alchemists, Indicating a Method of Discovering the True Nature of Hermetic Philosophy; and Showing that the Search after The Philosopher's Stone had not for its Object the Discovery of an Agent for the Transmutation of Metals...*, Boston, Crosby, Nichols, and Company, 1857; Cf. I. Bernard Cohen, "Ethan Allen Hitchcock: Soldier-Humanitarian-Scholar, Discoverer of the 'True Subject' of the Hermetic Art," *Proceedings of the American Antiquarian Society*, 61 (1951), pp. 29-136.

CHAPTER 2. GEORGE STARKEY AND EIRENAEUS PHILALETHES

George Starkey was the most prominent of all the American alchemists. Starkey practiced alchemy with the assistance of his close friend John Winthrop, Jr. while residing in New England, where Starkey also befriended Dr. Robert Child, who was imprisoned for sedition and forced to return to England in 1647; he would subsequently introduce Starkey to the members of the Hartlib circle in London.[43]

In 1650, the young Starkey gave away most of his medicines and left behind a highly successful medical practice to emigrate to London,[44] where he quickly joined up with Robert Boyle and his circle of utopian intellectuals.[45] George Starkey, or Stirk, was the son of Elizabeth and George Stirk,[46] a Puritan minister from Scotland who accompanied Governor John Bernard to the Bermudas in 1622[47] (which at that time were called Somers Islands and considered to be part of New

[43] George Lyman Kittredge, "Dr. Robert Child the Remonstrant," *Transactions of the Colonial Society of Massachusetts*, 21 (1920), pp. 1–146; George H. Turnbull, "Robert Child," *Transactions of the Colonial Society of Massachusetts*, 38 (1947), pp. 21-53.

[44] Hartlib, *Ephemerides*, [1650/51], A-B3; in Ronald Sterne Wilkinson, "The Hartlib Papers and Seventeenth-Century Chemistry, Part II: George Starkey," *Ambix*, 17 (1970), p. 87.

[45] Ronald Sterne Wilkinson, "George Starkey, Physician and Alchemist," *Ambix*, 11 (1963), pp. 125–133.

[46] Referring to Starkey, Child informed Hartlib: "The young Physitian of 21 or 22. y[ears] of age in N[ew] E[ngland] so famous already for curing the Palsy and other incurable diseases writes his name Stirke. Hee is a Presbyterian and of Scots-parents borne in Bermudes." Hartlib, *Ephemerides*, [summer 1650], G-H3; in Wilkinson, "Hartlib Papers II," p. 87.

[47] George Lyman Kittredge, "George Stirk, Minister," *Transactions of the Colonial Society of Massachusetts*, 13 (1912), pp. 16–17, 52-55.

England).[48] Starkey was born in the Bermudas on 8 or 9 June 1628.[49] After his father's death in 1637, Starkey was sent to Harvard following a letter written by the minister Patrick Copeland to John Winthrop, Sr. in 1639, requesting that Governor Winthrop find a good school for the fatherless boy.[50] While residing in the Massachusetts-Bay Colony, Starkey married the daughter of Colonel Israel Stoughton,[51] who along with the elder and younger Winthrop belonged to the first Board of Overseers, the governing body that established Harvard College.[52] Starkey's brother-in-law William Stoughton would later serve as judge in the Salem witch-trials and as the governor of Massachusetts, and Starkey's wife Susanna, almost certainly the only wife he took during his lifetime, died in England in 1662.[53]

Reverend Henry Dunster was the first president of Harvard College, and he strongly supported the pursuit of chemical medicine at Harvard, petitioning the Commissioners of the United Colonies in 1647 for books on various subjects, including physick, but he received no response to his request.[54] While under the tutorage of Reverend Dunster, Starkey began his alchemical studies in 1644 with the physician Richard Palgrave, who accompanied the elder Winthrop to America in 1630, and Starkey's classmate John Alcock joined them in these pursuits the following year.[55] Starkey writes in his autobiographical notes:

> In the year 1644 I was first invited to this study by Mr. Palgrave, physician of New England, while I was living at Harvard College,

---

[48] Jantz, "America's First Cosmopolitan," p. 7.

[49] John Gadbury, *Collectio Geniturarum...*, London, Printed by James Cottrel, 1662, p. 130; Jantz, "America's First Cosmopolitan," p. 23; Transcriber's insertion in a note of George Starkey, 1660, Harvard University, Houghton Library Autograph File, fol. 3r; in William R. Newman and Lawrence M. Principe (eds.), *George Starkey: Alchemical Laboratory Notebooks and Correspondence*, Chicago, University of Chicago Press, 2004, p. 331, (hereafter as *Starkey Notebooks and Correspondence*).

[50] Kittredge, "George Stirk," p. 53.

[51] George H. Turnbull, "George Stirk, Philosopher by Fire," *Publications of the Colonial Society of Massachusetts*, 38 (1949), p. 221.

[52] Samuel Eliot Morison, *The Founding of Harvard College*, Cambridge, Harvard University Press, 1935, pp. 193, 327-328, 408-409.

[53] [George Starkey], *The Dignity of Kingship Asserted, By G. S...*, W. R. Parker (ed.), New York, Columbia University Press, 1942, p. xix n. 7.

[54] C. Helen Brock, "The Influence of Europe on Colonial Massachusetts Medicine," *Publications of the Colonial Society of Massachusetts*, 57 (1980), p. 105.

[55] George Starkey, *Pyrotechny Asserted and Illustrated, To be the surest and safest means for Arts Triumph over Natures Infirmities. Being full and free Discovery of the Medicinal Mysteries studiously concealed by all Artists, and onely discoverable by Fire*, London, Printed by *R. Daniel*, for *Samuel Thomson* at the *Whitehorse* in *S. Pauls* Church-yard, 1658, p. 76; Starkey, *A Smart Scourge for a Silly, Sawcy Fool. Being An Answer to a Letter, at the End of a Pamphlet of* Lionell Lockyer..., London, s.n., 1664, pp. 6-7.

> under the tutorship and presidency of Henry Dunster. At the time I was between my sixteenth and seventeenth year...I was then assiduously studying philosophy, and I began this pursuit at the end of that very summer. But in the next year, namely 1645, I began to work—with God's assistance—on the true matter with a certain comrade, John Alcocke...I continued thus with great labor up to the year 1654, at which time the good God taught me the whole secret. This was the tenth year after my initial introduction to this art, which years I spent in incredible perseverance, with constancy of spirit, in highly erroneous labors. Around the end of the year 1654 the whole secret was revealed to me by divine grace.[56]

John Alcock became a physician and married Palgrave's daughter Sarah after he and Starkey graduated Harvard in 1646 with two others; his wife was also skilled in "physique and chirurgery,"[57] a practice that also appears to have been passed on to their son George, who graduated Harvard in 1673.[58] Following his graduation, Starkey established a successful medical practice in the Boston area and continued his alchemical pursuits,[59] receiving his M.A. from Harvard before his departure from New England,[60] arriving in London in November 1650.[61]

George Starkey was a remarkably successful chemist and physician whose writings were held in the highest esteem, and one Newton scholar has conferred upon Starkey's alter ego Eirenaeus Philalethes the honorable distinction as the "last great philosophical alchemist of the seventeenth century."[62] Starkey was almost certainly the "last great philosophical alchemist." Starkey's greatest works were published under the pseudonym Eirenaeus Philalethes, literally meaning, "A Peaceful Lover of Truth," who was also known as Cosmopolitan, "A Citizen of the World." So effective was Starkey in this deception that the identity of Anonymous Philalethes has

[56] [Starkey], 1660, Harvard University, Houghton Library Autograph File, ff. 3^r-4^r; in *Starkey Notebooks and Correspondence*, pp. 331-332.

[57] Sibley, *Graduates of Harvard*, vol. I, pp. 124–140.

[58] Samuel Eliot Morison, "The Library of George Alcock, Medical Student, 1676," *Transactions of the Colonial Society of Massachusetts*, 28 (1933), pp. 350-357.

[59] Kittredge, "George Stirk," pp. 53, 55 n. 2; Wilkinson, "George Starkey," p. 126.

[60] John Ferguson, *Bibliographica Chemica: A Catalogue of the Alchemical, Chemical and Pharmaceutical Books in the Collection of the late James Young of Kelly and Durris*, Glasgow, J. Maclehose and Sons, 1906, vol. II, p. 404.

[61] [Starkey], 1660, Harvard University, Houghton Library Autograph File, fol. 4^r; in *Starkey Notebooks and Correspondence*, p. 332.

[62] Betty Jo Teeter Dobbs, *The Foundations of Newton's Alchemy or "The Hunting of the Greene Lyon"*, Cambridge, Cambridge University Press, 1975, p. 52.

remained shrouded in mystery until the present generation.[63] Starkey produced the enormously influential writings of the "American philosopher" Philalethes while residing in London, claiming to have studied under this enigmatic figure in New England, whom Starkey supposedly persuaded to write the manuscripts that were in his trust.[64] The Philalethes compositions circulated in manuscript and only two of these were published during Starkey's lifetime: the two-part *Marrow of Alchemy* in 1654 and 1655, and *Sir George Riplye's Epistle to King Edward unfolded*,[65] published in 1655 without Starkey's consent and to his dismay in the *Chymical, Medicinal, and Chyrurgical Addresses* dedicated to Samuel Hartlib.[66]

Sometime between 1651 and 1654 Starkey composed the *Introitus Apertus*,[67] in which Philalethes declares that by the blessing of God he discovered the Philosophers' Stone in 1645 at the age of twenty-three, which he claimed to be the panacea that could cure all disease.[68] Although there is no hard evidence that Starkey was ever in mortal danger in New England,[69] Philalethes also claims that he was compelled to flee for his life after a rumor circulated that he was in possession of the treasured "Elixir of life."[70] Starkey's reasons for creating Philalethes can be found in the *Christianae Societatis Pactum*, dated 18 August 1652 and signed by Samuel Hartlib, John Dury, and

[63] Kittredge, "Robert Child," p. 146; Jantz, "America's First Cosmopolitan," pp. 9-24; Ronald Sterne Wilkinson, "Further Thoughts on the Identity of 'Eirenaeus Philalethes'," *Ambix*, 19 (1972), pp. 204-208; Wilkinson, "Some Biographical Puzzles Concerning George Starkey," *Ambix*, 20 (1973), pp. 235-244; Wilkinson, "Starkey, George," *Dictionary of Scientific Biography*, Volumes I-XIV, C. C. Gillispie (ed.), New York, Charles Scribner's Sons, 1970–1976, vol. XII, pp. 616-617; William R. Newman, "Prophecy and Alchemy: The Origin of Eirenaeus Philalethes," *Ambix*, 37 (1990), pp. 97–115.

[64] Philalethes, *Marrow Of Alchemy* [Part I], pp. A3-A4.

[65] [Eirenaeus Philalethes] (pseud.), "Sir George Riplye's Epistle to King Edward unfolded," *Chymical, Medicinal, and Chyrurgical Addresses: Made to Samuel Hartlib, Esquire*, London, Printed by *G. Dawson* for *Giles Calvert* at the *Black-spread Eagle* at the west end of *Pauls*, 1655, pp. 19-47.

[66] Jantz, "America's First Cosmopolitan," p. 20; Wilkinson, "Hartlib Papers II," p. 97; [George Starkey], Preface to *Sir George Riplye's Epistle to King Edward unfolded*, Sloane Collection, British Library, MS 633, fol. 3^r; in *Starkey Notebooks and Correspondence*, p. 313.

[67] William R. Newman, "The Authorship of the *Introitus Apertus ad Occlusum Regis Palatium*," *Alchemy Revisited: Proceedings of the International Conference on the History of Alchemy at the University of Groningen 17–19 April 1989*, Z. R. W. M. von Martels (ed.), Leiden, E. J. Brill, 1990, pp. 139–144; Newman, "Prophecy and Alchemy," pp. 99, 106.

[68] Eirenaeus Philalethes (pseud.), "Introitus Apertus, *ad* Occlusum Regis Palatium," *Bibliotheca Chemica Curiosa*, Volumes I-II, J. J. Manget (ed.), Genevæ, Sumpt...De Tournes, 1702, vol. II, p. 661.

[69] Child made an unsubstantiated claim to Hartlib that Starkey was "[con]fined for 2 y[ears] in N[ew] E[ngland] upon Suspition to bee a Spie or Jesuit." Hartlib, *Ephemerides*, 11 December 1650, K-L7; in Wilkinson, "Hartlib Papers II," p. 87.

[70] Philalethes, "Introitus Apertus," *Bibliotheca Chemica Curiosa*, vol. II, p. 666.

Frederick Clodius, in which the signatories pledge to assist Starkey and protect the greater secrets revealed by God:

> The greater secrets [*Arcana maiora*] that the divine benignity deigns to reveal to us must be dispensed with great caution: this also goes for those whom we admit to the notice of our society, lest that which should have remained secret be made public, and we expose ourselves thereby to the jealousy, malice, suspicion, and tyranny of ambitious men, or to the avarice and power of important persons...Above all as regards Mr. Starkey, not only will we see to it that [our endeavors] are not lacking on his behalf, but we will also work so that [they] serve his profit and honor.[71]

Jonathan Brewster, the son of Elder William Brewster of Plymouth, also practiced alchemy on the New England frontier while maintaining a close association with John Winthrop, Jr. and Gershom Bulkeley.[72] Brewster expressed a similar fear of exposure in a letter to Winthrop, which helps to explain Starkey's motivation in creating Philalethes:

> Sir, I intreate once again, spare to use my name, and lett my letters I send, either be safely kept, or burnt, that I write about it, for inde[e]d, Sir, I am more then before sensible of the evill effectes that will arise by the publishing of it. I should never be at quiett, neither at home, nor abroad, for one or other that would be enquiring & se[e]king after knowledg[e] thereof, that I should be tyared out, & forced to leave the place ; naye, it would be blas[t]ed abroad into Europ[e].[73]

George Starkey was an exceptionally skillful chemist and physician, and while residing in New England he was successfully curing "Agues,"[74] "Feaver," "Falling Sicknes[s]," "Dropsy,"[75] "Rickets,"[76] and "Palsy and other incurable diseases."[77] Immediately following his arrival in London, Child told Hartlib in December 1650 that Starkey had "already 20 good patients."[78] Robert Boyle submitted himself and

[71] G. H. Turnbull, *Hartlib, Dury and Comenius: Gleanings from Hartlib's Papers*, London, University Press of Liverpool, 1947, pp. 122–123.
[72] Morison, *Colonial New England*, p. 250.
[73] Brewster to Winthrop, 31 January 1656; in "The Winthrop Papers," *Collections of the Massachusetts Historical Society*, Fourth Series, 7 (1865), p. 80.
[74] Hartlib, *Ephemerides*, [September 1653], MM-MM4; in Wilkinson, "Hartlib Papers II," p. 103.
[75] Ibid., 11 December 1650, K-L7; in Ibid., p. 87.
[76] Hartlib to Boyle, 28 February 1653/54; in Thomas Birch (ed.), *The Works of the Honourable Robert Boyle*, Volumes I-VI, London, Printed for J. and F. Rivington [etc.], 1772, vol. VI, p. 80, (hereafter as *Works*); Wilkinson, "George Starkey," p. 130 n. 58.
[77] Hartlib, *Ephemerides*, [early 1650], E-F1, [summer 1650], G-H3; in Wilkinson, "Hartlib Papers II," pp. 86-87.
[78] Ibid., [December 1650], L4; in Ibid., pp. 87-88.

members of his aristocratic family to Starkey's chemical medicines,[79] and records from this time reveal that Starkey was curing patients after repeated attempts by other physicians had failed.[80]

During his brief life, Starkey struggled with mounting financial debts related to the high costs of his research and his charitable service to the poor,[81] which resulted in his imprisonment for debt in early 1654.[82] Following the publication of his polemical *Natures Explication* in 1657, Starkey suffered a ten-month confinement,[83] his defenders maintaining that this was the outcome of his bitter struggles with the London medical establishment.[84] In November of that year, John Beale told Samuel Hartlib that he had "heard of Starkeys distresse, & did expect, That his foule language would beget strong adversaryes."[85]

According to Starkey in *Natures Explication*, the attacks upon chemical medicine by the Galenists initiated the bitter conflict that continued for the duration of Starkey's life: "[T]he discourse is indeed

[79] Starkey to Boyle, 3 January and 16 January 1651 [Old Style, i.e., 1652], Royal Society Library, Boyle Letters, vol. V, ff. 129^v, 132^r; in Michael Hunter, Antonio Clericuzio, and Lawrence M. Principe (eds.), *The Correspondence of Robert Boyle*, London, Pickering & Chatto, 2001, vol. I, pp. 110, 116, (hereafter as *Correspondence of Boyle*); R. E. W. Maddison, *The Life of the Honourable Robert Boyle*, London, Taylor and Francis Ltd., 1969, pp. 77-78, 219.

[80] Hartlib, *Ephemerides*, 11 December 1650, K-L7, [1650/51], A-B3; in Wilkinson, "Hartlib Papers II," pp. 86-87, 93; Hartlib to Boyle, 28 February 1653/54; in Boyle, *Works*, vol. VI, p. 80.

[81] George Starkey, *Natures Explication and Helmont's Vindication. Or A short and sure way to a long and sound Life: Being, A necessary and full Apology for Chymical Medicaments, and a Vindication of their Excellency against those unworthy reproaches cast on the Art and its Profesors (such as were Paracelsus and Helmont) by Galenists...*, London, Printed by *E. Cotes* for *Thomas Alsop* at the two Sugar-loaves over against St. *Antholins* Church at the lower end of *Watling-street*, 1657, pp. 224-225.

[82] Hartlib to Boyle, 28 February 1653/54; in Boyle, *Works*, vol. VI, pp. 79-80.

[83] Starkey may have suffered detention again in the 1660s. In Lionel Lockyer's *Advertisement* (1664) for his poisonous "Pill extracted from the rays of the sun," which earlier that same year Starkey had denounced as fraudulent, an anonymous author attached a letter at the end of the pamphlet attacking Starkey: "He can Transmute Metals, if you will beleive him ; and yet is a pittiful Fellow, or else he would not have been so often in Prison for his Cousenage, in so much that he is as well known in *Newgate*, as most of the Common Rascals." A version of this same letter is also attached to the beginning of *Aut Helmont* (1665), another anonymous attack upon Starkey, perhaps by the same author who signs the letter with the initials G. S.: "I have known some of your fellow-prisoners at *Newgate*," and "How a man should have been in danger of hanging for being the Doctors Voucher to sell a horse, I know not, but these hard words, Sir, are a Juglers dialect; perhaps you are afraid to explain your self, lest you should be made to sing another *palinodium* in *Newgate* ; yet I have heard it is fatal to be some mens Vouchers, witness your Surety who was hanged for a Coiner, or rather your Disciple it may be, to whom you had taught the use of the new-fashioned Philosophers stone." Also, in a letter to Phillip Fryth at Rye in 1663, Starkey writes: "So long as I want my liberty, I am uncertaine where to reside." Lionel Lockyer, *An Advertisement Concerning those most Excellent Pills Called Pillulæ Radiis Solis Extractæ. Being An Universal Medicine...*, London, s.n., 1664, [p. 17]; *Aut Helmont, Aut Asinus: Or, St. George Untrust; Being a Full Answer to his Smart Scourge*, London, Printed for *R. Lowndes*, at the White Lion in S. *Pauls* Churchyard, 1665, pp. A3^v, B1^r, B2^v; Starkey to Fryth, 17 March 1662/63, Lewes, East Sussex Record Office, Frewen MS 5702; in *Starkey Notebooks and Correspondence*, p. 335.

[84] Turnbull, "George Stirk," pp. 241-244.

[85] Beale to Hartlib, 3 November 1657, Hartlib Papers, LII; in Wilkinson, "Hartlib Papers II," p. 108.

polemical, but the first that entered the list were themselves, whom because they bid defiance to the truth here asserted, with heaps of reproaches on such who were eminent in this Art (here defended) I was bold to meet, & to ingage conflict withal, and let them not complain, if they meet with shot for shot, and blow for blow."[86] These confrontational remarks are followed by an unbridled assault upon the Galenists, during which Starkey claims to have discovered the "mysteries of nature" and credits his talents to God:

> O foolish Doctors ! who hath bewitched you, that you will not see, nor abide the truth ? O silly and blind followers of these perverse blind guides ! how long will you be deceived?...Wo[e] is me, that I am and must be in this thing a Sonne of Contention, and must contend with almost all the earth...since it is truth that is to be defended, to betray which in a cause of so high concernment (as the lives of thousands) were so high an ingratitude to God, who hath discovered the mysteries of nature to me, (blessed be his name) that I might justly fear not only the deprivement of this Talent, but the other doom of the unprofitable Servant, the dread of whose exemplary punishment doth compell me thus to...expose what God hath discovered to me, to the view and censure of a captious generation, of whom I expect reproach, disdain and contumelie full measure, and heaped, yet is there a certain number of the sonnes of Wisdom, from whom I shall receive both thanks and encouragement.[87]

Starkey was so confident in his medicines that in *Natures Explication* he issued an open challenge to the Galenists of the College of Physicians: divide a group of patients, and for every one that Starkey cured his opponent would pay him a sum, and for each one cured by his opponent Starkey would pay twice the sum. Openly rejecting the methods employed by the Galenists, Starkey promised: "*I will engage to perform all my cures without bloud-letting, purging by any promiscuous Purge, or vomiting by any promiscuous Vomit, that is, which will work on all indifferently sick or no, without Vesication, or Cautery, without making any issue, or curious rules of diet, without Clyster or Suppository.*"[88] To this Starkey added: "Therefore according to what we know, we come to your own doors, and dare you to combat, we defie your Clysters as ridiculous, your Purges, and Vomits, and Bloud letting, as dangerous ; your Issues,

[86] Starkey, *Natures Explication*, p. a2^{v}.
[87] Ibid., pp. 236-237.
[88] Ibid., pp. A7^{v}-A8^{r}.

Cauteries, Blistering, *&c*, as cruell and needlesse ; and in a word, your whole Method we have impugned."[89] Starkey also writes:

> Our Doctors, (I mean the major part of them) maintain a method of medicine, which I impugn ; the Controversie concernes the way of restoring diseases safely, speedily, and certainly...and the reason why more cannot be expected from that Method, is because it is erroneous and defective, dangerous and impotent, partly lame and ridiculous, partly lamentable and desperate : To this Method as a remedy of its defects, I have opposed the way of curing and restoring diseases by powerful Medicaments, which are adequate remedies to the causes of the same, and have hazarded the cause in hand, and my reputation on the trial, if they dare to take me up.[90]

The Galenists had nothing to gain by accepting. They ignored Starkey's challenge, a regrettable decision since this would have been the first clinical trial in medical history; but the opportunity to conduct a similar medical experiment involving Starkey and the Galenists later emerged with the appearance of the Great Plague in 1665. The Plague ravaged the poorest inhabitants of London, destroying the lives of tens of thousands of inhabitants, and lacking an effective treatment to combat the disease and fearing for their own well-being, the majority of the College of Physicians abandoned the city *en masse*,[91] but Starkey courageously remained behind to care for the afflicted. He died after succumbing to the disease during one of the most fatal weeks of the epidemic.

According to the story that circulated among respected physicians in Europe and New England following his death, Starkey was the only physician in London who was able to cure the dreaded Plague, and it was the improper administration of the medication by his attendants that caused him to die almost immediately, taking with him the knowledge of the contents of his secret remedy.[92]

It was the opinion of George Thomson, a physician and close friend who was twice cured of the Plague by Starkey, the second time that same day, that Starkey's excessive drinking beforehand

[89] Ibid., p. b4^{r}.
[90] Ibid., p. A4^{r-v}.
[91] George Starkey, *An Epistolar Discourse...*, London, Printed by *R. Wood*, for *Edward Thomas*, at the *Adam* and *Eve* in *Little Brittain*, 1665, p. 56; Walter George Bell, *The Great Plague in London in 1665*, Second Edition, London, Bodley Head, 1951, pp. 62-63.
[92] Sibley, *Graduates of Harvard*, vol. I, p. 134.

caused the failure of the medicine.[93] Thomson writes, "[Dr. Starkey] had an extraordinary Gift bestowed on him of Curing others in far worse condition...that being very sensible of the impiety, hypocrisie, dishonesty, the imposture, subtile Frauds, disrespect of Real worth, odious Ingratitude, and other notorious Crimes of the Times, he was willing to resign himself to death."[94]

However, Philalethes did not depart this life with Starkey that fateful day but continued to exist as the mysterious Cosmopolitan, so legendary that through the following decades many claimed to have encountered Starkey's famous phantom *persona* in Europe and New England, and as late as the 1740s there was an assertion that the aged "American philosopher" was the functioning president of the Hermetic Society in Europe.[95]

Yet like the shameful treatment of Paracelsus and van Helmont, the unrestrained criticism and political attacks on Starkey would continue long after his death, at times by those who praised his eminent *Doppelgänger*, the great "American philosopher" Eirenaeus Philalethes.[96] Even Starkey's friend Robert Boyle praised Philalethes but rebuked the famous alchemist for his secrecy,[97] and after hearing a rumor that the Adept had been imprisoned in France, Boyle attempted to contact the American celebrity,[98] who remained America's most widely read and respected scientist until Benjamin Franklin nearly a century later.[99]

[93] George Thomson, *Loimotomia: Or The Pest Anatomized*, London, Printed for *Nath: Crouch*, at the *Rose* and *Crown* in *Exchange*-Alley near *Lombard-street*, 1666, pp. 3, 83, 86, 100.

[94] Ibid., pp. 100–101.

[95] Johann Heinrich Cohausen, *Hermippus Redivivus...*, E. Goldsmid (ed.), Edinburgh, Privately Printed, 1885, pp. 115–116.

[96] Cohausen calls Starkey "a vicious and extravagant Man," while Philalethes was "a Man of remarkable Piety, and of unstained Morals." Ibid., p. 116. According to Borrichius, Starkey was "a liar and given to wine," and Philalethes was "candid and frankly articulate." [Olaus Borrichius], *Olai Borrichii Itinerarium 1660–1665*, H. D. Schepelern (ed.), Copenhagen, Danish Society of Language and Literature, 1983, vol. III, p. 22; Borrichius, "Conspectus Scriptorum Chemicorum Celebriorum," *Bibliotheca Chemica Curiosa*, vol. I, p. 50.

[97] Royal Society Library, Boyle Papers, vol. XIX, ff. 187ᵛ–188ʳ; in [Robert Boyle], *Letters and Papers of Robert Boyle: From the Archives of the Royal Society*, M. Hunter *et al.* (eds.), Bethesda, Md., University Publications of America, 1990, [microform], (hereafter as *Letters and Papers*).

[98] Johann Michael Faustius (ed.), *Philaletha illustratus, sive Introitus apertus...*, Francofurti ad Moenum, Sumpt...Andreae, 1706, preface, [c2-3].

[99] Jantz, "America's First Cosmopolitan," p. 3; William R. Newman, *Gehennical Fire: The Lives of George Starkey, an American Alchemist in the Scientific Revolution*, Cambridge, Harvard University Press, 1994, p. 2.

Chapter 3. The Chymical Physicians Versus the Aristotelians

George Starkey was a Helmontian physician, a devout follower of the Flemish alchemist Jean Baptiste van Helmont,[100] whose collected writings were published posthumously in 1648 as the *Ortus Medicinæ* or the "Origins of Medicine."[101] Van Helmont was the most highly

[100] "Imagination and magnetism are linked in the work of Paracelsus...to whom 'The whole of heaven is nothing other than *imaginatio* influencing man, producing plagues, colds, and other diseases'...The magnet is a metaphor for the imagination...In the late sixteenth and early seventeenth centuries, after Paracelsus, magnetism became a central topic of natural philosophy, experimental science, and medical theory. The influence of the prominent authors Gilbert, Kepler, and Kircher on science and medicine until the eighteenth century can hardly be overestimated. William Gilbert (1544–1603) published his studies on magnetic phenomena under the title *De Magnete*...He was the first to distinguish 'electrics' from 'magnetics,' that is, the attraction caused by the amber effect from that caused by a lodestone...Magnetism turned out to be the world soul, and the magnetic force appeared to be a psychic force: it 'is animate or imitates a soul; in many respects in surpasses the human soul, while that is united to an organic body'...Kepler suggested replacing the word 'soul' by 'force,' that is, the force of magnetism... [Kircher] identified light with the 'attracting magnets of all things': it is connected ultimately with the heavens and works like the lodestone...[Van Helmont] considered the magnet to be a powerful instrument for magnetic (so-called sympathetic) cures...Van Helmont states that man also has a magnet, by which the plague poison is attracted from an infected person. The magnet thus pulls death into the body. But there is an antimagnet preventing the infection [*i.e.* the Philosophers' Stone]...van Helmont's elaborate theory of imagination as a 'magnetic' or 'sympathetic' power constructs a dynamic model of the disease process (and its cure) that includes physiology, pathology, psychopathology, and psychosomatic medicine, in terms of modern medicine." Heinz Schott, "Paracelsus and van Helmont on Imagination: Magnetism and Medicine before Mesmer," *Paracelsian Moments: Science, Medicine, & Astrology in Early Modern Europe*, G. S. Williams and C. D. Gunnoe, Jr. (eds.), Kirksville, Mo., Truman State University Press, 2002, pp. 137–145.

[101] "Van Helmont is consistent and definite in that disease is an affair of the *Archeus*. It is generated by it, takes place in it, and its outcome depends on it. For there is nothing that can happen in the organism—healthy or sick—without the *Archeus*. By *Archeus* Van Helmont understands the 'vital principle.' This governs the whole of the organism...The *Archeus* is a vital and spiritual force. It is not 'soul,' however. For it is not independent and does enter the body from outside in order to govern it. The *Archeus* is soul-like, psychoid. It is the psychic aspect of a specific living unit,

acclaimed physician of the seventeenth century, claiming intimate knowledge of the Philosophers' Stone and Alkahest that cured all disease.[102] Considering himself the "Reviver of Hippocrates"[103] and a "Philosopher by the Fire," van Helmont states in his "Modern Pharmacapolium and Dispensatory" that "Vegetables do only lay aside their juice and muscilage, by boyling in waters,"[104] and writes:

an individual. As such it is inseparably interlocked with its material and bodily aspect...Matter in itself is, according to Van Helmont, simply water, inert and indeterminate throughout. It is only through close interlocking with a disposing agent—the *Archeus*—that it becomes an object in nature. He believed that he had made evident the *Archeus* when he deprived an object of its course material cover 'through fire' and obtained a smoke with properties specific to the object. In this he saw its purest form, its divine kernel. He called the smoke with the 'new term' *Gas*, as different from such general media as air and water-vapour. This is indeed the gas of our modern textbooks of chemistry and Van Helmont is rightly remembered as its discoverer. To him, however, *Gas* meant much more—it was the *Archeus* that vitalized all objects, notably those of organic nature. Thus Van Helmont says: 'Archeus, qui est Gas spirituale'...In fact there are *Archei* not only in man and animals, but also in plants and minerals and indeed in any part of matter than functions even in the most primitive way. Up to Van Helmont these had been lumped together under such terms as 'occult qualities'...a certain attitude of the *Archeus* toward a certain, specific agent is required to produce certain effects. For example none but a certain emotional condition can produce plague...Nevertheless, even in the case of plague, it is a specific object (poison, virus) of which an image is formed and the reception of which is imagined...It does not act like a sword that indiscriminately wounds everybody hit. Plague becomes real...when its idea and image have been brought forth...In Van Helmont's spiritualist view each disease emerges as a specific *poison*." Walter Pagel, "Van Helmont's Concept of Disease—To Be or not to Be? The Influence of Paracelsus," *Bulletin of the History of Medicine*, 46 (1972), pp. 421-426.

[102] James Riddick Partington, *A History of Chemistry*, New York, St. Martin's Press, 1961, vol. II, pp. 209-242.

[103] "There is no area in which van Helmont's inspiration by and dependence upon Paracelsus is as evident as his ontological theory of disease." Walter Pagel, *Joan Baptista Van Helmont: Reformer of science and medicine*, Cambridge, Cambridge University Press, 1982, p. 149.

[104] "There seems little doubt that the accounts of *aer* and of *aither* found in the Ionian philosophers of the sixth century B.C. derived partly from a commonplace picture of the world already reflected in the poetry of Homer...Between the earth and the sky is *aer*, misty air. Beyond this *aer* is...the *aither*...it is akin to men's souls...Aristophanes has Euripides, in the *Frogs*, pray to *aither* as to a god...the status of *aither* remained very much unsettled in pre-Socratic science...The first theory of *aither* as such is probably not to be found until Aristotle...Although he himself usually called it 'the first body', Aristotle's new element was to be identified by later writers as *quinta essentia*, the fifth essence...*aither* has a representative, an analogue here on earth: *pneuma*. Not soul as such, but the instrument of animation in plants, animals, and men, *pneuma* is breath and is spirit...The Stoics integrated *aither* and *pneuma* even more fully...The medieval centuries saw deployment of the Greek ideas of *aither* and *pneuma* whenever...doctrines of spirit or of heaven were being given a physical interpretation...On Paracelsus's view, spirit of various kinds is the true object of the chemical and medical arts. For all minerals, plants, and animals embody and so imprison spirits ultimately of divine origin and celestial nature. From these hidden, invisible spirits they derive the secret powers elicited, concentrated, and coordinated by the scientific adept whose skill and wisdom arise likewise from heavenly powers implanted with the soul in his own body. The various metaphors and ancient teachings invoked by Paracelsans hardly cohere well enough to make a single, definite theory of *spiritus*; its constitution and relation to bodies, especially, remain obscure, not to say mysterious. For, it seems, no one spirit is common to all bodies, there being, rather, as many as distinct kinds of bodies. In any case the full powers of a body's spirit lie dormant until roused by the right agent. In the alchemists' work, fire is the great awakener; it can purify bodies by liberating their essences, their spirits, because it is itself a ubiquitous embodiment of spirit in nature. There are strong echoes here of the material Stoic *pneuma* and of the immaterial neo-Platonic *anima mundi*." G. N. Cantor and M. J. S. Hodge, "Introduction: Major themes in the development of ether theories from the ancients to 1900,"

> I praise my bountiful God, who hath called me into the Art of the fire [*Pyrotechnia*], out of the dregs of other professions. For truly, Chymistry, hath its principles not gotten by discourses, but those which are known by nature, and evident by the fire : and it prepares the understanding to pierce the secrets of nature, and causeth a further searching out in nature, than all other Sciences being put together : and it pierceth even unto the utmost depths of real truth...[105]

Van Helmont recounts his life and academic record in an introductory treatise to his "Origins of Medicine."[106]

> At the age of seventeen I had finished my course in philosophy, and it was then that I noticed that nobody was admitted to the examination who was not masked in his gown and hood, as if the robes warrant scholarship. The professors made a laughing stock of the academic youth that was to be introduced to the arts and learning, and I could not help wondering at the sort of delirium in the behaviour of professors, nay of everybody as much as the simplicity of the credulous youth. I retired into a deliberation in order to judge myself how much I was of a philosopher and had attained truth and science. I found myself inflated with letters and, as it were, naked as after partaking of the illicit apple—except for a proficiency in artificial wrangling. Then it dawned upon me that I knew nothing and that what I knew was worthless. I did astronomy, logic and algebra for pleasure, as the other subjects nauseated me, and also the Elements of Euclid, which became particularly congenial to me as they contained the truth...But I learned only vain eccentricities and a new revolution of the celestial bodies [Copernican theory], and what seemed hardly worth the time and labours I had spent...Having completed my course, I refused the title of Master of Arts since I knew nothing substantial, nothing true; unwilling to have myself made an arch-fool by the professors declaring me a Master of the Seven Arts, I who was not even a disciple yet.

Conceptions of ether: Studies in the history of ether theories 1740–1900, G. N. Cantor and M. J. S. Hodge (eds.), Cambridge, Cambridge University Press, 1981, pp. 3–10.

[105] Jean Baptiste van Helmont, "A modern Pharmacapolium and Dispensatory," *Oriatrike or, Physick Refined...*, J. Chandler (trans.), London, Printed for Lodowick Loyd, 1662, p. 462.

[106] "Having acquired a medical degree in 1599 at Louvain, frustrated and disgusted with the sham of academic life, Van Helmont embarked on the grand tour of Europe. He visited Switzerland and Italy between 1600 and 1602, and France and England between 1602 and 1605. He mentions a visit to London in 1604 as well as a previous occasion, possibly late in 1602, when he joined the Court at the Palace of Whitehall 'in the presence of the Queen herself'. Nothing he saw or learnt on his travels alleviated his frustration; he found only laziness, ignorance, and deceit. A period of practising medicine during a plague epidemic at Antwerp in 1605 made him even more conscious that useful knowledge and truth had eluded him. When offered a rich canonry he had refused to live on the sins of his fellow-men; he now declined to practise medicine, unwilling to grow rich on their misery. Nor would he accept such alluring calls as were extended to him by Ernest of Bavaria, Archbishop of Cologne, or the Emperor Rudolph II...Concentrating on chemical research did not prevent Van Helmont from following his own principles by ceaselessly curing the sick and devising and dispensing his own medicines free of charge." Pagel, *Joan Baptista Van Helmont*, pp. 6-7.

Seeking truth and science, though not their outward appearance, I withdrew from the university. I was promised a wealthy canonry on condition that I would only devote myself to theology; but St. Bernard warned me against living on the sins of the people. I prayed, however, to the Lord that he might make me worthy to receive my vocation as would please him best. This was the year in which the Jesuits began to teach philosophy at Louvain...One of their professors, Martinus del Rio...expounded his Disquisitions on Magic. I attended both courses with great zeal. But what I reaped in the end was empty straw and poor senseless prattle. In the meantime, in order to let no hour pass in vain, I became engrossed in Seneca, who pleased me immensely, and so even more did Epictetus. I thus seemed to have found the marrow of truth in moral philosophy...I prayed to the Prince of Life for the stamina to contemplate the naked truth and love it *per se*—a desire that was intensified by Thomas à Kempis and later by Tauler. Then I ardently hoped to attain Christian perfection by following Stoicism. Some time later, being fatigued by these exercises, I fell into a dream in which I saw myself as an empty bubble extending from earth to heaven. Above me hovered a tomb; below, however, in the place of earth was an abyss of darkness. Immensely frightened I lost consciousness of everything, including myself. When I recovered, I realized at once, that we live, move and exist in Christ Jesus...I became certain that, except by special grace, nothing but sin awaits us in any activity. Then I recognized that my Stoicism kept me an empty and inflated bubble, between the abyss below and the necessity of imminent death. It struck me that my studies were making me arrogant, though affecting to be moderate. Confident of the freedom of my will, I would readily forego divine grace, as if we are the arbiters of our destiny. Away with such monstrous blasphemy, I said. I renounced Stoic philosophy as hateful and beneath the dignity of a Christian, and tired and disgusted with too much reading, I amused myself by browsing in Mathiolus and Dioscorides—in the belief that nothing is so necessary to mankind as God's grace, admiring how in the plants it provides their appropriate needs and ripens their fruit. I soon noticed that botany has in no way progressed since the days of Dioscorides; that even to-day his pictures, nomenclature and descriptions of plants are after all these years the recognized basis of every discussion, but that no advance had been added, about their virtues, properties and use, except for the fictions of the later authors as to the grades of elemental qualities to which the whole composition of the plant was ascribed. But I knew that about two hundred plants, though identical in quality and grade, are quite different in virtues, and that a number of others, different in quality and grades, act synergically. So not the herbs (the seals of divine love), but the herbalists fell into disrepute with me. I enquired whether there was a text-book of the axioms and rules of medicine, for I believed that

> this could be taught like any art or science and was not merely a gift of grace...I was told that all particulars of the effects of plants, from the cedar to the hyssop, were to be found in Galen or Avicenna. But as I was not credulous nor found the desired certainty in the books, I well nigh suspected that it lay in the truth that He who created medicine remained its continuous distributor. Anxious and uncertain about which profession I should choose, I studied the customs and laws of nations, and the decisions of sovereigns. I saw that law is made up of human traditions, uncertain, unstable and devoid of truth. As there was no stability in human things and no core of certainty, I saw that my life would be useless were I to base it on human decisions...On the other side I was impressed by the misery of human life and the will of God, whereby everybody must sustain himself as long as he can. With a singular avidity I turned to the most gentle study of nature, and as the soul is enslaved by its inclinations I lapsed unconsciously into natural science. I read the institutions of Fuchs and Fernal which led me through the whole of medicine by way of a survey, and found myself amused. Is this the way medicine is taught—without theory, and a teacher who has received the gift of healing from a master? Is it not in this way that the whole of natural history is obstructed by qualities and elements? I therefore, read Galen's works twice, once Hippocrates, whose aphorisms I knew almost by heart, the whole of Avicenna and about a total of six hundred Greek, Arabic and modern authors, seriously and attentively, and compared and abstracted them. Then I re-read the collection of my notes and recognized my poverty, and the labours and years I had consumed angered me.[107]

Van Helmont rejected the logical-mathematical methods of the Aristotelians and embraced the Study of Nature, which he expounds upon in his treatise entitled "The ignorant Natural Phylosophy of *Aristotle* and *Galen*":

> [L]et them come to the Study of Nature, let them learn to know and seperate the first Beginnings of Bodies...Let the History of extractions, dividings, conjoynings, ripenesses, promotions, hinderances, consequences, lastly, of losse and profit, be added. Let them also be taught, the Beginnings of Seeds, Ferments, Spirits, and Tinctures, with every flowing, digesting, changing, motion, and disturbance of things to be altered. And all those things, not indeed by a naked description of discourse, but by handicraft demonstration of the fire. For truly, nature measureth her works by distilling, moystening, drying, calcining,

[107] Jean Baptiste van Helmont, "Studia Authoris," *Ortus Medicinæ...*, Amsterodami, Apud Ludovicum Elzevirium, 1648, pp. 16–18; in Walter Pagel, "The Reaction to Aristotle in Seventeenth-Century Biological Thought," *Science Medicine and History: Essays on the Evolution of Scientific Thought and Medical Practice written in honour of Charles Singer*, Volumes I-II, E. A. Underwood (ed.), London, Oxford University Press, 1953, vol. I, pp. 491-492.

> resolving, plainly by the same meanes, whereby glasses do accomplish those same operations. And so the Artificer, by changing the operations of nature, obtains the properties and knowledge of the same. For however natural a wit, and sharpness of judgement the Philosopher may have, yet he is never admitted to the Root, or radical knowledge of natural things, without the fire. And so every one is deluded with a thousand thoughts or doubts, the which he unfoldeth not to himself, but by the help of the fire. Therefore I confess, nothing doth more fully bring a man that is greedy of knowing, to the knowledge of all things knowable, than the fire. Therefore a young man at length, returning out of those Schooles, truly it is a wonder to see, how much he shall ascend above the Phylosophers of the University, and the vain reasoning of the Schooles.[108]

Van Helmont pronounced as ridiculous the use of metals and precious stones in medicine, and writes in his "Treatise of Fevers": "*Physitians* also, are wont to brag of their exhilarating Cordials, and restoring remedies prepared of Gold, and gems or precious stones, surely from a like stupidity with the rest."[109] Van Helmont reiterated this view in the "Modern Pharmacapolium and Dispensatory":

> Wherefore as I do (in general) pity the Compositions and Corrections of the Shops ; so I do as yet more detest the precipitatings, glassyfyings, and preparations of Mercury, Antimony, Tuttie, Sulphur, &c. And likewise, the adulterations of Spirits out of Spices, hot Seeds, Vitriol, Sulphur, &c. For they are prepared for gain, by our fugitive servants, and purchased by the Shops, rather to the disgrace of the Art of the Fire, than for the defect of the sick. I likewise bewail the shameful simplicity of those, who give men, leaf-gold, and bruised or powdered precious stones, to drink, with great hope, selling their ignorance, if not deceit, at a great rate...And therefore the more subtile error of those, is more to be bewailed, who corrode Gold, Silver, Corals, Pearls, and the like, by sharp liquors, and seem to dissolve them, and think that by this means, they are to be admitted within the veins...[110]

Van Helmont also challenged the Galenists to a clinical trial to determine the efficacy of their methods, which they avoided for obvious reasons:

> Oh ye Schooles...Let us take out of the Hospitals, out of the Camps, or from elsewhere, 200, or 500 poor People, that have Fevers, Pleurisies, &c. Let us divide them into halfes, let us cast lots, that one halfe of them may fall to my share, and the other to yours. I will cure them

[108] Helmont, "The ignorant Natural Phylosophy of *Aristotle* and *Galen*," *Oriatrike*, p. 45.
[109] Helmont, "A Treatise of *Fevers*," Ibid., pp. 970-971.
[110] Helmont, "Pharmacapolium," Ibid., pp. 466-467.

> without blood-letting and sensible evacuation...we shall see how many Funerals both of us shall have: *But* let the reward of the contention or wager, be 300 Florens, deposited on both sides: Here your business is decided.[111]

Jean Baptiste van Helmont was a Paracelsian,[112] a disciple of the Swiss-German physician Theophrast Bombast von Hohenheim, also known as Paracelsus,[113] the famous alchemist regarded as the "Father of Modern Chemistry"[114] who is gradually becoming credited by

---

[111] Helmont, "The Author Answers," Ibid., p. 526.

[112] "[Van Helmont], a scholar and a wealthy man, lived in Vilvoorde, near Brussels. In contrast to Paracelsus, who traveled restlessly all his life throughout Europe, van Helmont stayed at home and worked continuously in his chemical laboratory. The most important figure in the Paracelsian movement, van Helmont was in fact the founder of the so-called chemical philosophy. He was a critical follower of Paracelsus, principally sharing his alchemical approach and his religious attitude as a philosopher and doctor. However, van Helmont rejected the medical astrology of Paracelsus, and his analogy of microcosm and macrocosm." Schott, "Paracelsus & van Helmont," p. 142.

[113] "Chief among Paracelsus' contribution to medical theory was his new concept of disease. He demolished the ancients' notion of disease as an upset of humoral balance...Paracelsus completely reversed this concept, emphasizing the external cause of a disease...He considered each of these agents to be a real *ens*, a substance in its own right...He thus interpreted disease itself as an *ens*, determined by a specific agent foreign to the body, which take possession of one of its parts, imposing its own rules on form and function and thereby threatening life. This is the parasitistic or ontological concept of disease—and essentially the modern one. It was substantially elaborated by Helmont. The significance of specific disease-semina, its connection with imagination, ideas, and passions, and the bodily manifestation of spiritual impulses, as inculcated by Helmont, is clearly anticipated in Paracelsus' concept of disease...Paracelsus' general natural philosophy is spiritualist. The important forces in nature are the invisible 'spirits,' such as the *quintae essentiae*; these are the life substances of objects, and the *magus* may know how to extract them, particularly from herbs and chemicals. These 'spirits'...[are] *arcana* and primordial seeds (*semina*) that emanate immediately from God and direct and inform nature. Each contains its own *archeus*...Of the Paracelsian principles...Sulfur is also called the soul, and forms the link between the spirit (*geist*, mercury) and the solid body (salt)...Paracelsus was a great doctor and an able chemist...Paracelsus is basically consistent in theory and practice. A number of apparent contradictions in his work disappear when they are considered in their proper context, while others are clearly the result of developmental changes in his life and general outlook. What is in the end most remarkable in Paracelsus' work is that he achieved real advances in chemistry and medicine through the revival and original development of lore that had been kept alive only at a very low level (or had, indeed, been suppressed as heresy). This lore—alchemy, astrology, and the 'prohibited arts'—can be traced to Hellenistic and Oriental Neoplatonism, gnosticism, and syncretism; in Paracelsus' hands it became, if not scientific, at least protoscientific. It is difficult to overrate the effect of Paracelsus' achievement on the development of medicine and chemistry" Water Pagel, "Paracelsus, Theophrastus Philippus Aereolus Bombastus Von Hohenheim," *Dictionary of Scientific Biography*, vol. X, pp. 307-311.

[114] "[A]t least from the time of his brief official appointment in Basle in 1527 he was sarcastically labeled as the Luther of medicine. Paracelsus occasionally mentioned Luther, and he seems to have taken seriously the parallel between their roles...Paracelsianism constituted one of the vital ingredients of the so-called Scientific Revolution...The shorter religious, social and ethical writings of Paracelsus involved frequent reference to medicine...The prevalent anticlericalism of these prospective pamphlets is well-illustrated by the *De septem punctis idolatriae Christianae*. This attack on the corruption of the clerical hierarchy provided a model for his assault on academic medicine and the medical profession. Paracelsus painted a portrait of a Church almost totally overtaken by idolatry and seemingly vanquished by Satan...Especially in *De felici liberalitate*... The papacy was attacked for embodying the institution of corruption...Paracelsus aimed to replace the prevalent 'theory' of medicine derived from the ancients with a new theory, which

scholars in the history of science as the figure responsible for initiating the Scientific Revolution.[115]

Paracelsus[116] openly rejected the medical treatments of the Galenist orthodoxy in the University,[117] and at the instigation of the medical faculty of Leipzig the printing of his manuscripts on the "French Disease" was forbidden by the city council of Nürnberg in 1530.[118] In 1566,[119] the powerful Parliament and Faculty of Medicine in Paris

---

he also called the 'religion' of medicine. Galenic medicine is dismissed as a redundant scholastic exercise...The 'highest religion' of medicine required the intensive investigation of stones, roots, plants and seeds, in order to reveal their powers...Paracelsus acknowledged having learned from more than 80 peasants how to identify the signatures of plants and perform remarkable cures with them." Charles Webster, "Paracelsus: medicine as popular protest," *Medicine and the Reformation*, O. P. Grell and A. Cunningham (eds.), London, Routledge, 1993, pp. 57-70.

[115] Partington, *History of Chemistry*, vol. II, pp. 113–151; Allen G. Debus, *Man and Nature in the Renaissance*, Cambridge, Cambridge University Press, 1978, p. 15; Debus, *Paracelsus, Five Hundred Years: Three American Exhibits*, Bethesda, Md., Published by the Friends of the National Library of Medicine..., 1993, pp. 3–12; Debus, "Chemists, Physicians, and Changing Perspectives on the Scientific Revolution," *Isis*, 89 (1998), pp. 66-81.

[116] "In addition to what one might call the low medical culture of popular handbooks and plague tracts, the writing of Paracelsus reflects the high medical culture of Italian Renaissance medicine as represented by Marsilio Ficino (1433–1499)—a name...that drew a rare word of praise. Before Paracelsus, Ficino had addressed the cause of epidemic pestilence, explaining its spread by a supernatural power of the imagination...Paracelsus adopted this theory from the Ficino he admired as the 'best of Italian Physicians.'" Andrew Weeks, *Paracelsus: Speculative Theory and the Crisis of the Early Reformation*, Albany, State University of New York Press, 1997, pp. 57, 71; Cf. Walter Pagel, *Paracelsus: An Introduction to Philosophical Medicine in the Era of the Renaissance*, Basel, S. Karger, 1958, pp. 172–182.

[117] "Paracelsus' whole life and work seems to be an attempt at implementing the ideal of Ficino's priest-physician...It is from Ficino as the exponent of Neo-Platonism that Paracelsus derives his inspiration." Pagel, *Paracelsus*, p. 223. "Marsilio Ficino and Agrippa of Nettesheym. Both these can be safely regarded as fixed points in the literary sources through which neo-Platonic and magic tradition came to Paracelsus. There are, however, other probable sources of inspiration, notably Trithemius (1462–1516), one of the learned ecclesiastic teachers of Paracelsus...Paracelsus is rightly regarded as an exponent of the Renaissance...The influence of mediaeval neo-Platonism may have reached Paracelsus in various ways. He knew Arnald of Villanova and Hildegard of Bermersheim...He also knew the alchemical Lullists and John of Rupescissa." Walter Pagel, "Paracelsus and the Neoplatonic and Gnostic Tradition," *Ambix*, 8 (1960), pp. 158–159.

[118] Lynn Thorndike, *A History of Magic and Experimental Science*, Volumes I-VIII, New York, Columbia University Press, 1923–1958, vol. V, p. 380; Pagel, *Paracelsus*, p. 24.

[119] "[Paracelsus's] importance in the history of medicine is due to the fact that he undermined Galen's authority, an authority which had been valid for over a thousand years. It took a revolutionary character to question and sweep aside the burden of century-old scholastic teachings...He spent his whole life wandering from town to town...always with the news of his healing success having arrived before him. Over and over again, he was in trouble with the authorities and other local people of rank, and once again he was forced to flee in order to escape imprisonment or even punishment by death...Heavy attempts to suppress the science of Paracelsus took place...In 1530, the city of Nuremberg would not permit Paracelsus' work to go to print. In Paris in 1566, parliament prohibited the use of his medicines. By 1599, all his written books were listed in the Catholic church's index for prohibited books. 1603 saw the prohibition of his medicine by the faculty of medicine in Paris, and anyone ignoring the decree lost his licence to practice. The ban was not lifted until 1666. Graduates of Heidelberg's medical faculty were made to swear on oath that they would never administer Paracelsian medicines and it was not until 1655, when medical students declared that they were not prepared to graduate at Heidelberg under these conditions, that this passage was omitted from the doctors' oath...the spreading of the teachings of Paracelsus...by Alexander von Suchten...[who] was merciless in his

condemned the "dangerous innovations" of Paracelsus regarding the use of "antimony,"[120] and in 1615 this body issued a decree forbidding the sale of chemical medicines in the Kingdom of France.[121]

Along with the deep philosophical rift between the Galenists and Chemical Physicians,[122] this conflict also possessed sectarian religious overtones.[123] Martin Luther embraced the practice of alchemy,[124] and in

criticism of the traditional medicine based on book-learning. He declared that the most famous universities turned out nothing but prattlers and swindlers. It is probable that his indignant colleagues were responsible for von Suchten losing his post as a personal physician to the king of Poland. He was accused of being a heretic and atheist and two of his adversaries demanded nothing less than that he be subjected to torture." Herbert Breger, "*Elias Artista*—A Precursor of the Messiah in Natural Science," *Nineteen Eighty-Four: Science Between Utopia and Dystopia*, E. Mendelsohn and H. Nowotny (eds.), Dordrecht, D. Reidel Publishing Company, 1984, pp. 55-56.

[120] The use of terms like *antimony*, *mercury*, *sulphur*, and *zinc* by Paracelsus does *not* constitute evidence that he grasped anything about the properties of the substances commonly known by these names. Paracelsus also attacked the "impostures" that misused his alchemical remedies. W. P. D. Wightman, *Science in a Renaissance Society*, London, Hutchinson University Library, 1972, p. 41; Weeks, *Paracelsus*, pp. 134–135.

[121] Allen G. Debus, "Paracelsianism and the Diffusion of the Chemical Philosophy in Early Modern Europe," *Paracelsus: The Man and his Reputation, his Ideas and their Transformation*, O. P. Grell (ed.), Leiden, Brill, 1998, pp. 232-234.

[122] "[I]n the German Renaissance...[Agrippa] assailed alchemists...Agrippa denounced academic knowledge as vain and fruitless in comparison with the magical arts...Paracelsus ('Cacophrastus' and the 'Lutherus medicorum')...held one must unite theory and practice, put down books and dirty one's hands. Traditional learning is full of shit, and 'shit [*Dreck*] is its best part'...Paracelsus adopted as positive metaphor that form of manual labour perhaps most abhorrent, by its proximity, to academics: cooking." William Clark, "The Scientific Revolution in the German Nations," *The Scientific Revolution in National Context*, R. Porter and M. Teich (eds.), Cambridge, Cambridge University Press, 1992, pp. 94-95. The author's contention that "alchemy had anti-Christian and anti-social overtones" is distorted and inaccurate.

[123] "Writing in 1538 Paracelsus stated 'I do not here write out of speculation, and theorie, but practically out of the light of Nature, and experience, lest I should burden you and make you weary with many words.' The Paracelsian chemical physician is a man who is not afraid to work with his own hands. He is a pious man who praises God in his work and who lays aside all those vanities of his Galenist competitor and instead finds his delight in a knowledge of the fire while he learns the degrees of the science of alchemy...[In] the *Labyrinthus Medicorum* of Paracelsus (1538) we find no written books discussed at all. Instead we are told to seek out God through nature...As Descartes and Bacon were appalled by the universities, so too were these chemical philosophers. The teaching of Paracelsus at the University of Basel in 1527 had been an unhappy experience remembered largely for his conflict with the students and faculty alike as well as his insistence on lecturing in Swiss-German and his burning of the *Canon* of Avicenna. The Paracelsian disillusionment with the schools is reflected throughout the sixteenth and seventeenth centuries...Aristotelians and Paracelsians alike believed that heat would reduce a substance to its elements. Chemistry then seemed to be a basic method of getting to fundamentals in nature." Allen G. Debus, *The Chemical Dream of the Renaissance*, Cambridge, W. Heffer & Sons Ltd., 1968, pp. 11–12, 15.

[124] "The science of alchymy I like very well, and indeed, 'tis the philosophy of the ancients. I like it not only for the profits it brings in melting metals, in decocting, preparing, extracting, and distilling herbs, roots; I like it also for the sake of the allegory and secret significance, which is exceedingly fine, touching the resurrection of the dead at the last day. For, as in a furnace the fire extracts and separates from a substance the other portions, and carries upward the spirit, the life, the sap, the strength, while the unclean matter, the dregs, remain at the bottom, like a dead worthless carcass; even so God at the day of judgment, will separate all things through fire, the righteous from the ungodly." Martin Luther, *Table Talk*, 805; in Stanton J. Linden, "Alchemy and Eschatology in Seventeenth-Century Poetry," *Ambix*, 31 (1984), p. 102; Cf. Martin Luther, *Colloquia Mensalia*, H. Bell (trans.), London, Printed by *William Du Gard*..., 1652, p. 480.

Continental Europe most of the Chemical Physicians were Protestant, while the Galenists in the University were generally Roman Catholic, and the Catholic Church also embraced and violently defended Aristotelian philosophy.[125] In England, an intellectual backwater in the early seventeenth century,[126] the Cambridge Platonist movement emerged in the mid-1620s[127] yet there existed no English translations of any genuine works by Paracelsus before 1640 and his impact before the Puritan Revolution was very much less marked than on the Continent, after which Paracelsus became a dominant influence in English medicine.[128]

Van Helmont also suffered severe mistreatment which even surpassed that endured by Paracelsus.[129] He was persecuted by his enemies in the University and the Catholic Church after his work *On the Magnetic Cure of Wounds*[130] was submitted to the tribunal of

[125] Debus, "Paracelsianism and Chemical Philosophy," p. 233.

[126] John Henry, "The Scientific Revolution in England," *The Scientific Revolution in National Context*, p. 178.

[127] "Even before visiting England, he was probably inspired by the philosophical works of Francis Bacon, most of which were published during Hartlib's formative years. These writings presented the vista of the dramatic potentialities of empirical knowledge when organized efficiently and adequately patronised. Hartlib's first English visit, from 1625 to 1626, to complete his studies at Cambridge, must have convinced him that the puritan movement offered an ideal combination of patronage, religious enlightenment and sympathy with the new learning. Cambridge in 1626 was at the height of its influence as the focal point of English puritanism, and was witnessing the genesis of the Cambridge Platonist movement. The celebrated preacher, Richard Sibbes, had newly returned as Master of St Catharine's Hall and perpetual curate of Trinity parish. John Preston was at the end of his famous career as Master of Emmanuel College. His wealthy supporters, the Fiennes, Rich and Greville families, became the patrons of Hartlib; his pupils—the 'Spiritual Brotherhood'—became Hartlib's collaborators. Joseph Mede, one of Hartlib's early correspondents and supporter of Bacon, was at Christ's College completing his influential analysis of the apocalypse, *Clavis apocalyptica* (1627). John Milton and Henry More, students of Christ's, inherited Mede's erudition; they and other Cambridge Platonists were supporters of the Hartlib circle." Charles Webster (ed.), *Samuel Hartlib and the Advancement of Learning*, Cambridge, Cambridge University Press, 1970, pp. 6-7.

[128] Charles Webster, *The Great Instauration. Science, Medicine and Reform 1626–1660*, New York, Holmes & Meier Publishers, 1975, pp. 273-288; Webster, "Alchemical and Paracelsian medicine," *Health, medicine and mortality in the sixteenth century*, C. Webster (ed.), Cambridge, Cambridge University Press, 1979, p. 317.

[129] "Unfortunately, the enemies Van Helmont had made among his medical colleagues found strong allies in Jesuit circles and, through them, among the ecclesiastical authorities. Already in 1621 his 'magical' tract on the *Magnetic Cure of Wounds* had been published in Paris, without, as he asserted, his knowledge and with malicious intent. Religious prosecution followed, lasting for some twenty years...in 1623 members of the Louvain Medical Faculty denounced Van Helmont's tract as a 'monstrous pamphlet'. In 1625 the General Inquisition of Spain declared twenty-seven of its propositions as suspect of heresy, as impudently arrogant, and as affiliated to Lutheran and Calvinist doctrine." Pagel, *Joan Baptista Van Helmont*, pp. 8–12

[130] "[In] the 'magnetic cure of wounds', he developed theories which were the forerunners of modern immunological thinking." Pagel, "Reaction to Aristotle," p. 494. For the text, see Jean Baptiste van Helmont, *A Ternary of Paradoxies: The Magnetick Cure of Wounds...*, W. Charleton (trans.), London, Printed by *James Flesher* for *William Lee*..., 1650, pp. 1-93; Helmont, "De Magnetica Vulnerum Curatione," *Ortus Medicinæ*, pp. 746-780; *Oriatrike*, pp. 756-793.

Malines-Brussels and transferred to the Spanish Inquisition, which condemned and prohibited the writing as heresy and prohibited van Helmont from publishing for twenty years.[131] The faculty of the University of Reims also censured van Helmont, as did the theological faculties of Louvain and Cologne, the faculty of medicine of Louvain, and the College of Physicians of Lyon.[132] Van Helmont affirmed his innocence in 1627 and submitted to ecclesiastical discipline before the curia of Malines, which referred the matter to the Louvain Theological Faculty. Van Helmont repudiated his work in 1630. He was condemned by the Louvain Theological and Medical Faculties in 1633–1634 for "perverting nature by ascribing to it all magic and diabolic art and for having spread more than Cimmerian darkness all over the world by his chemical philosophy."[133] They proclaimed van Helmont guilty of circulating "monstrous, superstitious nonsense of the school of Paracelsus, that is, of the Devil himself."[134] Van Helmont was again arrested in 1634 and imprisoned by the Inquisition after repeated examinations of his views led to the accusation that he was a follower of Paracelsus and chemical medicine.[135] Along with van Helmont's writings, the works of Paracelsus and Michael Maier were also included in the Spanish *Index of Prohibited Books*,[136] and Paracelsus's name was placed in the "first class of authors of damned memory."[137]

Paracelsus repeatedly rejected the idea of making gold from base metals,[138] and in his medical practice he was often successful

[131] Thorndike, *History of Magic*, vol. VII, pp. 228-229.

[132] Ibid., vol. VII, p. 228; Allen G. Debus, *The Chemical Philosophy: Paracelsian Science and Medicine in the Sixteenth and Seventeenth Centuries*, Volumes I-II, New York, Science History Publications, 1977, vol. II, pp. 306-311; Debus, *Man and Nature*, p. 127.

[133] Pagel, *Joan Baptista Van Helmont*, p. 12.

[134] Walter Pagel, "Johann Baptist van Helmont," *Aufgang der Artzney-Kunst*, C. K. von Rosenroth (trans.), München, Kösel-Verlag, 1971, vol. II, Appendix XIII; Pagel, "Helmont, Johannes (Joan) Baptista Van," *Dictionary of Scientific Biography*, vol. VI, p. 254.

[135] Allen G. Debus, *Chemistry and Medical Debate: van Helmont to Boerhaave*, USA, Science History Publications, 2001, pp. 34-38.

[136] Franz Heinrich Reusch, *Der Index der Verbotenen Bücher*, Volumes I-II, Bonn, Verlag von Max Cohen & Sohn, 1883–1885, vol. I, p. 497, vol. II, pp. 178–179.

[137] Allen G. Debus, "Paracelsus and the Delayed Scientific Revolution in Spain: A Legacy of Philip II," *Reading the Book of Nature: The Other Side of the Scientific Revolution*, A. G. Debus and M. T. Walton (eds.), Kirksville, Mo., Truman State University, 1998, p. 149.

[138] Moritz of Hessen-Kassel possessed a large number of works by Paracelsus and was dedicated to the discovery of chemical medicines, assembling in his castle at Kassel the finest laboratory in Europe. Among his reforms at the University of Marburg was the appointment in 1609 of Johannes Hartmann to the first chair of *Chemiatria*, or chemical medicine. In what was originally spoken in his acceptance speech before the faculty and student body of the University of Marburg and later published, Hartmann said of Moritz: "He has given advice on the manner in which the other disciplines, including the school of medicine, are to be established according to a certain curriculum of lecture and practice wherein there is less room for thorny altercations

in producing cures in patients after other physicians had failed.[139] Paracelsus calls *alchimia* a pillar of medicine in his *Vom Terpentin*.[140] He distinguished this from the transmutation of metals, which he terms *alchimei*, using the German suffix -*ei* derisively to indicate the sense of *useless repetition*.[141] The Paracelsians followed German folk medicine and the Hippocratic homeopathic doctrine of curing by similitude,[142] and wholly rejected the dubious methods and dangerous medications employed by the Galenists.[143] The "Chymical Physicians" produced

concerning precepts than for work involved with the theory and practice of medicine...He wished this *Xustus* [*i.e.* the closed colonnade of a Greek gymnasium] to be opened...so that, beyond lectures and disputations, students of medicine may be taught not only about herbs and plants philosophically, but may proceed, in imitation of nature, out into the fields and gardens... And let them [he ordered] be well practiced both in dissection and in the hermetic philosophy, that is, the science of secrets of making good and useful medicines." Johannes Hartmann, *Opera Omnia Medico Chemica...*, C. Johrenio (ed.), Francofurti ad Maenum, Balthas. Christophori Wursti, 1684, vol. IV, p. 6; in Bruce T. Moran, *The Alchemical World of the German Court: Occult Philosophy and Chemical Medicine in the Circle of Moritz of Hessen (1572–1632)*, Stuttgart, Franz Steiner Verlag, 1991, pp. 59, 62-63, 116–118.

139 Ferguson, *Bibliographica Chemica*, vol. II, p. 171.

140 "[Paracelsus] says that 'the first was with God (*bei Gott*), the beginning, that is *ultima material*; this *ultima materia* He made into prime matter'...In another passage God, Prime Matter, Heaven and finally the soul (*Gemüt*) of man are juxtaposed as eternal and imperishable...[the *Arcana*] are direct emanations from divinity...This is the teaching of a treatise which has always enjoyed the reputation of authenticity. Here it says: 'all natural things flow from God and no other source... the things are His, the herb He created, not however the virtue that is in it. For each virtue is uncreated ; that is, God is without beginning and not created. Thus all virtues and powers were in God, prior to heaven and earth, and before all things were created, when God was a spirit and hovered penalty for the devil, first the element of air that is the *chaos* or heaven and thereafter the other elements'...we do have one outspoken piece of testimony for its creation in the Paracelsian *Corpus*: 'Thus God the Father created through His word things not in their ultimate state, but He only created *Prima Materia confusa*, that is the matrix, in which all Nature of the whole world was mixed together....called abyss and earth or a thing in which all things lie hidden...and this *Prima Materia* was the water on which the spirit of the Lord had hovered...*Materia prima* made from nothing and hence called *Abyssus*...'" Pagel, "Paracelsus and Neoplatonic Tradition," pp. 142–146.

141 Theophrast von Hohenheim gen. Paracelsus, *Sämtliche Werke. I. Abteilung: Medizinische naturwissenschaftliche und philosophische Schriften*, Volumes I-XIV, K. Sudhoff (ed.), München und Berlin, Druct und Verlag von R. Oldenbourg, 1922–1933, vol. II, p. 187.

142 "Toward the end of the medieval period serious alchemists, inspired and guided by... Paracelsus, concentrated their attention upon the medical aspects of alchemy, which earned them the name 'iatrochemists'. From that time on alchemy increasingly served medical practice by preparing elixirs and compounds intended not only to cure symptoms but above all to cure, and in fact to perfect, the whole man." Will H. L. Ogrinc, "Western society and alchemy from 1200 to 1500," *Journal of Medieval History*, 6 (1980), p. 103.

143 "Of the few authors who were singled out for praise by Paracelsus Hippocrates stands first in rank and admiration...In several places Van Helmont holds the deviation from Hippocrates through Galen responsible for the secular decline of medicine. He has even no hesitation in recognizing Hippocrates as superior to Paracelsus." Walter Pagel, *The Smiling Spleen: Paracelsianism in Storm and Stress*, Basel, S. Karger, 1984, pp. 23, 26-27.

simples and distilled various herbal remedies[144] in the belief that poisons could be altered to produce beneficial medicines.[145]

The practice of distillation in alchemy originated in Hellenistic Alexandria, and along with the ancient scientific and philosophical texts of the Greeks this technology was transmitted to the West[146] around the beginning of the last millennium by the cultures of the Near East.[147] Consequently, in medieval Europe the practice of distillation

[144] "The Paracelsians were also very interested in applying chemistry to the extraction of the virtues of plants through distillation. This application of the chemical philosophy has yet to be investigated in detail." Debus, "Perspectives on the Scientific Revolution," p. 74 n. 17.

[145] Allen G. Debus, "The Chemical Philosophers: Chemical Medicine from Paracelsus to Van Helmont," *History of Science*, 12 (1974), pp. 237-238, 242; Debus, *Chemical Philosophy*, vol. I, pp. 21-24, 116; Debus, *Man and Nature*, pp. 18, 39-53, 146–147.

[146] "The recovery of ancient science and philosophy in the twelfth and thirteenth centuries marks an epoch in the history of European intelligence. 'The introduction of Arabic texts into the studies of the West,' says Renan, 'divides the history of science and philosophy in the Middle Ages into two perfectly distinct periods. In the first the human mind has, to satisfy its curiosity, only the meager fragments of the Roman schools heaped together in the compilations of Martianus Capella, Bede, Isodore, and certain technical treatises whose wide circulation saved them from oblivion. In the second period ancient science comes back once more to the West, but this time more fully, in the Arabic commentaries or the original works of Greek science for which the Romans had substituted compends'—Hippocrates and Galen, the entire body of Aristotle's writings, the mathematics and astronomy of the Arabs. The full recovery of this ancient learning, supplemented by what the Arabs had gained from the Orient and from their own observations, constitutes the scientific renaissance of the Middle Ages. The most important channel by which the new learning reached western Europe ran through the Spanish peninsula...The science of mediaeval Spain was, of course, an importation from the Mohammedan East. It was not specifically Arab, save for the Arab power of absorbing rapidly the older culture of the Byzantine Empire, Egypt, Syria, and the lands beyond. Fundamentally it was chiefly Greek, either by way of direct translation or through the intermediary of Syriac and perhaps Hebrew versions of Aristotle, Ptolemy, Euclid, Hippocrates, and the rest, but developed in many fruitful ways by elements from the farther East and by a certain amount of specific observation and discovery under the caliphs. The men of science were from all parts of Islam, few indeed being Arabs, but they shared the speech and culture which gave the several caliphates their common civilization." Charles Homer Haskins, *Studies in the History of Mediaeval Science*, Second Edition, Cambridge, Harvard University Press, 1927, pp. 3-6.

[147] "The first important body of alchemical texts are those written in Greek at dates between c. A.D. 100 and A.D. 600...The Greek alchemists speak of distilling sulphur, *θείον*, but every chemist must regard the processes described by them and the design of their apparatus as being quite inconsistent with the distillation of a substance of such a high boiling-point. Furthermore, these distillations were, in Greek alchemy at least, said to produce a liquid, the *divine water*, and not a solid such as would result from distilled sulphur. There is scarcely sufficient evidence to enable us to conclude what was distilled...It does not appear that any of the well-known Arab pharmacists employed distilled alcohol as a drug, nor that it was known to the Arabs before the thirteenth century...It is thought that the first mention of alcohol as a drug is in the work of Salernus, about 1150...An early reference to *aqua vitae* is found in the Acts of the Dominican provincial chapter at Rimini in 1288...This suggests that the process of distilling *aqua vitae* was not new in 1288...Arnaldus da Villanova, writing about 1309–12, states that it can be obtained from wine or lees of wine by distillation...He does not, however, treat it as a substance of a spiritual character, or as differing materially from the general class of drugs...Conrad Gesner tells us downrightly that Ramon Lull 'was the first man to write of the quintessence, though it was unknown to all the physicians of his time, nor written in any book nor tried in practice'... It may perhaps be said that the central idea of the writings of Paracelsus is the preparation of what we may term 'spiritual medicines', and the influence of the works we have been discussing, for example, the Lullian treatises and those of Rupescissa, is easily recognizable. His spiritual medicines are classed as *arcana* and *quintessences*, but the distinction between his usage of these

became associated with many famous and influential individuals, including the thirteenth and fourteenth century figures Raymond Lull,[148] Arnald de Villanova,[149] and John of Rupescissa,[150] who are among the earliest philosophers of the West to distill alcohol for use as a drug.[151] During this period the monasteries played a medical

terms is not obvious...Paracelsus does not, however, acknowledge his debt to the earlier authors who wrote on the quintessence, for in his work *On the correction of impostures*, he condemns the treatises of Arnald da Villanova, of Rupescissa...as utterly untrue; and it would appear that he did not identify the spirit distilled by their process with his ideal quintessence. It is noteworthy that he does not seem to use the word 'quintessence' to represent alcohol, and it is he who first gave the latter substance the name of *alcool vini*...After the time of Paracelsus we note a double movement of opinion about the Quintessence. The alchemical authors no longer tend to regard it as the physical substance, *aqua vitae*, then becoming familiar to all as a medicine or beverage, but rather to treat it as something far more subtle and hard to discover. The writers on distillation, on the other hand, tend to write plainly and simply about spirit of wine as a physical substance, and to drop the idea that it was a substance of a different order from other liquids. An example of the former type of work is Leonhard Thurneysser's *Quinta essentia*, which treats the quintessence with a degree of obscurity that puts it as far out of reach as the philosopher's stone." F. Sherwood Taylor, "The Idea of the Quintessence," *Science Medicine and History*, vol. I, pp. 251-263.

[148] Raymond Lull (*ca.* 1235–1315) was born at Palma (Majorca) and led an unrestrained life until his conversion (*ca.* 1263–1266), after which he attempted to conduct the Muslims back to the Christian Church. Thorndike writes: "[Lull's] famous Art came to him as a sudden inspiration in the midst of long study and reflection and was, he and his followers believed, received by direct divine illumination. Hence his title, 'the illuminated Doctor.'" Lull often derided the alchemists and rejected the transmutation of metals, and in the mid-sixteenth century the Inquisition also placed his works in the *Index Expurgatorius*. R. J. Forbes, *Short History of the Art of Distillation: From the Beginnings up to the Death of Cellier Blumenthal*, Leiden, E. J. Brill, 1948, pp. 59-60; Thorndike, *History of Magic*, vol. II, pp. 862-873; J. N. Hillgarth, *Raymon Lull and Lullism in Fourteenth-Century France*, Oxford, Clarendon Press, 1971, p. 134 n. 369; Michela Pereira, *The Alchemical Corpus Attributed to Raymond Lull*, London, The Warburg Institute, University of London, 1989, pp. 1-2.

[149] Arnald de Villanova (*ca.* 1240–1311) was a Catalan who treated popes and kings and was often entrusted with diplomatic missions, but also had his troubles with the French Inquisition, who had him arrested and later forbade the possession or reading of his books. Villanova also rejected the transmutation of metals and is said to have sojourned in Naples with Raymond Lull, and while shipwrecked on the coast of Africa he wrote *On Wine*, which describes medicinal wines and their uses. R. J. Forbes, *Studies in Ancient Technology*, Leiden, E. J. Brill, 1955, vol. III, p. 123; Forbes, *History of Distillation*, p. 61; Joseph Ziegler, *Medicine and Religion c. 1300: The Case of Arnau de Vilanova*, Oxford, Clarendon Press, 1998, p. 40 n. 120.

[150] John of Rupescissa (*ca.* 1300–1365) was a Catalan and Franciscan chemist and prophet also imprisoned several times. Rupescissa writes under the influence of Lull, and his principal work *De consideratione quintæ essentiæ* contains the doctrine of the fifth essence, the *Aqua vitae*, containing alcohol. Pereira writes: "[T]he fifth essence of wine...[was] introduced into Western alchemy by John of Rupescissa...For Rupescissa the fifth essence was a medicine. Even when he mentioned gold, it was only to reinforce the recognized medical virtues of the fifth essence." Forbes, *History of Distillation*, pp. 64-65; Pereira, *Alchemical Corpus Attributed to Lull*, pp. 11-20.

[151] "The early history of the art of distillation goes back to the beginnings of alchemy, for distillation was an operation of fundamental importance in the practice of alchemy...At first sight it appears somewhat surprising that the discovery of alcohol should have been made as late at the eleventh century. The art of distillation had been known for several centuries before and had been applied to a most remarkable variety of substances. Wine and other alcoholic beverages were known and widely used from a much more remote period. The reason for its comparatively late discovery probably lies in the imperfection of the distilling apparatus used by the Greeks and Arabs...The early use of alcohol in Italy supports the belief that it was discovered in that country. The cultivation of the vine was carried out on a very large scale in Southern Italy, and alchemy was extensively practiced there from the eleventh century, the school of Salerno doubtless being an important influence. As alchemy was largely practiced by ecclesiastics,

role in European society, and these institutions generally tolerated the fermenting and distilling of alcohol, and many also had botanical gardens that provided the ingredients for pharmaceutical products.[152]

The importance of botany to early medicine is reflected in the two massive volumes that comprise the *Herbarium* of John Locke, containing specimens collected from the Oxford Botanic Garden, or the Physick Garden.[153] Locke was committed to Helmontian medicine[154] and owned three of Starkey's works.[155] Commenting on the *Historia* of the Paracelsian Leonhard Thurneysser,[156] Locke writes:

---

it is probable that the discovery of alcohol was kept secret for some time through fear of an accusation of magic or heresy...[Villanova] was the first to apply to alcohol the name 'aqua *vitæ*' or 'eau de vie,' which has since come into common use. Previously it had denoted the elixir of life or any liquid having marvelous properties...The somewhat fantastic eulogies of alcohol which are found in the writings of Arnold of Villanova and Raymond Lully have not been without their echoes in much later years...During the fourteenth and fifteenth centuries alcohol was gradually ceasing to be an expensive pharmaceutical product and was becoming a beverage in general use. Its use as a specific remedy against plague probably led to its becoming widely known during the 'Black Death' of 1348. It is obvious also that a process by which sour wine and stale beer and cider could be converted into 'aqua *vitæ*' would attract many practitioners. The discovery that alcohol could be made from grain probably took place in the fourteenth or fifteenth century...In England distilling was apparently confined for many years to the monasteries, which were the centres of medical science in those days. The prohibition of alchemists or 'multipliers' by Henry IV in 1404 probably prevented any distilling outside the monasteries. With the dissolution of the monasteries in the sixteenth century, the monks took up various professions as a means of subsistence, and some became doctors and apothecaries and others became brewers and distillers, and distilling developed into an industry." Underwood, "Historical Development of Distilling," pp. 35-42; See also Forbes, *History of Distillation*, pp. 57-60, 101; T. Fairley, "Notes on the History of Distilled Spirits, Especially Whiskey and Brandy," *Analyst*, 30 (1905), pp. 300-304; James Comer, "Distilled Beverages," *The Cambridge World History of Food*, K. F. Kiple and K. C. Ornelas (eds.), Cambridge, Cambridge University Press, 2000, vol. I, pp. 653-656; Robert G. W. Anderson, "The Archaeology of Chemistry," *Instruments and Experimentation in the History of Chemistry*, F. L. Holmes and T. H. Levere (eds.), Cambridge, MIT Press, 2000, pp. 5-34.

[152] Anderson, "Archaeology of Chemistry," p. 21.

[153] John Harrison and Peter Laslett, *The Library of John Locke*, Oxford, Oxford University Press, 1965, pp. 26-27, 152–153 nos. 1427, 1428.

[154] "[Paracelsus] adapted chemistry to medicine and tried to find the medicinal value of chemical compounds, devising new methods of preparing and purifying them. The iatrochemical school counts many scientists among its followers who play a part in the story of distillation, such as [Jean Baptiste Van Helmont]...They pave the way of the new science of the balance, of analytical chemistry...Apart from the true chemical textbooks of this era we find a second series of books on practical technology, the so-called *Berg, Probier- und Kunstbüchlein*, mostly tracts on mining, treatment of ores, metallurgical and similar practical problems...They too play a part in the history of distillation but our main sources are a third class of texts called the *Arznei-, Kräuter- und Destellierbücher* mostly written by physicians, apothecaries and botanists, who discuss not only the distillation methods and apparatus but also the treatment of different herbs and flowers and the medicinal value of their decoctions, 'oils', and 'spirits'." Forbes, *History of Distillation*, p. 100.

[155] Locke possessed Philalethes's *Enarratio Methodica* (1678), *Introitus Apertus* (1667), and *Secrets Reveal'd* (1669). The *Enarratio* is spurious, although it contains two genuine works by Starkey: *Experimenta De Praeparatione Mercurii Sophici* and *Vade-Mecum Philosophicum*, the latter a dialogue between a master, Agricola Rhomaeus, a pseudonym for Starkey, and his pupil Eirenaeus Philoponos Philalethes. Harrison and Laslett, *Library of Locke*, pp. 208, 238; Newman, *Gehennical Fire*, pp. 268, 296-297 n. 47.

[156] Leonhard Thurneysser, *Historia sive Descriptio Plantarum omnium*, Berlini, Michael Hentzske, 1578.

"In this first booke he gives the discription & vertue of several plants... He gives also up & downe several ways of prepareing subtilitates plantarum as he calls them, or chymical p[re]parations of them... There is also in him the explication of severall of Paracelsus's termes & names of plants."[157] In a correspondence of Samuel Hartlib in 1649, Benjamin Worsley writes: "Spiritts of herbes & simples; drawne by or destilled with wine; have beene things by all, both physicians, Chymists, & Philosophers, much cried up; and magnifyed, since the first that destillation came in practise!"[158]

In 1651, Worsley informed Hartlib of Starkey's experiments involving the decoction of medicinal plants,[159] and Hartlib also recorded Starkey's production of a "Balsam of Vegetables" which he gave to Robert Boyle.[160] In a letter to Hartlib, Child made reference to Starkey's distillations of "oyles" from medicinal plants,[161] and Clodius told Hartlib that the "Essence of Oil of Roses as Stirk makes it, is a very fine gentle Purge being 3 or 4 drops taken inwardly."[162] Drawing a clear distinction between the pharmacopoeias and the application of simples (i.e., powdered matter of a single plant) by the Chemical Physicians and the Galenists, Starkey writes in *Natures Explication*: "As the *Galenists* may and do use Minerals, so we do use both Vegetables and Animals, only we differ in our preparations, and in our intentions in application."[163] Referring to the Galenists' use of minerals and other dangerous substances, Starkey writes, "They who desire to read more particularly concerning the folly and futility of vulgar Medicaments, I recommend them to the noble *Helmont* his *Pharmacopolium ac Dispensatorium modernum*, where this subject is handled *ad nauseam*."[164] Starkey then mentions the plant called "Hellebore" in pre-Linnaean botany:[165]

[157] British Museum, Add. MS 32544, ff. 118ᵛ–119ʳ, 121ʳ; Bodleian Library, MS Locke c. 42 (first part), p. 12; in Patrick Romanell, *John Locke and Medicine: A New Key to Locke*, Buffalo, N.Y., Prometheus Books, 1984, pp. 98-99, 207-209; Partington, *History of Chemistry*, vol. II, pp. 152–155.

[158] Worsley to Hartlib, Hartlib Papers 26/33/9–10; in Antonio Clericuzio, "The Internal Laboratory. The Chemical Reinterpretation of Medical Spirits in England (1650–1680)," *Alchemy and Chemistry in the 16th and 17th Centuries*, P. Rattansi and A. Clericuzio (eds.), Dordrecht, Kluwer Academic Publisher, 1994, p. 54.

[159] Hartlib, *Ephemerides*, [April 1651], E-E2; in Wilkinson, "Hartlib Papers II," p. 94.

[160] Ibid., [April 1653], HH-HH5; in Ibid., p. 102.

[161] Child to Hartlib, 8 April 1653, Hartlib Papers, XV; in Ibid., p. 100.

[162] Hartlib, *Ephemerides*, [summer 1655], 31-311; in Ibid., p. 105.

[163] Starkey, *Natures Explication*, p. 108.

[164] Ibid., p. 213.

[165] "From the start of my studies devoted to the medical art, the drugs of the Galenists seemed absurd to me, and Pyrotechny was more satisfactory for obtaining a store of medicaments. From the iatrochemists it was Mathias Untzer who first came into my hands, then Quercetanus along

> [I]n many vegetable Simples under the mask of virulency, great and noble virtues are hidden, which are kept by the poisonous appearance from rash hands, as the apples of the *Hesperides* were feigned to be kept by a watchful Dragon ; or as the passage to the Tree of life, was guarded by a flaming sword in the hand of Cherubims. Thus in Hellebore under the churlish vomitive poyson caused with convulsion both of stomach and nerves, is hidden a most noble remedy...[166]

On the subject of the poisonous "Hellebore," Starkey writes in *Pyrotechny*:

> And here ingenuous Reader observe the rottenness of the *Galenical structure*, who in Herbs of excellent virtue only look to the *vomitive* or *laxative* vemone, which may well be compared to *the flaming sword in the hand of the Cherub, that guards the passage to the tree of life*, So this face of *Venome* oft hides most noble and admiral endowments in many *simples*, by reason of which Poisonous outside they cannot get admittance into the more retired closets of Nature...But this veil being taken away, then appears the true, noble, and specifick virtues of it, and consequently of any other churlish *Vegetable*, which the *Galenists* by reason of their misty method, cannot endure to behold, with full view, and open eies, but they are discovered unto us, and taught us by the means and through the discipline of the fire, which is our so much commended *Pyrotechny*.[167]

In his *Usefulnesse of Experimental Philosophy*, Boyle also acknowledges the efficacy of simples employed by Paracelsus and van Helmont:

> [I]t is known there are divers chymists, and others, that practise physick, who so doat upon the productions of their furnaces, that they will scarce go about to cure a cut finger...And methinks those, that practise, as if nature presented us nothing worth the accepting, unless it be cooked and perfected by *Vulcan*, might consider, that *Paracelsus* himself oftentimes employeth simples for the cure even of formidable diseases. And though for particular reasons I be inclinable enough to think, that such searching and commanding remedies, as may be so much of kin to the universal medicine, as to cure great numbers of differing diseases, will be hardly obtained without the help of chymical preparations, and those perhaps of minerals...*Helmont* himself, a person more knowing and experienced in this art, than almost any of the chymists, scruples not to make this ingenious confession..."I believe

with the writings of Basilius, then Croll with Hartmann, and with many others, and finally that best interpreter of Paracelsus, Helmont the Bruxellian." [George Starkey], "Praxis chemiatria," Sloane Collection, British Library, MS 631, fol. 198^r; in *Starkey Notebooks and Correspondence*, p. 319.

[166] Starkey, *Natures Explication*, p. 281.

[167] Starkey, *Pyrotechny*, pp. 101–102.

> simples in their own simplicity are sufficient for the curing of all diseases."[168]

In his "Modern Pharmacapolium and Dispensatory," van Helmont writes:

> The Art of Healing is every where drawn into the Tragedies and scorn of the vulgar. Because Physitians will not be wise, but according to the custom of the Schools...I cannot but be angry at the describers of simples : For although there be no field more spacious, plentiful, and delightful in the face of the whole earth, and where the mind is more delighted, than in Herbarism ; yet there hath scarce been a less progress made in any other thing. For truly the *Arabians*, *Greeks* or *Gentiles*, *Barbarians*, wild country People, and *Indians*, have observed their own Simples much more diligently than all the *Europeans*. For even in the dayes of *Plato* (wherein *Diascorides*, a man of War, lived) nothing almost hath been added to Herbarism : but much diminished. *Galen*, from a desire of robbery, wrote this study of another, his name being suppressed. He being plainly a non-Diascorian, snatched up the words of *Diascorides*. The which, in the mean time, *Pliny* hath besprinkled with many trifles : Because, as its very likely, he being of a mean judgement, not being able to distinguish between truth and fals[e]hood, scraping many things together, on every side, hath described them, that he might equalize his name into the greatness of his Section. But even into this day, the more learned part of Physitians do as yet carefully dispute only about the faces and names of Herbs : As if the vertues could not speak before their countenance were known ; the virtues I say, being first delivered by *Diascorides*...God out of the eternal providence of his goodness and wisdom, hath abundantly provided for future necessities. He himself hath made and endowed Simples for the appointed ends of all necessities. Therefore, I believe, that the Simples, in their own simplicity, are sufficient for the healing of all Diseases. Therefore we must more study about the searching into the virtues, than about disputing any hard questions : Seeing that in Simples there is a perfect cure, and healing of all Diseases...I also think, that God hath perfectly, and sufficiently composed in Simples, compleat remedies of any Diseases whatsoever...Lastly, I believe, that God doth give the knowledge of Simples, to whom he will, from a supernatural grace : but not by the signes of nature...Therefore I have laughed at *Paracelsus*, because he hath erected serious trifles into the principles of healing...*Mathiolus*, *Tabernomontanus*, *Brasavolus*, *Ruellius*, *Fuchsius*, *Tragus d' Allichampius*, and other observers of Herbes, are hitherto buried, only about the faces, and visual knowledge of Plants : but their virtues, they all as one, describe out of *Diascorides*...There is none amongst them all,

[168] Robert Boyle, *Some Considerations touching the Usefulnesse of Experimental Naturall Philosophy*, Oxford, Printed by Hen: Hall..., 1663; in Boyle, *Works*, vol. II, p. 133.

> who hath knowingly described the properties of Simples, even as he, who has described all things, from the Hyssop, even unto the Cedar of *Libanus*. As a sure token, that true knowledges or Sciences are not elsewhere to be fetched, than from the *Father of Lights*...The *Father of Lights* therefore is to be intreated, that he may vouchsafe to give us knowledge, such as once he did unto *Bezaleel* and *Aholiab*, for the glory of his own Name, and the naked charity towards our Neighbours. For so, the Art of Medicine should stand aright in us, under every weight. But it is to be feared, lest he who hath suffered the Books of *Salomon* to perish, may reserve this knowledge of Simples for the age of *Elias* the Artist.[169]

The followers of Helmontian chemistry and medicine were violently opposed to Aristotelian philosophy and Galenic medicine that controlled the academic environment throughout Europe, a monopoly they regarded as arbitrary and illegal,[170] which had condemned Paracelsus and were equally intolerant of his philosophical partisans.[171]

Paracelsus was the leading figure of a long line of anti-Aristotelian reformers[172] including Francis Bacon, René Descartes, J. B. van

[169] Helmont, "Pharmacapolium," *Oriatrike*, pp. 457-459.

[170] Charles Webster, "English Medical Reformers of the Puritan Revolution: A Background to the 'Society of Chymical Physitians,'" *Ambix*, 14 (1967), pp. 16-41.

[171] "In 1476, Ficino asserted the superiority of Plato's ethics over Aristotle's; in 1477 he accused Aristotle of kicking his teacher Plato; and in another letter that same year, he remarked that even Aristotle could not tolerate the false calumnies uttered against Plato, a comment that only made sense if one presumed that Aristotle was normally critical of Plato. Even earlier, in his commentary on the *Philebus* of 1469, Ficino bragged of destroying the pettifogging criticisms of the Aristotelians against Plato concerning the highest good and accused Aristotle of slandering Plato. About the same time, in his commentary on the *Symposium*, he also made sure to correct deceptive Aristotelian interpretations of Plato concerning the soul...Ficino did enter into the Plato-Aristotle controversy, but on his own terms and with a minimum of effort...to win the goodwill of contemporary Aristotelians, he decided on essentially a rhetorical policy. He insinuated and on occasion baldly asserted harmony between Plato and Aristotle...Ficino shared Pico's desire for a *concordia philosophica*...For Ficino, *concordia philosophica* was never a goal in itself; it was only an instrument in achieving his real goal, which was the victory of the one true pious philosophy, the *Theologia Platonica*." John Monfasani, "Marsilio Ficino and the Plato-Aristotle Controversy," *Marsilio Ficino: His Theology, His Philosophy, His Legacy*, M. J. B. Allen, V. Rees, and M. Davies (eds.), Leiden, Brill, 2002, pp. 190–196.

[172] "I am not Luther, I am Theophrastus, the Theophrastus you called "Cacophrastus" in Basel... And if I am not match enough for you as Cacophrastus, I tell you, my shoestrings know more than you and all your schoolmasters, Galen and Avicenna, and all your high schools. If you can't accept that as true, lay both cures on the scales and see how they tip.' Theophrastus was of course not Luther. But the two men were contemporaries, and no one but Luther is so appositely other for the medical reformer [Paracelsus]...Luther and Paracelsus both rejected the nonscriptural foundations of Scholasticism, objecting strongly to Aristotelian philosophy...Paracelsus cast the Aristotelian philosophy of medieval Scholasticism aside in favor of a nature that he viewed almost as a second bible, complementary to the written one; and in favor of what he called *experientia*...The traditional recourses of the Galenists were as discredited for Paracelsus as the commerce in indulgences for the Lutherans: either recourse, Galenic prescriptions no less than indulgences, was thought to be motivated by greed and venality...[Alchemy] might be regarded as the nominalism of the Paracelsian medicine...Could scholars...have relished his execrations to

Helmont,[173] George Starkey, and Robert Boyle,[174] who felt that the absolute control of the University by the Aristotelians could not be broken without the use of vituperative attacks,[175] and neither side demonstrated any mercy in the ensuing battles.[176] The astronomers of the Scientific Revolution also came into conflict with the Aristotelians in the University and with Church authorities that also embraced Aristotelian philosophy.[177] A contemporary of Paracelsus, Copernicus saw his heliocentric theory replace the geocentric theory of Ptolemy

'shit on your Pliny, Aristotle, your Albertus, Thomas, Scotus...'?" Weeks, *Paracelsus*, pp. 1, 13, 46. Quoted from Paracelsus, *Sämtliche Werke*, vol. VIII, pp. 43, 47, 138.

[173] "Van Helmont's earliest experience of academic studies had left him almost obsessed with the defects of tradition Galenic medicine...His intention was the demolition of the whole ancient system of the elements, the humours and their qualities, their mutual conflict and contrariety in mixture, their excess or deficiency...Just as 'rotten' as the theories of the ancients, was their therapy. In this blood-letting was pre-eminent. Galen relied upon it in all fevers with the exception of hectic fever. Van Helmont's response to this is, if venesection cools, why not use it in hectic fever? He himself forbids it absolutely...Purges and laxatives are similarly criticized and rejected." Pagel, *Joan Baptista Van Helmont*, pp. 19, 157–158.

[174] Newton and Locke possessed the writings of all of these anti-Aristotelian reformers. John Harrison, *The Library of Isaac Newton*, Cambridge, Cambridge University Press, 1978, pp. 93, 131–132, 209-210, *et passim*; Harrison and Laslett, *Library of Locke*, pp. 152, 202, 238, *et passim*.

[175] "Paracelsus had died in 1541, a solitary figure with few friends, no immediate pupils, but many adversaries. These were united in refusing him credibility, if not summarily dismissing his brusquely proffered new ideas against the ruling syllabus as the ravings of a madman or the oracles of a demonic magician. His awkward personal behaviour, his vagrancy in life, style and speculation, his coarseness in expressing perpetual dissent remained fresh in the memory of contemporaries, as indeed it has persisted in history. Few of his numerous medical works and nothing of his philosophy, his naturalistic and mystic cosmology, his religious and exegetic outpourings had been published in his lifetime...By 1560 an ever increasing number of first editions in the vernacular and Latin translations became accessible through the mass medium of printing..." Pagel, *Smiling Spleen*, p. 143.

[176] P. M. Rattansi, "Paracelsus and the Puritan Revolution," *Ambix*, 11 (1963), pp. 24-32; Rattansi, "The Helmontian-Galenist Controversy in Restoration England," *Ambix*, 12 (1964), pp. 1-23; Debus, *Chemical Philosophers*, pp. 238-244; Harold J. Cook, *The Decline of the Old Medical Regime in Stuart London*, Ithaca, Cornell University Press, 1986, pp. 122–132, 145–160.

[177] "On March 5, 1615, the Congregation of the Index banned Copernicus's *De revolutionibus* on the grounds that it defended 'the false Pythagorean doctrine that the earth moves and the sun is motionless', a doctrine 'altogether contrary to Holy Scripture'. In February 1632, Galileo published his *Dialogue on Two Chief World Systems*, in which he strongly defended the Copernican world-view. In April 1633, he was brought to trail before the Holy Office in Rome...On June 22, Galileo was condemned, 'on vehement suspicion of heresy'...The *Dialogue* was banned; Galileo was forced to abjure the heresy that 'the sun is the center of the world and motionless and that the earth is not the center and moves'. He was sentenced to perpetual house-arrest." Richard S. Westfall, *Essays on the Trial of Galileo*, Vatican City State, Vatican Observatory Publications, 1989, p. v.

based on Aristotle,[178] while Tycho was a Paracelsian,[179] and Kepler[180] and Galileo were also anti-Aristotelians.[181]

Starkey's position on natural philosophy is articulated in *Natures Explication*:

> Are they Physicians by profession ? so am I, educated in the Schools as well as they, graduated as well as they, nor was my time idly spent, but in the Tongues and course of Philosophy usually taught, in Logick and other Arts read in the Schools...'Tis not because I never read the usual Philosophy, that I do not embrace it, nor because I am a stranger to the usual Method of Medicine, that I speak and write against it, and rather choose the Chymical way then it. For the vulgar Logick and Philosophy, I was altogether educated in it, though never satisfied with it ; at length *Aristotles* Logick I exchanged for that of *Ramus*, and found my self as empty as before : and for Authors in Medicine *Fernelius* and *Sennertus*, were those I most chiefly applied my self to, and *Galen*, *Fuchsius*, *Ayicen*, and others I read, and with diligence noted, what I could apprehend useful, and accounted this practical knowledge a great treasure, till practical experience taught me, that what I had learned was of no value, and then was I to seek for a new path, in which I might walk with greater certainty, and by Gods blessing, by the tutorage of the fire, I attained true Medicines taught obscurely by *Paracelsus*, but only explained by labour and diligence in the Art of Pyrotechny.[182]

Following the Hippocratic tradition, it was van Helmont's view that disease was of divine origin, and that the source of chemical and medical knowledge was the intellect through illumination and experience,[183] opposing the conventional Galenic-Aristotelian

[178] Edward Grant (ed.), *A Source Book in Medieval Science*, Cambridge, Harvard University Press, 1974, p. 823; Wilbur Applebaum (ed.), *Encyclopedia of the Scientific Revolution from Copernicus to Newton*, New York, Garland Publishing, Inc., 2000, pp. 162–168; Robert S. Westman, "The Copernicans and the Churches," *God and Nature: Historical Essays on the Encounter between Christianity and Science*, D. C. Lindberg and R. L. Numbers (eds.), Berkeley, University of California Press, 1986, pp. 76–113.

[179] Ole Peter Grell, "The Acceptable Face of Paracelsianism...," *Paracelsus and his Reputation*, pp. 254-256, 261-262.

[180] Charles Webster, *From Paracelsus to Newton: Magic and the Making of Modern Science*, Cambridge, Cambridge University Press, 1982, pp. 6, 30; Job Kozhamthadam, *The Discovery of Kepler's Laws: The Interaction of Science, Philosophy, and Religion*, Notre Dame, University of Notre Dame Press, 1994, pp. 51-63.

[181] Grant, *Source Book in Medieval Science*, p. 816; Applebaum, *Encyclopedia of the Scientific Revolution*, pp. 245-255; William R. Shea, "Galileo and the Church," *God and Nature*, pp. 114–135.

[182] Starkey, *Natures Explication*, pp. a3^{r}-a4^{r}.

[183] The English Helmontians did not reject the study of anatomy; they rejected the dangerous methods employed by the Galenists. George Thomson, *Galeno-pale: Or, A Chymical Trial of the Galenists...*, London, Printed by *R. Wood*, for *Edward Thomas*, at the *Adam* and *Eve* in *Little Britain*, 1665, p. 26.

approach in the University based on theory, reason, and physiology.[184] This echo of the voice of Plato philosophically appealed to Puritan sensibilities. With the wide support of eminent figures in Charles II's court, including the king,[185] George Starkey and other Helmontian physicians launched a fierce attack upon Galenic medicine and the University curricula in the mid-1660s.[186]

Although both sides attempted to gain his support, Boyle, who was not a physician, committed to neither in the dispute, likely fearing of the wrath of the Galenists. In a treatise that was suppressed by Boyle for this reason among others, he strongly criticizes the methods and intellectual rationale of the Galenists, and until his death Boyle remained sympathetic and committed to the empirical approach of the Helmontian physicians.[187]

---

[184] Alice Browne, "J. B. van Helmont's Attack on Aristotle," *Annals of Science*, 36 (1979), pp. 575-591; Antonio Clericuzio, "From van Helmont to Boyle: A study of the transmission of helmontian chemical and medical theories in seventeenth-century england," *British Journal for the History of Science*, 26 (1993), pp. 303-334.

[185] "There is no more amazing scene in history, if it be true, than that related of Charles the Second's deathbed, when fourteen doctors were said to have stood and wrangled round the monarch's bed before they first bled and then administered an emetic to the dying man. John Freind (1675–1728) who wrote a history of medicine, may have had scenes like this in mind when he wrote that 'contemporary doctors were mountebanks whose impudence was equal to their guilt in tormenting people in their last hours!'" Edith Grey Wheelwright, *The Physick Garden: Medicinal Plants and their History*, Boston, Houghton Mifflin Company, 1935, p. 154.

[186] Thomas, Sir Henry, "The Society of Chymical Physitians: An Echo of the Great Plague of London, 1665," *Science Medicine and Society in the Renaissance: Essays to Honor W. Pagel*, A. G. Debus (ed.), New York, Science History Publications, 1972, vol. II, pp. 56-71; Webster, *Great Instauration*, pp. 281-282; John Morgan, *Godley Learning: Puritan Attitudes Towards Reason, Learning, and Education, 1560–1640*, Cambridge, Cambridge University Press, 1986, pp. 41-61; Harold J. Cook, "The Society of Chemical Physicians, the New Philosophy, and the Restoration Court," *Bulletin of the History of Medicine*, 61 (1987), pp. 61-77; J. Andrew Mendelsohn, "Alchemy and Politics in England 1649–1665," *Past and Present*, 135 (1992), pp. 30-78.

[187] "Boyle planned a sequel to *Usefulness*, developing the passages in that work that were implicitly critical of the orthodox therapy of his day in a more overtly reformist direction; this treatise was to have been called 'Some Considerations & Doubts about the Vulgar Method or Practice of Physick'. Though only synopses and fragments of its text survive, enough is extant to give a sense of its content and approach, which was undoubtedly more aggressive than its predecessor, excoriating the shortcomings of Galenic medical practice and its intellectual rationale. In the end, however, Boyle suppressed it for various reasons; that to which he gave most prominence was the hostility of professional doctors, who 'were not well pleased, that a person not of their profession should offer to meddle with it, though with a design of advancing it." Michael Hunter, "The Reluctant Philanthropist: Robert Boyle and the 'Communication of Secrets and Receits in Physick,'" *Religio Medici: Medicine and Religion in Seventeenth-Century England*, O. P. Grell and A. Cunningham (eds.), Aldershot, Eng., Scolar Press, 1996, p. 251. For more on this manuscript, including a transcript of the extant passages from it, see Hunter, "Boyle Versus the Galenists: A Suppressed Critique of Seventeenth-Century Medical Practice and its Significance," *Medical History*, 41 (1997), pp. 322-361.

In what also belongs to the suppressed history of the Scientific Revolution,[188] Robert Boyle was undoubtedly an anti-Aristotelian,[189] a position that he held largely as a result of the influence of George Starkey.[190] In his *Origine of Formes and Qualities*, Boyle refers to the "Dark and Narrow Theories of the Peripateticks [Aristotelians]" and "*the Lazy* Aristotelian *way of Philosophizing*," and consequently "*that in about* 2000 *years since* Aristotles *time, the Adorers of his Physicks, at least by Vertue of* His peculiar Principles, *seem to have done little more*[191] *then Wrangle, without clearing up (that I know of) any mystery of Nature, or producing any useful or noble Experiments*."[192] In his *Free Enquiry*, Boyle writes: "I take divers of *Aristotle's* opinions relating to religion, to be more unfriendly, not to say pernicious, to it, than those of several other heathen philosophers."[193]

In his "Essay of the Holy Scriptures," Boyle makes reference to completed but unpublished papers that contain his refutation of Aristotle's philosophy (which unfortunately have not survived), and he praises and defends Copernicus, Francis Bacon, René Descartes, and the "Chymists" among others in an attack upon "Aristotle's

[188] "Boyle was also much more involved in the study of alchemy than successors like Birch were happy to admit. Birch seems to have felt obliged to protect Boyle against accusations of credulity on this score, and, in an attempt to defend Boyle's reputation, he and his collaborator, Henry Miles, tampered with the historical record, suppressing key documents among Boyle's papers. This is especially apparent in their treatment of Boyle's letters, for whereas Boyle had corresponded quite assiduously with alchemists and other occultists, letters to Boyle from such figures were to a disproportionate extent omitted from the selection of correspondence that Birch printed, or worse, were actually destroyed. This is revealed by surviving inventories made by Miles, which refer to many letters on alchemical topics which are now lost, while the suggestion that this was due to a conscious policy on Miles's part is supported by the fact that a number of such items are dismissed as of 'No Worth' or 'unintelligible'...Particularly vexing was their suppression of the profuse evidence which once existed concerning Boyle's interest in casuistry, in other words the salving of his conscience concerning difficult moral dilemmas, on which he sought the advice of leading churchmen. Far more material relating to Boyle's casuistical interests survived when Birch was writing his *Life* than is now the case, and the fact that this is due to Birch and Mile's disapproval is made clear by a letter in which the latter explained to the former how 'a very judicious friend' to whom he had shown the documents considered it 'better to omit [th]em as not suited to the genius of the present age'." Michael Hunter, "Robert Boyle (1627-91): a suitable case for treatment?" *British Journal for the History of Science*, 32 (1999), p. 265.

[189] In his *Things Above Reason* (1681), "Boyle as usual loses no opportunity to attack the Aristotelians, and there are frequent gibes at the Schoolmen who slavishly follow his teachings." John F. Fulton, *A Bibliography of the Honourable Robert Boyle; Fellow of the Royal Society*, Second Edition, Oxford, Clarendon Press, 1961, p. 98.

[190] William R. Newman and Lawrence M. Principe, *Alchemy Tried in the Fire: Starkey, Boyle, and the Fate of Helmontian Chymistry*, Chicago, University of Chicago Press, 2002, pp. 222-223.

[191] The word *more* is printed twice.

[192] Robert Boyle, *The Origine of Formes and Qualities*..., Oxford, Printed by H. Hall..., 1666, pp. B4^{v}, [269].

[193] Robert Boyle, *A Free Enquiry Into the Vulgarly Receiv'd Notion of Nature*..., London, Printed by H. Clark..., 1685/6; in Boyle, *Works*, vol. V, p. 163.

Adorers" and "Aristotles Schoole ... which hath of latter Ages bred most of my Antagonists." Boyle openly reveals his anti-Aristotelian sentiments throughout this diatribe, criticizing the frequent "Mistakes" of Aristotle, "whose Ability's I yet reverence, ev'n when I doe not his Opinions," and then praising "our Helmont" and other anti-Aristotelian figures,[194] whom "I may boldly oppose...to any Galenicke, or Peripateticke Names."[195]

> And tho Aristotles School be that which hath of latter Ages bred most of my Antagonists; yet he that shall reade Telesius, Bassa, Veralum, Copernicus & (not to mention the Chymists) Gassendus[,] Regius & the rest of the Atomists, & other Free Philosophers, shall find his soveraignty & Axioms not undisputed:[196] to omit ‹that› the Acute Campanella, hath vigorously invaded him in all the severall Provinces of Phylosophy's Dominion; & in each Particular Discipline treated of by Aristotle, hath introduc'd new Theorems destructive to the Old ones; & that as many thinke, Alwayes with equall Probability, & often with more Truth. And, not to make Instances too wide of our discourse, that Ingenious Gentleman Monsieur Des Cartes, whome sure None that know him take for a Bigot, who hath (in my Opinion) in his Treatise Des Passions,[197] giv'n us the Luckiest Conjectures of divers things relating to the soule that I have hitherto met with; & who hath

[194] "In England both the Elizabethan Statutes for Cambridge (1570) and the Laudian Code for Oxford (1636) maintained the authority of the ancients, and this spirit also was carried over to professional societies. When Dr. John Geynes had the temerity to suggest that Galen was not infallible (1559), he was forced to sign a recantation before being received back into the company of the Royal College of Physicians. It was surely acceptable—even commendable—for the humanist to criticize the vulgar annotations and emendations foisted on the works of the ancients by the Arabs and the Scholastic philosophers, but for many the original and pure texts seemed like an impregnable fortress of truth not to be added to or altered in any way...at his examination for the M.A. in 1536, Petrus Ramus began his attack on Aristotle in defending the thesis 'Everything which Aristotle states is false.' Similarly—in Italy—Telesio launched a strong attack on Medieval Aristotelianism in his Academy at Cosenza. Both Ramus and Telesio stressed the observation of Nature as a new foundation of knowledge rather than the constant repetition of the views of the ancients...Descartes tells us that after completing the entire course of study at one of the most celebrated Schools of Europe, 'I found myself embarrassed with so many doubts and errors that it seemed to me that the effort to instruct myself had no effect other than the increasing of my own ignorance'...when we look at Ramus, Telesio, Gilbert, Harvey, Descartes and Bacon—all of whom who have been referred to as founders of modern scientific method, we sense a disillusionment with the formal training of the period and its emphasis on the infallible truths of antiquity." Debus, *Chemical Dream*, pp. 8–11.

[195] Robert Boyle, "Essay of the holy Scriptures," ff. 29-31, 44; in Michael Hunter and Edward B. Davis (eds.), *The Works of Robert Boyle*, London, Pickering & Chatto, 2000, vol. XIII, pp. 190–191, 197, (hereafter as *Works of Boyle*).

[196] The natural philosophers mentioned by Boyle are Bernardino Telesio (1508–1588); Sebastiano Basso (fl. 2nd half 16th cent.); Francis Bacon, Ist Baron Veralum (1561–1626); Nicholas Copernicus (1473–1543); Pierre Gassendi (1592–1655); Henricus Regius (1598–1679); Tommaso Campanella (1568–1639); René Descartes (1596–1650); Athanasius Kircher (1602–1680); and Daniel Sennert (1572–1637).

[197] This and the following reference are to Descartes's *Passions de l'âme* (1649) and *Lettre à Pere Dinet* (1642).

> Question'd All ‹which› men have thought unquestionable, this Famous Philosopher ‹I say› whose Parts have already gayn[e]d him a sect: asserts (how Truely, it were foreine to my Argument to determine) the Immortality of the Soul to be naturally more Demonstrable by Reason, then any Physicall Assertion of any sect of Philosophers. And this very Gentleman in a Printed Epistle adress't to noe lesse (& no lesse concern'd) a Schooleman then ‹the› Pere Dinet, then Provinciall of the French Jesuites, boldly casts the Aristotelians a ‹yet› unaccepted Gantlet; & openly challenges all Aristotle's Adorers, to name the solution of any one Naturall Question, by the Principles Peculiar to the Peripateticke Philosophy, which he cannot demonstrate to the illegitimate or false. What I thinke of the Philosophy of Aristotle, (whose Ability's I yet reverence, ev'n when I doe not his Opinions) having in some Philosophicall Papers largely enough express'd, I shall now onely say, that I should be very sorry there could be as much Reason brought against the Three Persons in the Trinity, as I thinke I could bring against the Foure Elements in Mixt Body's: & that his own Mistakes have not infrequently borne Witnesse to that Ingenuous Confession[198] of his wherein he sayth...that as the Eyes of Batts are to the Day-Lights brightnesse, to is our Soules intellect, to the most manifest or discernable things of All.[199]

In his repudiation of Aristotle and the Aristotelians, Boyle also invokes the authority of van Helmont and "the Adepti" in this polemic against the Aristotelians that controlled the University:

> I might add to my former Instances, most of the Adepti, some of whose writings Nature her selfe seemes to have endited; & they to have but lent their Pens & Names to the Bookes that have ennobled them. And tho remoter Times had bin unable to furnish me with the Instances I have (out of numerous others) chosen to mention; yet our own Age were capable sufficiently to supply me; since I may boldly oppose our Kircherus, our Campanella, our Helmont[,] our Sennertus, & our Veralum (to mention noe others of the Noble Number of as Eminent Votarys to Gods written, as Doctours in his Created Booke) to any Galenicke, or Peripateticke Names...[200]

[198] Boyle here gives the original Greek text and translation of Aristotle's *Metaphysics* (I), 993^{b}9–11.
[199] Boyle, "Essay of the holy Scriptures," ff. 29-31; in *Works of Boyle*, vol. XIII, pp. 190–191.
[200] Ibid., fol. 44; in Ibid., vol. XIII, p. 197.

CHAPTER 4. ROBERT BOYLE AND GEORGE STARKEY

Robert Boyle's committed pursuit of the Philosophers' Stone extended over forty years[201] and received strong momentum when Robert Child,[202] who had befriended George Starkey in Boston through the younger Winthrop, introduced Boyle to Starkey soon after Starkey's arrival in England.[203]

In what appears to be the inception of Philalethes and the ensuing campaign of misinformation, Starkey almost immediately told Boyle

[201] According to Principe and Hunter, Boyle showed little interest in science prior to 1649, and was almost exclusively involved with theological and ethical pursuits. Lawrence M. Principe, "Style and Thought of the Early Boyle: Discovery of the 1648 Manuscript of *Seraphic Love*," *Isis*, 85 (1994), pp. 247-260; Principe, "Newly Discovered Boyle Documents in the Royal Society Archive: Alchemical Tracts and his Student Notebook," *Notes and Records of the Royal Society of London*, 49 (1995), p. 63; Michael Hunter, "How Boyle Became a Scientist," *History of Science*, 33 (1995), pp. 59–103. For another view, see Malcolm Oster, "Biography, Culture, and Science: The Formative Years of Robert Boyle," *History of Science*, 31 (1993), pp. 177-226; Antonio Clericuzio, *Elements, Principles and Corpuscles: A Study of Atomism and Chemistry in the Seventeenth Century*, Dordrecht, Kluwer Academic Publishers, 2000, pp. 109–110.

[202] "In the early 1650s, Boyle was intensely active in alchemy...and read the works of...van Helmont and Paracelsus...Boyle's alchemical activity seems to have slackened around the end of the decade, and remained less intense through the 1660s. There are, nevertheless, still a fair number of alchemical laboratory operations in the Papers and a few pieces of alchemical correspondence. By the 1670s, however, Boyle's alchemy was again vigorous, as marked by his composition of the *Dialogue on Transmutation* (early 1670s), the publication of its conclusion as the anti-elixir tract (1678) the revelation of the incalescent mercury (1675/6), the alchemical sentiments of *Producibleness* (1677), and the commencement of extensive correspondence with continental alchemists, which continued well into the 1680s. Voluminous laboratory records from the 1680s attest to Boyle's continued alchemical interest throughout that decade, at the end of which he successfully moved to have the Act against Multipliers repealed (1689)." Lawrence M. Principe, "Boyle's alchemical pursuits," *Robert Boyle Reconsidered*, M. Hunter (ed.), Cambridge, Cambridge University Press, 1994, p. 98.

[203] Starkey, *Pyrotechny*, preface, [xv]; Kittredge, "Robert Child," p. 101; Wilkinson, "Hartlib Papers II," p. 96.

of "a filius Hermetis in N[ew] E[ngland] who had the elixir."[204] Preparing the way for Starkey's arrival, in the previous summer Child told Samuel Hartlib that Starkey was already famous in New England for curing "Palsy and other incurable diseases,"[205] and in late November of that year Worsley brought Hartlib "the first news of young Dr. Starkey come hither out of New England with a ful[l] confidence of the Althahest [Alkahest]."[206] After meeting Hartlib in December and Boyle around this same time,[207] Starkey was given immediate entrance into the "Invisible College" or the "Philosophical College,"[208] the forerunner of the Royal Society,[209] established by Boyle, Worsley, and Hartlib with the purpose of creating a system of rationalizing alchemy.[210] The discovery of the "immortal liquor"

[204] Hartlib, *Ephemerides*, [January 1650/51], A3; in Wilkinson, "Hartlib Papers II," p. 90.

[205] Ibid., [summer 1650], G-H3; in Ibid., p. 87.

[206] Ibid., 29 November 1650, K-L3; in Ibid., p. 87.

[207] Ibid., 11 December 1650, K-L7; in Ibid., p. 87.

[208] "The period from 1654 to 1600 [1660] was one of relentless disappointment for the Hartlib circle, who saw their imaginative and idealistic schemes for social reconstruction ignored by parliament and opposed by the educational establishment...Hartlib's last two years coincide with the opening phase of the restoration (1660–1662). For him this was a time of poverty, illness and isolation...Even in 1662, his death went virtually unnoticed. Indeed, after 1660 he was treated with studied indifference by most of his former acquaintances. His name and associations evoked too many memories of the puritan revolution. His carefully preserved correspondence contained embarrassing information, at a time when few intellectuals wished to draw attention to their careers under the protectorate. Thus Hartlib's voluminous papers, petitions and reform tracts were consigned to oblivion...it is difficult to access Hartlib's direct influence, because of the sharp discontinuity at the restoration, creating a strong ideological bias against association with him...Most significant in this context is Hartlib's lack of mention in the records of the Royal Society, founded in 1660...Hartlib was well acquainted with many of the leading founders of the Society...It was obviously impolitic for a society seeking royal patronage to associate with such a potent reminder of the protectorate. Hartlib was a man under suspicion, his correspondence being intercepted by government agents. Furthermore his Baconianism, practical and unsystematic, was too close to the 'mechanics' for the taste of the Royal Society. Nevertheless, Hartlib exercised more indirect influence on the early Royal Society than is generally recognised...He corresponded regularly until his death with Boyle and Oldenburg, two of the most important members. Boyle had probably developed his enthusiasm for natural philosophy, particularly chemistry, through association with Hartlib's circle, in which he had long been one of the central figures. Boyle, more than any other figure, created the scientific image of the early Royal Society. He emancipated the Baconianism of Hartlib's circle from its mystical and alchemical elements, making it a viable tool for the rational and experimental study of chemistry." Webster, *Samuel Hartlib*, pp. 61-64, 69-70.

[209] "In his utopian *New Atlantis* Bacon went on to describe 'Saloman's House' with its provisions for workshops, scientific instruments and laboratories—all of the requirements for the proper collection of scientific data according to Bacon's scheme for a new science. This was to become a major source of inspiration for the founders of the Royal Society of London." Debus, *Chemical Dream*, p. 10.

[210] Marie Boas [Hall], *Robert Boyle and Seventeenth Century Chemistry*, Cambridge, Cambridge University Press, 1958, pp. 6-7, 24-25; Douglas McKie, "The Origins and Foundation of the Royal Society of London," *The Royal Society: Its Origins and Founders*, H. Hartley (ed.), London, Royal Society, 1960, pp. 21-23; R. E. W. Maddison, "Studies in the Life of Robert Boyle, F.R.S.: Part VI. The Stalbridge Period, 1645–1655, and the Invisible College," *Notes and Records of the Royal Society of London*, 18 (1963), pp. 104–124; Charles Webster, "New Light on the Invisible College:

occupied the majority of Hartlib's associates,[211] and motivated by the attainment of useful medicines,[212] Boyle and Starkey joined efforts to produce various medicinal preparations.[213] Their collaboration lasted about eighteen months, and their interactions continued throughout the decade, and may have continued into the mid–1660s.[214]

The Social Relations of English Science in the Mid-Seventeenth Century," *Transactions of the Royal Historical Society*, Fifth Series, 24 (1974), pp. 19-42; Webster, *Great Instauration*, pp. 57-67.

[211] In early 1649, Hartlib was working on a process "to turne Iron into Gold," and referring to "the Experiment of Iron and Antimony" mentioned in a lost letter of Boyle, he adds: "there is nothing to bee gotten by it." Hartlib, *Ephemerides*, [1648/9], A-B8 to B-C1; in Ronald Sterne Wilkinson, "The Hartlib Papers and Seventeenth-Century Chemistry, Part I," *Ambix*, 15 (1968), p. 61.

[212] "*Vulcan* has so transported and bewitched me, that as the delights I taste in it make me fancy my laboratory a kind of *Elysium*, so as if the threshold of it possessed the quality the poets ascribed to that *Lethe*, their fictions made men taste of before their entrance into those seats of bliss, I there forget my standish and my books, and almost all things..." Boyle to Lady Ranelagh, 31 August 1649; in Boyle, *Works*, vol. VI, pp. 49-50.

[213] In a passage dated 21 March 1693, Robert Hooke wrote in his Diary: "I saw neer 100 of Mr Boyles high Dutch Chymicall bookes ly exposed in Moorfeilds on the railes." Fulton, *Bibliography of Boyle*, p. v.

[214] Wilkinson, "George Starkey," p. 129; Clericuzio, "Helmont to Boyle," p. 315; Lawrence M. Principe, *The Aspiring Adept: Robert Boyle and his Alchemical Quest*, Princeton, Princeton University Press, 1998, p. 160; Newman and Principe, *Alchemy*, pp. 213, 234-236, 265-266.

A man of affluence who was born[215] into one of the most privileged aristocratic families[216] in England,[217] Robert Boyle financially supported Starkey's research and his development of various medicines, an activity in which Starkey was unrivaled by his contemporaries.[218] Starkey exerted a strong and lasting influence upon Boyle,[219] who

[215] "Not only did Boyle have to endure a separation at birth, followed by the death of his mother, but he then found himself despatched across the waters to attend Eton College, with his elder brother Francis, at the age of eight years...His academic work proved so vital to the young student that by his own reckoning 'his Master would be sometimes necessitated to force him out to Play ; on which, & upon Study, he look'd as if their natures were inverted'. The accumulation of traumatic experiences ranging from loss of the family home, wet-nursing and swaddling, maternal bereavement and abandonment to boarding school, exacerbated by one or more homosexual seductions by 'gown'd Sodomites' with 'goatish Heates' during his fourteenth year, contributed to Robert Boyle's growing incapacity to tolerate a fully intimate relationship during adulthood. To cap it all, the Earl of Cork then died, leaving Robert Boyle an orphan at the age of sixteen years. His close association with his sister Lady Ranelagh...with whom he lived for some twenty-three years, may have come to represent a sublimated oedipal spouse." Brett Kahr, "Robert Boyle: a Freudian perspective on an eminent scientist," *British Journal for the History of Science*, 32 (1999), p. 283.

[216] "Privilege brings obligations—*noblesse oblige*...Boyle's childhood was beset by tragedy...His mother died of consumption when he was three...The maternal role in his life was therefore a gap, a lack, a hole, a vacuum. Fresh air, or air, became a leitmotif in his life...Air became a unifying theme in Boyle's life, the harnessing of air in his air pump, for example, filled the vacuum mentioned above...His sensitivity was both his weakness and his strength. It led him to pursue interests in his chosen fields of science and intellectual advancement...It took him to the leading edge of discovery and exploration. Had he been living today he might well have been a candidate for the Nobel Prize...In Florence the recorded sexual encounter was a homosexual approach which he firmly rejected. It seems to have deterred him from sex for life...the maternal role in Boyle's life was a lack. The nearest anyone came to filling for him was his sister, Katherine... Although she was twelve years his senior, they were alike in many ways: sensitive, interested in ideas, exploration, and becoming part of the emerging climate of opinion of their age...By being close to her, he was able to avoid confronting his own sexual identity. In its place he developed this quasi-incestuous relationship with his sister...His sister Katherine was a remarkable woman in her own right. Good-looking, she was high-spirited, intelligent and handsome...She had to contend with a bad marriage, arranged by her ambitious, dynastically minded father... After the predictable collapse of her marriage...He may have replaced her husband to some extent, and she may have been keen to keep him for herself...The evidence suggests that Boyle defended himself against adult love, just as he had defended himself as an adolescent against what he called 'Goatish Heates' such as might be found in Florence...We can detect a note of triumph in his description of this Florence episode...Such a triumph was something of a Pyrrhic victory and meant his emotional ties stayed at a rather shallow level, lacking in real intimacy." John Clay, "Robert Boyle: a Jungian perspective," *British Journal for the History of Science*, 32 (1999), pp. 285-292.

[217] "Boyle is interesting in this respect, in that evidence of deep internal conflict lies on the surface of a mind dedicated to rational argument and persistent enquiry into natural phenomenon...He refused to accept the presidency of the Royal Society of London, because he would have been legally bound to swear an oath. Since he could not be sure of his capacity in the future to honour the oath, he could not in good conscience swear it...Boyle was deeply conflicted, and he seems to have led an austere life, always attending to, and overcoming, temptations and a proneness to guilt. He was a man driven by conscience, but that is too general a notion." Karl Figlio, "Psychoanalysis and the scientific mind: Robert Boyle," *British Journal for the History of Science*, 32 (1999), pp. 307-311.

[218] Starkey to Boyle, 16 January 1651 [Old Style, i.e., 1652], Royal Society Library, Boyle Letters, vol. V, fol. 131^r; in *Correspondence of Boyle*, vol. I, p. 117; [George Starkey], *George Starkey's Pill Vindicated...*, London, s.n., 1660[?], p. 6.

[219] Antonio Clericuzio, "Robert Boyle and the English Helmontians," *Alchemy Revisited*, p. 192; Newman and Principe, *Alchemy*, pp. 207-272.

was also heavily influenced by Paracelsus and van Helmont,[220] whom Boyle considered to be the "Learn'd expositors"[221] and "Patriarchs" of alchemy.[222] Boyle became intimately acquainted with and adopted the theories of van Helmont predominantly through George Starkey.[223] Boyle actively sought out Starkey's knowledge of chemical medicine, and the theories and experiments of van Helmont occupied an important role in Boyle's early chemical studies, particularly before the publication of his most famous work *The Sceptical Chymist*,[224] in which he praises and defends van Helmont and quotes Paracelsus at length.[225]

Robert Boyle's alliance with George Starkey was a critical step in Boyle's chemical career and his pursuit of the "grand Elixir," which Boyle makes reference to in his *Usefulnesse of Experimental Philosophy*.[226] Starkey makes similar references to this collaboration in his *Pill Vindicated*,[227] as well as in *Pyrotechny*, a work the "Philosopher by Fire" dedicated to "*Robert Boyl* Esq; My very good Friend."[228] Referring to Starkey as the "industrious chymist" and addressing the work

[220] Forbes, *History of Distillation*, pp. 100–109, 164–166, 182, 264.

[221] Robert Boyle, *Some Motives and Incentives To the Love of God* [*Seraphic Love*], London, Printed for *Henry Herringman...*, 1659; in Boyle, *Works*, vol. I, p. 262.

[222] Robert Boyle, *The Sceptical Chymist...*, London, Printed by F. Cadwell for F. Crooke, 1661; in Boyle, *Works*, vol. I, p. 560.

[223] Newman and Principe, *Alchemy*, pp. 222-223.

[224] "Signs of Paracelsus's (and of the Paracelsians') influence can be traced in Boyle's early chemistry. At the beginning of the 1650s, Jean Baptiste van Helmont's iatrochemistry, which...was widely known in the Interregnum, became a major source of Boyle's chemical investigations. *The Usefulnesse* and the so-called first draft of *The Sceptical Chymist* are largely based on Helmontian theories and experiments. Though in *The Sceptical Chymist* Boyle criticizes van Helmont's theory of water and *semina* as the principles of natural bodies, crucial Helmontian views are still adopted there and in subsequent works. Boyle's diary of January 1649 (old style) contains notes on ferments and on chemical spirits, two topics which played a prominent part in Paracelsian and Helmontian iatrochemistry. Part I of *The Usefulnesse*, largely written between 1648 and 1650, shows evidence of Boyle's commitment to iatrochemistry and to Paracelsianism. In the first essay Boyle extols Paracelsus and explicitly criticizes Galenic physicians. Unlike most of the latter, Paracelsus—despite his 'many extravagances'—improved our knowledge of nature and produced useful remedies. Boyle commends both Paracelsus's inquisitive attitude and the Paracelsian view that investigations of nature disclose God's power in the world. Like other members of the Hartlib Circle, Boyle was convinced that Paracelsianism contributed both to the advancement of learning and to the promotion of the Christian religion. The first essay of *The Usefulnesse* also contains a short account of Boyle's presumably extensive investigations of the nature and preparation of antimony—a substance which Paracelsians had introduced into medicine. When discussing the relationships of chemistry to medicine, Boyle evidently adopted the Paracelsians' view, namely that chemistry provides the foundation to medicine." Clericuzio, *Elements, Principles and Corpuscles*, pp. 109–110.

[225] Boyle, *Sceptical Chymist*; in Boyle, *Works*, vol. I, pp. 485-486, 495-500, 523, 542, *et passim*.

[226] Boyle, *Usefulnesse of Experimental Philosophy*; in Boyle, *Works*, vol. II, pp. 97-98, 135, 151–152, 215-216.

[227] [Starkey], *Pill Vindicated*, pp. 6-7.

[228] Starkey, *Pyrotechny*, preface, [xv].

to *Pyrophilus* (i.e., Boyle), Boyle recounts in *Usefulnesse* a particular experiment that involved the reluctant participation of Starkey:

> The first of these is the same powder, which passeth under the name of *ens Veneris* ; which appellation we gave it, not out of a belief, that it equals the virtues ascribed by *Helmont*, to what he calls the true *ignis Veneris*, but partly to disguise it a little, and partly upon the account of the occasion, whereon it was first found out ; which was, that an industrious chymist (whom you know) and I, chancing to look together upon that tract of *Helmont's* which he calls *Butler*, and to compare it somewhat attentively with other passages of the same author, we both resolved to try, whether a medicine, somewhat approaching to that he made in imitation of *Butler's* stone, might not be easily made out of calcined vitriol : and though upon trials we found this medicine far short of what *Helmont* ascribes to his, yet finding it no ordinary one, we did, for the mineral's sake it is made of, call it *ens primum Veneris*.[229]

Starkey was very independent and often showed a reluctance to engage in alchemical corroborations, and in his early letters to Boyle Starkey claims to have discovered the preparation of the liquor Alkahest, and his production of "six whole pounds of the liquor Alkahest, which are now in preparation" made Starkey a master without need of partners.[230] Starkey produced the *Ens Veneris* with Boyle's financial support but without Boyle's direct participation, who later received the medicine from Starkey, leaving Boyle to assume that the highly secretive Starkey had literally interpreted van Helmont's formula in his preparation of the medicine.[231]

Referring to Starkey's reluctant participation in transmutation experiments with the Hartlib circle, John Dury related to Hartlib that "hee complaines that it's a round Worke (to goe like a horse in a Mill about one and the same Work continually)."[232] In a letter to Boyle in the spring of 1651 containing Starkey's coveted transmutation formula, Starkey declares himself without need of "partners, & in a thing wch I Could Command as a Master, I would not work as an Amanuensis, Nor yet would I in such a way of lucre prostrate so great a secret...not willing to imbrace a life (in Exchange of a studious search of Natures

[229] Boyle, *Usefulnesse of Experimental Philosophy*; in Boyle, *Works*, vol. II, p. 135.

[230] Starkey to Boyle, 26 January 1651 and 3 February 1651/2 [Old Style, i.e., 1652], Royal Society Library, Boyle Letters, vol. V, ff. 133^r^–134^r^, 135^v^; in *Correspondence of Boyle*, vol. I, pp. 118–126, 129; Maddison, *Life of Boyle*, p. 77.

[231] [Starkey], *Pill Vindicated*, p. 6.

[232] Hartlib, *Ephemerides*, [April? 1651], E-E7, 8; in Wilkinson, "Hartlib Papers II," p. 91.

mysteryes) wch might be Compared with y^t of a Mil[l]horse running round in a wheele today, that I may doe ye same tomorrow."[233]

In *Usefulnesse*, Boyle provides further details of Starkey's production of the *Ens Veneris*:

> [A]n industrious chymist (of our acquaintance) and I, chancing to read one day together that odd treatise of *Helmont*, which he calls *Butler*... we at length concurred in concluding, that either the *lapis Butleri* (as our author calls it) or at least some medicine of an approaching efficacy, might (if *Helmont* did not mis-inform us) be prepared by destroying (as far as we could by calcination) the body of copper, and then subliming it with sal armoniack...But the person I discoursed with, seeming somewhat diffident of this process, by his unwillingness to attempt it, I desired, and easily persuaded him, at least to put himself to the trouble of trying it with the requisites to the work, which I undertook to provide, being at that time unable to prosecute it my self, for want of a fit furnace in the place where I then chanced to lodge. And though at first we did not hit upon the best and most compendious way, yet during the sublimation, he being suddenly surprized, as both himself and his domesticks two days after told me, with a fit of sickness, attended with very horrid and seemingly pestilential symptoms, was reduced to take some of this medicine out of the vessels before the due time, and upon the use of it found, as he told me, an almost immediate cessation of those dreadful symptoms, but not of the paleness they had produced. This first prosperous experiment emboldened us to give our remedy the title of *Primum Ens Veneris*, which, for brevity's sake, is wont to be called *Ens Veneris* ; though I am far from thinking, that it is the admirable medicine, to which *Helmont* gives that name, at least, if his *Ens Veneris* did really deserve half the praises by him ascribed to it.[234]

The "Stone" of the Irish alchemist Butler produced the *Ignis Veneris* of van Helmont, who in his often-cited treatise entitled *Butler* writes:[235]

> There was a certain *Irish-man*, whose name was *Butler*, being sometime great with *James* King of *England*, he being detained in the prison of the Castle of *Vilvord*, and taking pitty, on *Baillius* a certain Franciscan monk, a most famous Preacher of *Gallo-Brittain*, who was

[233] Starkey to Boyle, [April/May 1651], Royal Society Library, Boyle Letters, vol. VI, fol. 100; in Boyle, *Letters and Papers* [microform]. The full text is reproduced in William R. Newman, "Newton's *Clavis* as Starkey's *Key*," *Isis*, 78 (1987), pp. 564-574. This letter is also in *Correspondence of Boyle*, vol. I, pp. 90–103.

[234] Boyle, *Usefulnesse of Experimental Philosophy*; in Boyle, *Works*, vol. II, pp. 215-216.

[235] In December 1652 Starkey became acquainted with Henry Carey, Lord Dover, described by Hartlib as "a great Chymist. A kinswoman of his should have married Butler with whom he was very wel[l] acquainted." Referring to Starkey, Hartlib informed Boyle in February 1654 that he had heard of "secret transactions between him and my lord *Dover*." Hartlib, *Ephemerides*, [December 1652], EE-EE4; in Wilkinson, "Hartlib Papers II," p. 99; Hartlib to Boyle, 28 February 1653/54; in Boyle, *Works*, vol. VI, p. 80.

> also imprisoned, having a formidable Erysipelas in his arme ; on a certain evening when as the sick Monk did almost Despaire, he swiftly tinged a certain little Stone in a spoonful of Almond Milk...Butler said unto the keeper of the Prison, reach this supping to that Monk, and how much soever he shall take thereupon, he shall be whole at least within a short hours space ; which thing even so came to pass with the greatest admiration of the Keeper, and the sick man not knowing from whence so sudden health had shined on him, seeing that he was ignorant that he had taken any thing...On the morning following, I being intreated by great men, came to *Vilvord* as a witness of his deeds : Therefore I contracted a friendship with *Butler*. Presently afterwards, I saw a poor old Woman[,] a La[u]ndress, who from sixteen years of age or thereabouts, laboured with an intolerable Megrim [Migraine], presently cured in my presence...I was amazed, as if he were become another *Mydas*; but he smiling on me said : *My most dear Friend, unless thou come thitherto, so as to be able by one only Remedy, to cure every Disease, thou shalt remain in thy Young Beginnings, however old thou shalt become.* I easily assented thereto, because I had learned that thing from the secrets of *Paracelsus*...[236]

Robert Boyle told Hartlib in September 1653 that Starkey's *Ens Veneris* was an excellent remedy for agues, fevers, headache, and other such diseases, and was a true "Medicina Pauperum because for 5 Sh[illings] so much may be prepared with it as may serve a 100 Poore People."[237] In *Pyrotechny*, Starkey mentions his preparation of calcined *Vitriol* named "Oil of *Venus*," of which Starkey writes: "Nature hath not a more soverain remedie, for most (not to say all) diseases : This is the true *Nepenthe* of Philosophers."[238] Starkey writes in his *Pill Vindicated* :

> I may say confidently that I was the first that made this [*Ens Veneris*] in *England* (that is known) in the year 1[6]52 I prepared it, for the Honourable *Robert Boyl* Esq; one of the Royal Society, who hath wrote of its excellency, as his extant Treatise thereof can testifie...I must assure those that are concerned, that what I made in 16[5]1 for the Honourable *Robert Boyl* Esq; and is by him commended, is so inferior a preparation to that as is made by me now, that the former deserves not the same name with this latter...This is the *Nepenthes verum* of the Phylosophers, the true *Ladanum* without *Opium*, lengthening the life by Gods permission, and conquering powerfully monstrous tragical maladies...[239]

[236] Helmont, "Butler," *Oriatrike*, p. 587.
[237] Hartlib, *Ephemerides*, [September 1653], MM-MM4; in Wilkinson, "Hartlib Papers II," p. 103.
[238] Starkey, *Pyrotechny*, pp. 32-33.
[239] [Starkey], *Pill Vindicated*, pp. 6-7.

Starkey's *Liquor Alchahest* was also dedicated to Robert Boyle, published posthumously in 1675,[240] the year of Boyle's famous public announcement of discovering alchemical "Mercury." Boyle firmly believed in the existence of the Philosophers' Stone and the "liquor Alkahest," which he considered to be a valuable medicine, as well as a vehicle to communicate with angels and rational spirits.[241] Starkey personally communicated to Boyle in early 1651 that while residing in New England, he had obtained a copy of the *Theatrum Chemicum*, "w[hi]ch being brought to him and hee falling asleep in the reading of it Hee was biden to arise and looke out such a place in it that speakes of the true Philosophical fire, w[hi]ch accordingly hee did, and turned to the very place and sleeping in the darke yet underlined it w[hi]ch hee hath to shew to this day."[242] In a letter to Boyle in January 1652,[243] Starkey also claimed that after falling asleep in his laboratory he received a visit from his Eugenius, or "Good Spirit," who revealed to Starkey the secret of the Alkahest.

> I bring new things to you...I now tell you that God...finally opened to me the gates of nature and given me not only the understanding but also the possession of that immortal liquor Ignisaqua...But it is otiose to expand on this joy, for you will say it is a miracle that you have obtained the thing in such short order. Cease marvelling, for it is a thing neither of one who flies nor of one who runs, but of the merciful God, before whom the entire world is as a fungus...I know well enough what the alkahest is, and I have gathered this in effect from Paracelsus as well as from Helmont, but it was from the Father of Lights that I brought about the preparation of the thing itself...And its preparation corresponds singularly to the description of Paracelsus...I could not have found it without the immediate finger of God...Therefore I will complete my history of finding the alkahest briefly. Having completed a new furnace (which I now have), being unoccupied with labours around the first hour of night, and fatigued by reading and meditating, after resting my head upon my forearm, a deep sleep came upon me.

---

[240] [George Starkey], *Liquor Alchahest, Or A Discourse Of that Immortal Dissolvent Of Paracelsus & Helmont...*, London, Printed by *T. R.* & *N. T.* for *W. Cademan* at the *Popes-Head* in the Lower Walk of the *New Exchange*, 1675. Starkey's friend Jeremiah Astell dedicated the work and identifies Starkey as the author in the preface.

[241] Royal Society Library, Boyle Papers, vol. VII, ff. 134^{v}–150^{v}; in Boyle, *Letters and Papers* [microform]; Michael Hunter, "Alchemy, magic and moralism in the thought of Robert Boyle," *British Journal for the History of Science*, 23 (1990), pp. 396-398; Principe, *Aspiring Adept*, pp. 190-201; Jan W. Wojcik, *Robert Boyle and the Limits of Reason*, Cambridge, Cambridge University Press, 1997, pp. 135, 141–144.

[242] Hartlib, *Ephemerides*, [1650/51], A6-A-B1; in Wilkinson, "Hartlib Papers II," p. 87.

[243] It was a common practice in England during this period to begin the year on 25 March, the feast of the Annunciation. Starkey's letter is dated 26 January 1651.

> And behold! I seemed intent on my work, and there appeared a man, entering the laboratory, at whose arrival I was astonished. But he greeted me and said, "May God support your labours." When I heard this, realising that he had mentioned God, I asked who he was, and he responded that he was my Eugenius; I asked whether there were such creatures. He responded that there were...Finally I asked him what the alkahest of Paracelsus and Helmont was, and he responded that they used salt, sulphur, and an alkalised body, and though this response was more obscure than Paracelsus himself,[244] yet with the response an ineffable light entered my mind, so that I fully understood. Marvelling at this, I said to him, "Behold! Your words are veiled, as it were by fog, and yet they are fundamentally true." He said, "This is so necessarily, for the things said by one's Eugenius are all certain [*scientifica*], and those just said by me are the truest of all."[245]

Starkey appears to have had a number of revelations with the Alkahest, for in an autobiographical note he writes in 1660: "the good God taught me the whole secret...About the end of the year 1654 the whole secret was revealed to me by divine grace."[246] Perhaps referring to this great secret, Starkey writes in early 1656: "God revealed to me the whole secret of the liquor alkahest; let eternal blessing, honor,

[244] "Paracelsus is known for introducing numerous variations in his diagnoses, analyses, and prescriptions. His theory of plague retains elements of constancy from the early fragmentary *Volumen Paramirum*...'You should know that all diseases are cured in five ways, and our medicine therefore begins with the cure and not with the causes, for the reason that the cure reveals to us the cause'...plague is 'four thousand years' old according to Paracelsus...[He asserts] the human being is made after the image of God: namely as body, soul, and spirit...[Paracelsus writes:] '[T]he same *prima materia* is nothing other than sulphur, salt, and mercury, which are the soul of the element, its spirit and its proper substance. And the three things here mentioned have in them all metals, all salts, all gems, etc...the *prima materia* is together in the mother, as if in a sack, namely composed of three parts. Now, however, as many as are the kinds of fruit, even that many are the kinds of sulphur, salt, and also mercury'...Paracelsus harks back to the insight of the *Volumen Paramirum*. Everywhere there is a poison: 'For there is no food that does not contain within it poison'...Applying the three things to the discussion of specific diseases such as 'St. Anthony's fire' (*brant*), the Basel lectures universalize the significance of the three [elements] in order to account for life in the organic being or the unity of the inorganic object...[Paracelsus writes:] 'It is not as they say, that alchemy makes gold, makes silver; here the project is; make *arcana* and direct them against the diseases; that is the end'...Like Christ at the Judgment, the alchemist separates good and evil: 'Who would contradict that in all good things a poison lies hidden and exists...mustn't one separate the poison from the good and take the good but not the bad?'...[Paracelsus writes:] 'Thus [what] is contained in alchemy, to be acknowledged here [as] medicine, is the cause of the great concealed virtue that lies in the things of nature, which are not revealed to anyone, unless alchemy makes them manifest and brings them forth.'" Weeks, *Paracelsus*, pp. 62, 73, 116–117, 124, 126, 153–154. Quoted from Paracelsus, *Sämtliche Werke*, vol. I, pp. 123, 165, vol. II, p. 41, vol. III, pp. 41-42, vol. VIII, pp. 185, 191, 197, 371.

[245] Starkey to Boyle, 26 January 1651 [Old Style, i.e., 1652], Royal Society Library, Boyle Letters, vol. V, fol. 133r; in *Correspondence of Boyle*, vol. I, pp. 119–121.

[246] [Starkey], 1660, Harvard University, Houghton Library Autograph File, fol. 4r; in *Starkey Notebooks and Correspondence*, p. 332.

and glory be to Him!"[247] Later that same year Starkey would write: "Today by the gift of God, an anonymous friend revealed to me the full practical knowledge of the grand Elixir under the sacrament of silence."[248] Starkey also writes in his autobiographical notes in 1658: "From the year 1647 up to this year and day, I have exerted myself in the search for the liquor alkahest with many studies, vigils, labors, and costs. Today for the first time it has been granted and conceded to my unworthy self by the highest Father of Lights, the best and greatest God, to attain complete knowledge of it and to see its final end. To Him let there be eternal praise both now and forever. Amen."[249]

In Philalethes's *Ripley Reviv'd*, Starkey makes a number of extraordinary claims concerning the alchemical Elixir:

> [O]f all Medicines in the World it is the highest, for it is the true *Arbor vitæ*, which doth answer the universal desires of them who have it...it kills all the venom of any disease or malady, so that those diseases which do astonish the beholders, are by this overcome even *ad miraculum* : for suppose a man dying with the Tokens of the Plague, so that he is upon the very point of departure, (and the decree be not past, for then there is no recovery) if he have but a drop of this *Elixir* poured down, so that he swallow it, he shall immediately recover, and in short time he will be restored to his former health...For although by Venery, or a *Tabes*, or Bleeding, or by any other way a man be debilitated, he may be restored by this *Elixir*, not only to perfect health, but also to such a measure of strength which he never had before. Yea and a man or woman who is born to hereditary weakness, may be changed into a more then ordinary strength by the use of our Medicine : or a man who by labour, sickness and years, is come to the Graves mouth, even to drop in it, may by use hereof be restored his hair, his teeth, and his strength, so that he shall be of greater agility then in his youth, and of greater strength, and may live many years, provided the period of the Almighties decree be not come.[250]

In *Ripley Reviv'd*, Starkey declares: "GOD only is the dispenser of these glorious Mysteries."[251] Starkey also makes reference to the "Lady the Sages have called *Juno*, or the Metallick Nature," and identifies

[247] George Starkey, 20 March 1655/56, Sloane Collection, British Library, MS 3750, fol. 19v; in Ibid., p. 175.

[248] George Starkey, 18 August 1656, Royal Society MS 179, fol. 3v; in Ibid., p.306.

[249] George Starkey, 20 September 1658, Harvard University, Houghton Library Autograph File, fol. 2r; in Ibid., p. 329; Starkey, *Natures Explication*, p. 295.

[250] Philalethes, "First Six Gates of Sir George Ripley's Compound of Alchymie," *Ripley Reviv'd*, pp. 243-247.

[251] Philalethes, "An Exposition upon Sir George Ripley's Epistle to King Edward IV," Ibid., p. 24.

her with the Shu'lammite sister-bride of Solomon's *Song of Songs.*[252] In what appears to be an account of another of Starkey's encounters with the supernatural, Starkey then reveals this figure to be Wisdom, the "Queen of Heaven," who is eulogized in Solomon's *Proverbs*: "Long life is in her right hand; in her left hand are riches and honor."[253]

> Then I lift up mine eyes, and behold I saw *Nature* as a Queen gloriously adorned, sitting upon her Throne, and in her hand a fair Book, which was called, *Philosophy Restored to its Primitive Purity* ; whom with low submission I did obeysance to, and she graciously took notice of me, and gave me this Book to eat it up, which I did, and straight-way she had another of the same in her hand : Then was my Understanding so enlightened, that I did fully apprehend all things which I saw and heard ; and when I approached to any Gate or Door, straight-way (as though they were acted by a sensitive Spirit) they opened of their own accord : And all in the House did fealty to me, and said that I was to be honoured as Lord of the place : *For*, say they, *the Queen and He are in love united, and she moreover hath plighted her troth to him.* Then I considered with my self, and behold the Book that I had devoured (like a Charm) had so commanded my Spirits, that I could think of nothing more than the enjoyment of this rare Beauty which I had beheld : And while I was full of these thoughts, behold I heard a Voice behind me, saying, *What wouldest thou in this World?* I was a little astonished at the Voice, but yet boldly answered, *Nothing but that I might once more see that admirable Perfection which once I beheld in a Nymph, which not long since I saw, who with seeming affection did salute me, and gave me a Book to eat ; which when I had eaten, my Intellectuals seemed as though the Candle of the Lord had been kindled in them : But since I could never see her whom my Heart longs for : Oh that I might only be so happy again!* Then said the Voice, *Thou art happy in that thou hast seen her, more happy in that she gave thee that Book, which few in an Age attain to ; most happy in that thou couldest and didst eat it, which every one that hath it cannot do : She therefore whom thou seekest for, is gone into her retired Solitudes, and as a Legacy hath left thee two great Treasures, the Treasure of Riches, and the Treasure of Long Lif[e]* : Then said I, *Ah Sir, this you tell me of, is nothing but an aggravation of my misery ; for all the wealth in this world I count but as a straw in comparison of the enjoyment of that most admirable Lady's presence, whose Service I should take for a greater happiness, than if I were Master of all the World besides. If then I may not see her again, my Life will be to me a burden, and to what then will Long Life avail?* Thus I sat bemoaning my self, and I heard a shrill Voice as if were close by me, and I looked suddenly, and behold an unspeakable Light, in comparison whereof the Sun it self seemed dark ; and close by me I saw a most secret place, and in it a secret Room of

[252] Philalethes, "First Six Gates of Sir George Ripley's Compound of Alchymie," Ibid., p. 98.
[253] Proverbs 3:16, RSV.

Diaphanous matter, and round, and within it this Lady whom I formerly had seen, upon her Throne ; and another in the person of a King, in most gay Raiment, as it were a Robe of beaten Gold, which reached from his shoulders to the ground, and a Crown of pure Gold on his head ; and a third person, who like a Water-bearer had a Pitcher on his shoulder, and in the midst of it there burned as it were a Lamp : The sight was excellent, yet I could not be pleased, for that I saw this Lady stark naked with this King, so in private ; and while I viewed the Room, I found it was exquisitely closed on every side, so that it seemed as if it were made of one intire piece of Crystal. I marveled at what I saw : for the House was but small, the Chamber less, and the Closet of Crystal to sight no bigger than a small Egg ; and the three Parties, with all the Accoutrements of them, might well have been inclosed in a Hazel Nut : Yet was their Delineaments so lively, that I might easily discern her intire shape, whom I could not but with distracted thoughtfulness and a sad countenance behold ; which she perceiving, said unto me, *Friend, Why art thou sad? I am not sad*, quoth I, *most Noble Lady, but am pensively meditating on what I behold, which doth not a little amaze me, the sight not being to be parallell'd in* John Tradescants *Chamber of Rarities, which is the System of the Novel Rarities of the known World : For Whom I lately beheld glorious upon a Throne in the Majesty of a Queen, I now see cloistered up in a small Diaphanous Pix, in a stature so small as is scarce credible : Moreover, whom I deemed so piously virtuous a Lady, to be so retiredly naked with a man, only attended with a Water-bearer, makes me very thoughtful what this thing should be. Moreover, it was my hopes so to have ingratiated my self into your favour, as to have been a Servant unto you, who I see are otherwise provided of a Lover.* Then said she, *My Friend, what you admire in this strange Metamorphosis of me, know that it is by a Magical Vertue, which is alone given to me from GOD, my immediate Lord and Ruler ; and for any Diabolical Art, which your Scruple seems to manifest your suspition of, it is because of your unexperience in these things ; and this your Ignorance is no way provoking unto me, for in these Affairs (though a man) yet you are but a Child ; and this liberty I allow all my Sons while they are Children, so to speak, so to think, and so to act ; and I love to hear and answer their childish prattle. Know then that the Devil is but one of my Servants, and in my Kingdom he doth serve GOD, his and my Lord : And though of all my Servants he is the worst, yet he can do nothing of himself, either without me, or against me, or above me : He for the most part is a deceitful Jugler, and doth make things appear, that are not ; but whatever is actually effected by him, is nothing but what is in my Power : He only applies Agents to Patients, and adds a little of his own villanous qualities, as a circumstantial aggravation of the horror of what he thus (by my virtue) brings to pass, and then his villanous mind attributes that to himself, which is my Act, that so he might arrogate the honour due to my Lord and his Master. Now I will tell you a strange thing, which yet is very true : I am obedient to all my Subjects, which are many, and they obey me ; I rule them, and they do as it were inforce me, for so my*

Lord hath pleased to ordain it : If they call me, I am straight at hand ; yea, in my Body which thou seest (which is no Body (but only representative) for I am all Spirit) I feel the Sympathies and Antipathies, the Actions and Passions of every thing in the World ; and I must be always present, for nothing is or can be well done without I be[*ing*] *present : I always work according to the subject and its disposition, which doth alter the effect wonderfully. In a word, whatever thou seest that I am, and more then thou canst see by far, though thou hadst the Eyes of* Argus. *My Rule is not as is the Rule of Princes among Men, but I am serviceable to all, yea to the least Worm in the World ; and because I am so serviceable, therefore my Master hath appointed that nothing can or may disobey me, or offer violence to me ; the Devil here hath no power, though malice enough : Therefore my Lord hath given me his own* Diploma *to make me the more Honourable ; first, An Omnisciency of all things which are done in the World, as touching the Being, Conservation, or Mutation of them ; and next, An Omnipresency, by which I am every where present at once, and I am seated in the Will of God, which is my Centre...Now as concerning your jealousie for that you see me naked with this King, know that this place and my Kingdom are in the State of Innocency, though we are by the Fall of* Adam *laid subject to Vanity ; and till the final Restitution of that Fall, I am forbidden to work any thing of my own accord beyond the state of fading corruptibility, though all things have an incorruptible Spirit, which when Heaven and Earth shall be renewed, shall cause an Immutable glory in all these things. Know then that this King is my Servant, and he hath many Brethren who in their passage to him are taken Prisoners, and kept in bondage, and there is no way to Redeem them, unless he give his Flesh and Blood for their Ransom, which cannot be effectual, unless he die and arise from the Dead : This I cannot perform alone my self, nor can any help me herein but Man alone ; for God hath here limited my power...If thou couldest but understand and believe, thy very Soul would command all Nature in the whole Fabrick of it : for if thou didst but know things as they are, thou wouldest withal clearly see the Dignity of thy Soul, being the Image of God ; and this would command Faith, and kindle Desire : Now Faith and a kindled Desire in the Soul is that extatical Passion which attracts the whole* Phænomena *of Nature. This is the Dignity of a Mental Man. Now then, my Friend, hearken to me, and what I advise, that do ; help me in what I cannot, and I will help thee in what thou canst not ; so shalt thou be (to GOD subordinate) Lord both of me and mine ; and the Blood of this King, which redeems his Brethren, will give thee a Medicine to command all the Imperfections of thy mortal body ; and though it be no Antidote against Death, the irrevocable Decree being past, yet it triumphs over all the Miseries of Life, both of Poverty and Sickness, and it possesseth a Man of the most incomparable Treasures of this World.* Then full of Admiration, with Tears for very Joy trickling down abundantly, I bespake her, and said, *Lady, I thank you for your so great favour to me, as so familiarly to discourse with me ; Now then, without any Complement, I am yours* (ad usque aras) *and whatever you please, that will I do.* Then said she, *Under this Chamber and Closet there is a Stove, put Fire into it, for this King must sweat to death. Ah sweet Lady,* said I, *and what will become of you? Care not you for that*, said she, *do you as I bid you : But yet farther to satisfie your curious mind, let me tell you, That I indure without hurt the most violent Fires which are or can be made, for I am in them all, and no less in the most frozen places.* Then I considered, and methought my Understanding it was inlarged, and I perceived the extent of Nature, and of a sudden she appeared not to my sight ; but where she was I saw a most exquisite Light, which took up an incredible small room, and methoughts my Head seemed as it were diaphanous : And while I considered these things, it came into my mind to wonder what was become of my Guide, for I miss'd him. While these thoughts perplexed me, an Answer was given, as if from an intelligent Spirit within the Glass, saying, *Let not thoughts fill your mind, he whom who seek is with us, for so it must be, this King is his Lord.* This straight made me view the complexion of the Water-bearer, and his countenance told me that he was my very Guide : Then I viewed his Pitcher well, and found that his Pitcher was clear as pure Silver ; and what was strange, the Bearer, and the Pitcher, and the Water in it were one ; and in the midst of the Water, as it were in the very centre, there was a most radiant twinkling Spark, which sent forth its Beams even to the very surface of the Water, and appeared as it were a Lamp burning, and yet no way distinguishable from the Water. The Voice then spake to me a second time, *Delay not to put Fire under us, and govern it as you shall hear the Voice direct you.* Then I put fire in at the open door at the top of the high Turret...And as I wistfully beheld it, I saw as it were a goodly Lady in the midst of it, which was no way resembling the former Beauty which I had discoursed withal, whose Name was *Nature* ; yet indeed very beautiful, even to the parallel of *Helena*. This Lady was naked, and of an admirable fair complexioned Skin, as bright as the finest Silver ; at first she appeared very small, and waxed bigger and bigger, until the Water appeared no more, but she her self had transmuted its whole substance into her shape. This sight I beheld with pity, for she (far unlike the first Lady) was wholly impatient of the heat which I had made, and yet was so inclosed in the Closet that she could not get out ; she sweat therefore even as though she would melt, and seemed as though continually fainting : Then the King (who seemed as it were glad of the heat) seeing her knew her to be his Sister, his Mother, and his Wife, and compassionating her estate, ran unto her and took her in his Arms, and she feeling him, did so strongly embrace him, that he could not shake her off, and with her sweat partly, and partly with her tears, she did so bestream his Kingly Robes, which shone like unto *Tagus* or *Pactolus*, that they were all suddenly changed into a colour Argent : the King loving her exceedingly, asked her what she desired? She answered, *That her desire was to have of him Conjugal Fealty* ; *for*, said she, *I cannot endure this heat, but I must die in it, and without me your Highness can have no Off-spring* : The King condescended,

> and granted her Request, and so soon as she conceived the Kings Seed, she said that she was better able to endure the Fire which did prevail upon her. Therefore not contended, she had a second, a third and fourth Benevolence, even to the eleventh time : Then said the King, *I am very faint and weak* : and trying to go, as formerly, his Legs and Feet failed him, his Flesh and Body wasted as it were to nothing, and so continued worse and worse, until at length his Body being thus wasted by Venery, began to sweat exceedingly, so long he sweated, till he was as it were wholly consumed ; and his Wife seeing what fell out, wept bitterly, and her tears mingling with her Husbands sweat, grew into a large stream, in which both she and the King were drowned...[254]

In the same work, Starkey identifies the immortal "liquor Alkahest" with Ripley's *Aurum Potabile*, the "Drinkable Gold" of the Adepts.

> And now I come to the second reward that Wisdom doth bring with her, and that is length of days ; and here I have transposed some few Verses of *Ripleys* concerning *Aurum Potabile...Paracelsus* the first Author of this, did name this dissolving Water his Alkahest, his *Ignis Gehennæ*, his *Corrosivum Specificum*, with many other names. This Medicine thus made of *Gold* by the Alkahest, as it is Philosophical and real, so it is very excellent, and known only to the *Adepti*.[255]

Paracelsus was said to have conceived the word Alkahest from the German *Allgeist*, meaning "All Spirit," a term that was popularized by van Helmont, which he also called *Ignis Gehennae*, the "Fire of Hell."[256] Starkey refers to the liquor Alkahest as "all spirit,"[257] and calls it "the ultimate torture of fire, for it burns more than common fire, and is called fire water, or the fire of Gehenna."[258] There is only one mention of the "liquor Alchahest" by Paracelsus, which he says is drunk with "the wine of life,"[259] and Paracelsus was also said to have received his

---

[254] Philalethes, "First Six Gates of Sir George Ripley's Compound of Alchymie," *Ripley Reviv'd*, pp. 103–117.

[255] Philalethes, "Sir George Ripley's Preface," Ibid., pp. 78-79.

[256] Ladislao Reti, "Van Helmont, Boyle and the Alkahest," *Some Aspects of Seventeenth-Century Medicine & Science: Papers Read at a Clark Library Seminar, October 12, 1968*, Los Angeles, University of California, 1969, pp. 6-7.

[257] In unpublished notes Starkey refers to the Alkahest as "'Agehest,' that is, all spirit." [George Starkey], "Praxis chemiatria," Sloane Collection, British Library, MS 631, fol. 199^r; in *Starkey Notebooks and Correspondence*, p. 325.

[258] George Starkey, "Vitriologia," Sloane Collection, British Library, MS 3750, fol. 12^v; in Ibid., p. 163.

[259] Paracelsus, *De Viribus Membrorum*, 1526–1527, vol. II, p. 5; in Partington, *History of Chemistry*, vol. II, p. 138.

knowledge of the "Elixir of life"[260] from a visiting angel.[261] In his "Tree of Life," van Helmont writes:

> I have written concerning long Life, what I know to be true ; not indeed for Young Beginners, as neither to be comprehended by readings...There shall at sometime be an Adeptist (in its own maturity of Dayes) who shall understand that I have spoken Truth...what things I have written of the Cedar, I have offered for a memorial of honour towards God, who hath been propitious or favourable unto me. But other things which there are concerning the Cedar, shall be buried with me ; for the World is not capable thereof...Indeed *Paracelsus* hath been silent (even as in most of his other Descriptions) as to the addition of the Liquor Alkahest, wherewith the whole matter is presently solved throughout its whole, and the Medicine succeeds according to his Description...for with one only small drop being given to drink in Wine, I have oftentimes so refreshed those that were desperate through a contagious Fever, that they have as yet dined with me at noon, who at midnight had received the last or extream[e] unction of holy Oyl. Truly through want of the Being of Cedar, the Elixir of Propriety doth relieve. But what shall I say ? The Alkahest is required ; which is not granted to thinkers, but only to knowers, and that indeed, to those on whom Knowledge is doubled.[262]

The alchemists were employed with the production of "alcohol," defined in this period as both "powder of antimony" and "spirit of wine."[263] Among the names assigned to alchemical "antimony" were the "Red Lion," the "Fiery Dragon," the "Fiery Satan," the "Son of Satan," and the "Ultimate Judge."[264] In 1651, Worsley related to Hartlib that Starkey was conducting "an exp[e]rim[ent] of making Powder w[hi]ch shall be kindled by the sun," and "lucriferous exp[e]rim[ents] of Antimony w[hi]ch hee should have fallen upon first and afterwards prepared his univ[ersal] or more aprooved medicines. Also

---

[260] "In according ferments a central position in his natural philosophy, Van Helmont would seem to have been indebted to alchemy in the first place and to Paracelsus in the second... Paracelsus had bracketed ferment with such powerful arcana as elixir and semen. These are seen as predominantly spiritual and it is through fermentation that corporality is lost and bodies 'ascend to their exaltations'. The 'highest ferment which is reserved in nature' brings about the maturing of fruit by means of digestion and growth. Elixir is 'medicine fermented from the seven metals'. Christ, the 'food of the soul', the divine word, is called *fermentum* in pseudo-Paracelsian tracts..." Pagel, *Joan Baptista Van Helmont*, pp. 79-81.

[261] Principe, *Aspiring Adept*, p. 194.

[262] Helmont, "The Tree of Life," *Oriatrike*, p. 81.

[263] Forbes, *History of Distillation*, p. 107.

[264] Joseph William Mellor, *A Comprehensive Treatise on Inorganic and Theoretical Chemistry*, London, Longmans, Green and Co., 1929, vol. IX, p. 341; John Read, *Prelude to Chemistry...*, New York, Macmillan Company, 1937, p. 310 n. 33.

an Exp[e]rim[ent] of preserving by way of decoctions the s[c]ent[,] colour[,] shape of plants or flowers."[265]

In 1652, Hartlib writes that Starkey's "Elixir Proprietatis" was "very fragrant and refreshing. He hath found out a kind of fermentation whereby he can prepare excellent Medicin[e]s and Cordials as good as if they were done w[i]th the Alcah[est] yet without the Alcahest."[266] In a letter to Johann Moriaen in 1651, Starkey claimed that from his philosophical "Mercury" he had prepared Paracelsus's "Mercury of life" that was also described by van Helmont.[267] In a letter to Boyle the following year, Starkey writes:

> I know the sulphur of Glaure of Augurelli[268] and Helmont and its extraction...Recently, when your brother's daughter suffered a plague (as they say) in the intestines, an endemic illness, a servant was sent to me...and I sent a little of the prepared golden sulphur. Nor do I doubt its success...it was added that if her illness continued they would send for me on the following day, and since I have heard nothing further thence, I gather for certain that the girl has convalesced.[269]

In his *Producibleness of Chymical Principles*, Boyle writes: "For *Raymond Lully*, whom I take to be one of the greatest chymical philosophers... speaks of mercury in a dark and allegorical sense," and "the word *sal armoniacum*...amongst Hermetic philosophers it often signifies, not common sal armoniac...but a very differing and much more noble and operative thing."[270] In what helps to explain the various references to "sulphur," "mercury," and "antimony" by Paracelsus, van Helmont, and Starkey, Boyle writes in his *Sceptical Chymist*:[271]

> I find that even Eminent Writers (such as *Raymund Lully*, *Paracelsus*, and others) do so abuse the termes they employ, that as they will now and then give divers things one name ; so they will oftentimes give

[265] Hartlib, *Ephemerides*, [April 1651], E-E2; in Wilkinson, "Hartlib Papers II," p. 94.

[266] Ibid., [October 1652], DD-DD4; in Ibid., p. 99.

[267] "A little while ago the secret preparation (from this mercury) of the mercurius vitae of the great Paracelsus (with his successor and best interpreter Joan Van Helmont describing it) was made clear to me." Starkey to Moriaen, 30 May 1651, Hartlib Papers, 17/7/2a; in *Starkey Notebooks and Correspondence*, p. 37.

[268] Starkey is referring to Giovanni Aurelio Augurello (1451–1524) and his poem *Chrysopoeia*.

[269] Starkey to Boyle, 16 January 1651 [Old Style, i.e, 1652], Royal Society Library, Boyle Letters, vol. V, ff. 131^v–132^r; in *Correspondence of Boyle*, vol. I, pp. 115–116.

[270] Robert Boyle, *The Producibleness of Chymical Principles*, Oxford, Printed by Henry Hall..., 1680; in Boyle, *Works*, vol. I, pp. 637, 643.

[271] "Divers of the Hermetic Books have such involv'd obscuritys that they may justly be compar[e]d to Riddles written in Cyphers. For after a Man has surmounted the difficulty of decyphering the words & terms, he finds a new & greater difficulty to discover ye meaning of the seemingly plain Expression." Royal Society Library, Boyle Papers, vol. XXV, fol. 295; in Boyle, *Letters and Papers* [microform].

> one thing many names...even in technical words or terms of art, they refrain not from this confounding liberty ; but will, as I have observed, call the same substance, sometimes the sulphur, and sometimes the mercury... And now I speak of mercury, I cannot but take notice, that the descriptions they give us of that principle or ingredient of mixt bodies, are so intricate, that even those, who have endeavoured to polish and illustrate the notions of the chymists, are fain to confess, that they know not what to make or [of] it either by ingenuous acknowledgements, or descriptions, that are not intelligible.[272]

Among his contributions to the medical sciences, George Starkey was known for his celebrated laudanum "Sope Pills," a remedy consisting of "antimony"[273] and opium,[274] the latter a medicine inherited by the Chemical Physicians from the Hippocratic medical tradition,[275] which many of them were known to prescribe for their patients and habitually consume themselves. George Wilson relied upon Starkey's formulas in his famous textbook, *A Compleat Course of Chymistry*, first published in 1691, in which he records the formula for "Dr. *Starkey's* Pill," containing "Extract of Opium," "Tincture of Antimony," and "the *Corrector*." Wilson writes: "This I had from the Ingenious Dr. *Starkey*'s own Mouth, in the Year 1665. a little before his Death, who then told me, He gave *Matthews* the former for a little Mon[e]y ; but this is it which he successfully made use of himself. It is both more Diaphoretic, and a greater Anodine than the former ; and I have hear'd it affirm'd by several Gentlemen, who have made use of it in their Practice, to be the best Laudanum they ever met with."[276] In a later edition Wilson listed the contents of the "Universal Anodine" as opium and the Secret Corrector in French brandy.[277]

---

[272] Boyle, *Sceptical Chymist*; in Boyle, *Works*, vol. I, pp. 520-521.

[273] Lockyer's Pill contained common antimony that poisoned those who took it. George Starkey, *A Brief Examination and Censure Of Several Medicines, of late years Extol'd for Universal remedies, and Arcana's of the highest preparation...Namely, Lockyers pill, Hughes pouder, Constantines Spirit of Salt, with several other of their kind, by which the Art of Pyrotechny is in danger of being brought into Reproach and Contempt...*, London, s.n., 1664, pp. 17-26; Newman, *Gehennical Fire*, pp. 192-200.

[274] George Kendall, *An Appendix to the Unlearned Alchimist; Wherein is contained the true Receipt of that Excellent Diaphoretick and Diuretick Pill...*, London, Printed for *Joseph Leigh...*, 1664[?], pp. 3, 10–15, 31-38; George Wilson, *A Compleat Course of Chymistry...*, London, Printed for *W. Turner...*, 1700, pp. 272-273.

[275] Ch. Rice, "Historical Notes on Opium," *New Remedies*, 5 (1876), pp. 229-232, 6 (1877), pp. 144–145, 194–195; John Scarborough, "Hermetic and Related Texts in Classical Antiquity," *Hermeticism and the Renaissance*, I. Merkel and A. G. Debus (eds.), Washington, D.C., Folger Shakespeare Library, 1988, pp. 27-37.

[276] Wilson, *Compleat Course of Chymistry* [1700], pp. 272-273.

[277] George Wilson, *A Compleat Course of Chymistry...*, London, Printed for John Bayley..., 1709, p. 310.

George Kendall also recorded the formula for Starkey's Pill, "given to me by the first author of it, Mr. *George Starkey*," which was formerly "com[m]itted to the custody of Mr. *Richard Mathews*, in the year 1655, by Mr. *George Starkey* the first that found it out."[278] According to Kendall, the ingredients of Matthew's Pill are the "Corrector," Hellebore, liquorish, and opium, and Wilson's formula for Starkey's Pill contains the same ingredients excluding liquorish, the ingredient of "antimony" recorded in Wilson's *Chymistry* and the "Hellebore" recorded by Kendall are clearly identical and refer to the same thing.[279] An invaluable medicine throughout the ages, tincture of opium belatedly entered the official *London Pharmacopœia* in 1721,[280] and a variation of Starkey's opium pills was later accepted by the pharmaceutical collection in 1746.[281] So famous were Starkey's opium pills[282] that "Starkey's Soap" remained a name used in soap making into the late nineteenth century.[283]

The eleventh century Iranian physician and alchemist Avicenna also wrote about the astounding effects of opium,[284] and he is said to have died from an overdose of the drug.[285] Paracelsus prepared his own laudanum and also possessed a reputation for frequent intoxication;[286] van Helmont also produced his own laudanum and was respectfully known as "Doctor Opiatus."[287] Starkey was preparing laudanum for

[278] Kendall, *Appendix to the Unlearned Alchimist*, pp. 1, 31.

[279] Ibid., pp. 3-4, 31-32.

[280] *Pharmacopœia Londinensis*, Londini, Typis G. Bowyer, Impensis R. Knaplock..., 1721, p. 154.

[281] *Pharmacopœia Londinensis*, Londini, Apud T. Longman, T. Shewell, et J. Nourse, 1746, p. 25.

[282] In 1687, Reverend Richard Blinman of New London bequeathed to his relatives "10 pound weight of Dr. Starky's Pill" and "a Quart Bottle full of the Tincture of Starky's Pill Diaphoretick," probably purchased in England. Isaac Greenwood, "Rev. Richard Blinman of Marshfield, Gloucester and New London," *New-England Historical and Genealogical Register*, 54 (1900), pp. 42-43.

[283] H. Dussauce, *A General Treatise of the Manufacture of Soap, Theoretical and Practical*, Philadelphia, Henry Carey Baird, 1869, pp. 333, 666.

[284] Péter Tétényi, "Opium Poppy (*Papaver somniferum*): Botany and Horticulture," *Horticultural Reviews*, 19 (1997), p. 400.

[285] Rice, "Notes on Opium," p. 232.

[286] "We learn from the contemporary account of Johannes Oporinus, and from Paracelsus's own hint, that he dictated many of his works to 'secretaries'...In 1555, Oporinus recalled in a letter to Johann Weyer...how the master, who had been dignified by a dual position at the university and as an official physician to the city [Basel], was in the habit of returning home late after garrulous nocturnal drinking bouts. Sometimes Paracelsus would fall into bed fully clothed and then wake up in a fit of rage, terrifying young Oporinus by banging against the wall the sword he carried with him at all times and claimed to have received from an executioner. But it could also happen that the master would arouse himself and his amanuensis from sleep, and, though still inebriated, dictate forth his philosophy with such fluidity and clear sense that, as Oporinus recollected, a sober person might not have improved upon it." Weeks, *Paracelsus*, pp. 3-4.

[287] Ibid., pp. 144, 232.

Boyle in early 1652,[288] and in 1653 Boyle told Hartlib that Starkey had prepared "a great store now" of laudanum, "w[hi]ch together with his Ens Veneris[289] and of Haimatinum[290] are excellent Medicin[e]s."[291] Referring to Starkey in a letter dated February the following year, Hartlib wrote to Boyle of "the strange virtues and effects, which his laudanum hath already produced among us here."[292]

Starkey also engaged in the production of alcoholic beverages, and he too had a well-known reputation for habitual intoxication.[293] In early 1653, Starkey attempted the large-scale production of alcoholic beverages in collaboration with another alchemist and brewer named Webb, and referring to this enterprise Hartlib wrote that Starkey had "also engaged one for making Wines out of Corne."[294] In September of 1653 Starkey told Hartlib that the "businesse of trebble fermentation begins to grow pretty common, wherby bier or ale becomes cleere as rock Water."[295]

In an undated letter of Starkey to Hartlib published in 1655, Starkey refers to "the excellencie of Honey...*by help of it and grain, may be made most excellent Wine, nothing inferiour to the richest Canary or Greek wines, and by the mixture of it with the Juyce of fruits, the best French or Rhenish Wines may be paralell'd, if not surpassed. Nor will any of the Specifick Odour, either of the Honey, or of the Corn, after a threefold fermentation remain.*"[296] In a letter to Boyle in February 1654, Hartlib reported that Starkey had been imprisoned for debt, and was now hiding out in the London suburb of Rotherhith:[297]

---

[288] Starkey to Boyle, 3 January 1651 [Old Style, i.e., 1652], Royal Society Library, Boyle Letters, vol. V, fol. 129^v; in *Correspondence of Boyle*, vol. I, p. 110.

[289] Ladanum, distinguished from Laudanum containing opium.

[290] Essence of Blood.

[291] Hartlib, *Ephemerides*, [September 1653], MM-MM4; in Wilkinson, "Hartlib Papers II," p. 103; Turnbull, "George Stirk," p. 235.

[292] Hartlib to Boyle, 28 February 1653/54; in Boyle, *Works*, vol. VI, p. 80.

[293] [George Starkey], "A Perfect Day Booke," Sloane Collection, British Library, MS 3711, fol. 68^r; in *Starkey Notebooks and Correspondence*, pp. 124–125; *Aut Helmont, Aut Asinus*, p. B4^r.

[294] Hartlib, *Ephemerides*, [March 1652/53], HH-HH2; in Wilkinson, "Hartlib Papers II," p. 102.

[295] Ibid., [autumn 1653], NN-NN1; in Ibid., p. 103.

[296] Starkey to Hartlib, n.d.; in [Samuel Hartlib], *The Reformed Common-Wealth of Bees. Presented in severall Letters and Observations to Sammuel Hartlib Esq...*, London, Printed for *Giles Calvert* at the *Black-Spread-Eagle* at the West-end of Pauls, 1655, p. 30.

[297] In a letter to Hartlib, Child described Starkey "as a bird who is flowen into ye world before fully feathered, or as a good vessell with much saile & little Ballast," and Starkey "wants as yet ye ballast of yeares & Experience of ye world, but if he have ye Alkahest as I hope he hath, he hath enough whithersoever he goes." Child to Hartlib, 2 February 1652/53, Hartlib Papers, XV; in Wilkinson, "Hartlib Papers II," p. 100.

> Dr. *Stirk*...is altogether degenerated, and hath, in a manner, undone himself and his family. I know not directly how many weeks he hath lain in prison for debt ; but after he hath been delivered the second time, he hath secretly abandoned his house in *London*, and is now living obscurely, as I take it, at *Rotherhith*. He hath always concealed his rotten condition from us ; nor hath there been any communication between him and my son [Clodius], as long as you have been in *Ireland*. Mr. *Webb* doth now rail at him and curse him, as having been most wretchedly seduced and deceived by him.[298]

In the summer of 1655, Hartlib recorded that Starkey had "gone to Bristol to assist the Worke of Refining there and to pr[actice] physick,"[299] where he remained for a year before returning to London.[300] In his "Philosophicall Diary" from the same year, Boyle recorded Starkey's recipes for pear and apple cider said to be equal to the best French wines, his "Wine of Corne" produced from grain, and the distilled liquor Starkey called "Spirit of Corne."[301]

In Henry Oldenburg's synopsis of Robert Boyle's "Essay of Poisons," composed in the late 1650s, Boyle refers to laudanum as an "excellent" medicine, and he repeats a story related by Paracelsus's disciple Johann Oporinus:[302]

> As for remedies drawne from vegetable poisons, I shall only mention Opiate medicin[e]s...I know a physitian, much cried up for the cures, he does with opiate remedies, who uses no other preparation of it, than the hiding it crude in a piece of bread dipt in good wine... And such an one Helmont mentions: and indeed I have seen and used a Laudanum, which the Communicator of it to me affirms to have been Helmonts, of whose strange efficacy I know many noble instances... Operinus saith in the life of Paracelsus, how that great boaster bray'd, that with his Laudanum he could almost revive the dead, and Operinus himself confesses his having performed incredible matters with it...The preparation of it I cannot yet get leave to communicate, but when I gaine it, I shall not only present it to you, but be willing, that all the world were acquainted with so usefull a medicin[e]; and in the mean time, having prepared it ourselves, we shall venture to inform you, that the most material of the process is extant in one of the opiat[e] preparations of Paracelsus, where his adding some things beyond our

[298] Hartlib to Boyle, 28 February 1653/54; in Boyle, *Works*, vol. VI, pp. 79-80.

[299] Hartlib, *Ephemerides*, [summer 1655], 34-346; in Wilkinson, "Hartlib Papers II," p. 105.

[300] Harvard University Archives, HUG 4486, Box 3, Folder 5; in Newman, *Gehennical Fire*, p. 320 n. 23.

[301] Robert Boyle, "A Philosophicall Diary. Begun this First of January 1654/5," Royal Society Library, Boyle Papers, vol. VIII, ff. 142^v–143^r nos. 35-37; in Boyle, *Letters and Papers* [microform].

[302] Boyle cites Oporinus's "Life of Paracelsus," published in Daniel Sennert, *De chymicorum cum Aristotelicis et Galenicis consensu et dissensu* (1619).

> skill to prepare, has made artists neglect a proces[s], which, without those additions, is of excellent use, tho not so great, as with them. And I shall presume to add yet further (which direction well considered may make you need no other) that the Laudanum, I have been commending to you, is made by repeated fermentations, digestions and filtrations of opium, with an idoneous juice, impregnated with a few vegetable Ingredients, for the most part usual in other Laudanums. Helmont somewhere says, that, Omne Narcoticum perit in Alcali ["Every narcotic perished in an alkali"]...[303]

Boyle then refers to the "Ingenious physitian" George Starkey and his vegetable "Correctors."

> But, though I have hinted to you in generall, that divers other noxious vegetables, besides opium, may be corrected by Alcalies, yet I must not yet acquaint you circumstantially with the manner of preparing and using those Alcalisate Correctors, for fear of injuring an Ingenious physitian, who, having taken great paines about the improving of poysons, do's now, by remedies prepared of them, yet the greatest part of his credit and subsistence.[304]

Boyle produced laudanum himself and often extolled the virtues of opium in his writings,[305] and in 1674 he published a paper devoted to van Helmont's *Laudanum* in the *Philosophical Transactions* of the Royal Society.[306] This legacy continued with the eminent British physicians Thomas Sydenham, William Cullen, and John Brown, known for their devotion to opium and chronic intoxication.[307] Celebrated as the "English Hippocrates" and considered the greatest epidemiologist of his time, Sydenham declared, "I would not be a doctor without

[303] [Robert Boyle], "Observations out of Mr B. Essay of turning Poisons into Medicins," ff. 84v-85v; in *Works of Boyle*, vol. XIII, pp. 252-254.
[304] Ibid., fol. 87v; in Ibid., p. 256.
[305] Clericuzio, "Helmont to Boyle," p. 327.
[306] Robert Boyle, "An Account of the two Sorts of the *Helmontian Laudanum*," *Philosophical Transactions*, 9 (1674), pp. 147–149; in Boyle, *Works*, vol. IV, pp. 149–150.
[307] Rice, "Notes on Opium," p. 195.

opium,"[308] and he also maintained a close relationship with Boyle[309] and Locke.[310]

In New England, John Winthrop, Jr. frequently dispensed herbal remedies in his practice, often in beer, and his success in brewing beer from corn was reported to Boyle along with several bottles as evidence.[311] An alchemical associate of van Helmont, Winthrop also administered laudanum in his medical practice, which he presumably prepared himself.[312] Winthrop's close associate Gershom Bulkeley also consulted van Helmont's esteemed laudanum formula for his opium preparation.[313] Reverend Bulkeley also attempted to produce

[308] Tétényi, "Opium Poppy," p. 400.

[309] "[B]eing the thirteenth or fourteenth child of a mother, that was not above 42 or 43 years old when she died of a consumption, it is no wonder I have not inherited a robust, or healthy constitution. Many also have said, in my excuse, as they think, that I brought myself to so much sickliness by over much study...the grand original of the mischiefs, that have for many years afflicted me, was a fall from an unruly horse into a deep place, by which I was so bruised, that I feel the bad effects of it to this day. For this mischance happened in *Ireland*, and I being forced to take a long journey before I was well recovered, the bad weather I met with...put me into a fever and a dropsy...for a compleat cure of which, I past into *England* and came to *London* ; but in so unlucky a time, that an ill-conditioned fever raged there, and seized on me among many others : and though, through God's goodness, I at length recovered, yet left me exceeding weak for a great while after ; and then for a farewel[l], it cast me into a violent quotidian or double tertian ague, with a sense of decay in my eyes...after such a train of mischiefs, which was succeeded by a scorbutick cholick [scurvy], that struck into my limbs, and deprived me of the use of my hands and feet for many months, I have not enjoyed much health, notwithstanding my being acquainted with several choice medicines ; especially since divers of these I dare not use, because by long sitting, when I had the palsy, I got the stone, voiding some large ones (as well as making bloody water) and by that disease so great a tenderness in my kidneys, that I can bear no diureticks, though of the milder sort ; and that I am forced to forbear several remedies for my other distempers, that I know are good ones..." Robert Boyle, "Medicinal Experiments," *Some Receipts of Medicines, For the most part Parable and Simple. Sent to a Friend in America*, London, s.n., 1688, [preface]; in Boyle, *Works*, vol. V, pp. 315-316. Boyle was in Ireland between June 1652 and July 1654, apart from two months in 1653. In August 1649, Boyle informed his sister Katherine that "These three or four weeks I have been troubled with the visits of a quotidian ague...and, in the intervals of my fits, I both began and made some progress in the promised discourse of *Public Spiritedness* : but now truly weakness, and the doctor's prescriptions, have cast my pen into the fire..." Boyle to Lady Ranelagh, 2 August 1649; in Boyle, *Works*, vol. VI, p. 48. In a letter dated 23 July 1649, almost certainly intended for Hartlib, Boyle writes: "Your last Commands are of such a quality, that I must confesse, that 'tis not without some Reluctancy that I obey them. Not so much because all the moments employ'd in that Duty, must be snatcht from my new-erected Furnaces, & that many of the lines Yow exact must be trac't by a hand benum[m]'d (as well as my Fancy) by the unwelcome visits of a Quotidian Ague..." Boyle Letters, vol. VI, fol. 3^r; in Boyle, *Letters and Papers* [microform]. Reproduced in R. E. W. Maddison, "The Earliest Published Writing of Robert Boyle," *Annals of Science*, 17 (1961), pp. 165–166.

[310] John F. Fulton, "Boyle and Sydenham," *Journal of the History of Medicine and Allied Sciences*, 11 (1956), pp. 351-352; Maddison, *Life of Boyle*, pp. 128–129, 133; Romanell, *Locke and Medicine*, pp. 17, 69-89.

[311] Birch, *History of the Royal Society*, vol. I, pp. 198, 205-206; Fulmer Mood, "John Winthrop, Jr., on Indian Corn," *New England Quarterly*, 10 (1937), pp. 131–133; George E. Gifford, Jr., "Botanic Remedies in Colonial Massachusetts, 1620–1820," *Publications of the Colonial Society of Massachusetts*, 57 (1980), pp. 266-267.

[312] Wilkinson, "Hermes Christianus," p. 232.

[313] Jodziewicz, "Bulkeley of Connecticut," pp. 17-23, 78-80 nos. 22, 35, 45, 47, 73.

"philosophical wine" consulting Weidenfeld's *Secrets of the Adepts*, a book dedicated to Robert Boyle that is devoted to the preparation of "Lully's Spirit of Wine."[314] This collection of various alchemical authors contains recipes for "mineral" and botanical extracts in wine, and Bulkeley narrowly escaped a disastrous accident in 1700 when his furnace exploded while he was preparing this concoction.[315]

[314] [Johann Seger Weidenfeld], *Four Books of Johannes Segerus Weidenfeld, Concerning the Secrets of the Adepts; Or, Of the Use of Lully's Spirit of Wine: A Practical Work*, London, Printed by *Will. Bonny*, for *Tho. Howkins* in *George-Yard* in *Lombard-Street*, 1685.

[315] Jodziewicz, "Bulkeley of Connecticut," p. 19 n. 44.

CHAPTER 5. ISAAC NEWTON, ROBERT BOYLE, AND GEORGE STARKEY

The large body of surviving manuscripts estimated to contain over a million words offers compelling testimony to Isaac Newton's prolonged and passionate interest in alchemy, an obsession that extended over three decades, and spanned most of his great achievements.[316] Newton's career in alchemy gained momentum in 1669 following his discovery of the law of gravity,[317] and his intensely devoted search for the "Universal Solvent" continued until 1696, when Newton moved from Cambridge to London after being appointed Warden of the Royal Mint.[318] According to his assistant Humphrey Newton, Sir Isaac isolated himself for weeks at a time in his live-

[316] "Probably he had begun to read alchemical literature in 1668; in 1669 he purchased chemicals, chemical glassware, materials for furnaces, and the six massive folio volumes of *Threatrum chemicum*, a compilation of alchemical treatises. He established a laboratory of his own at Trinity College, and the records of his subsequent laboratory experimentation still exist in manuscript. Each brief, and often cryptic, laboratory report hides behind itself untold hours with hand-built furnaces of brick, with crucibles, with mortar and pestle, and the apparatus of distillation, and with charcoal fires; experimental sequences sometimes ran for weeks, months, or even years... The manuscript legacy of his scholarly endeavor is very large and represents a huge commitment of his time, but to it one must add the record of that extensive experimentation, a record that involves an amount of time impossible to estimate but surely equally huge." Betty Jo Teeter Dobbs and Margaret C. Jacob, *Newton and the Culture of Newtonianism*, Atlantic Highlands, N.J., Humanities Press, 1995, p. 24.

[317] "[I]f you meet w[i]th any transmutations out of one species into another (as out of Iron into Copper, out of any metall into quicksilver, out of one salt into another or into an insipid body &c) those above all others will bee worth your noting being ye most luciferous & many times lucriferous experiments too in Philosophy...There is in Holland one—Bory, who some yeares since was imprisoned by the Pope to have extorted from him some secrets (as I am told) of great worth both as to medicine & profit, but hee escaped into Holland...pray enquire w[ha]t you can of him..." Newton to Aston, 18 May 1669; in H. W. Turnbull, J. F. Scott, A. R. Hall, and L. Tilling (eds.), *The Correspondence of Isaac Newton*, Volumes I-VII, Cambridge, Published for the Royal Society at the University Press, 1959–1977, vol. I, pp. 10–11.

[318] In 1699 Newton became Master of the Mint, a position that he held until his death in 1727. Richard S. Westfall, *Never at Rest: A Biography of Isaac Newton*, Cambridge, Cambridge University Press, 1980, pp. 526-531.

in laboratory at Trinity College in Cambridge,[319] where he soberly labored to expose the hidden identity of the mysterious Philosophers' Stone,[320] so absorbed in his work that he frequently lost track of the days of the week.[321] During this epic period in science,[322] Newton's years at Trinity[323] were spent pursuing alchemy, theology, and esoteric matters, while the subject of optics held Newton's attention only briefly in the late 1660s and early 1670s,[324] and mechanics and

[319] "About 6 weeks at spring, and 6 at the fall, the fire in the elaboratory scarcely went out, which was well furnished with chemical materials as bodies, receivers, heads, crucibles, etc., which was made very little use of, the crucibles excepted, in which he [Newton] fused his metals; he would sometimes, tho' very seldom, look into an old mouldy book which lay in his elaboratory, I think it was titled *Agricola de Metallis*, the transmuting of metals being his chief design, for which purpose antimony was a great ingredient..." Newton to Conduitt, 17 January 1727/28, King's College, Keynes MS 135; in Dobbs, *Foundations*, p. 8.

[320] "He [Newton] was turning Grey, I think, at Thirty, and when my Father observed yt to him as ye Effect of his deep attention of Mind, He would jest w[i]th ye Experim[en]ts he made so often w[i]th QuickSilver, as if from thence he took so soon that Colour." Wickins to Smith, 16 January 1728, King's College, Keynes MS 137; in Rob Iliffe, "Isaac Newton: Lucatello Professor of Mathematics," *Science Incarnate: Historical Embodiments of Natural Knowledge*, C. Lawrence and S. Shapin (eds.), Chicago, University of Chicago Press, 1998, p. 121. Nicholas Wickins was the son of John Wickins, Newton's roommate at Trinity. Westfall, *Never at Rest*, p. 74.

[321] Cambridge University Library, Add. MS 3975; in Richard S. Westfall, "The Role of Alchemy in Newton's Career," *Reason, Experiment, and Mysticism in the Scientific Revolution*, M. L. R. Bonelli and W. R. Shea (eds.), New York, Science History Publications, 1975, pp. 196, 309 n. 20.

[322] "In the beginning of the year 1665 I found the Method of approximating series & the Rule for reducing any dignity of any Binomial into such a series. The same year in May I found the method of Tangents..., & in November had the direct method of fluxions & the next year in January had the Theory of Colours & in May following I had entrance into ye inverse method of fluxions. And the same year I began to think of gravity extending to ye orb of the Moon & (having found out how to estimate the force with wch globe revolving within a sphere presses the surface of the sphere) from Keplers rule...I deduced that the forces wch keep the Planets in their Orbs must [be] reciprocally as the squares of their distances from the centers about wch they revolve: & thereby compared the force requisite to keep the Moon in her Orb with the force of gravity at the surface of the earth, & found them answer pretty nearly. All this was in the two plague years of 1665 & 1666. For in those days I was in the prime of my age for invention & minded Mathematicks & Philosophy more then at any time since." [Isaac Newton], Cambridge University Library, Add. MS 3968.41, fol. 85; in Richard S. Westfall, "Newton's Marvelous Years of Discovery and Their Aftermath: Myth versus Manuscript," *Isis*, 71 (1980), p. 109.

[323] "Newton told his future biographers that he had been an autodidact who as a very young man had proceeded by almost plodding hard work to his mastery over mathematical and scientific texts. He had discovered the law of universal gravitation 'by thinking upon it continually.' One set of narratives about his early life linked his extraordinary labors to a fit and powerful body. Conduitt recorded how Newton had physically worsted the school bully by smashing his face against the side of the church and his cunning use of the wind to outjump his schoolmates in a long-jump contest, while, after a wealth of stories concerning Newton's mechanical acumen, Conduitt concluded that 'Sr Isaac had the mechanicks hands as well as the head of a Philosopher.'" Iliffe, "Isaac Newton," pp. 131–132.

[324] "[I]n 1669 he gave his first set of lectures on his optical discoveries. He sent papers on optics to the Royal Society in the 1670s also, but the strong opposition his papers encountered from Hooke and others discouraged him from further publication to such an extent that he never published the full text of *Opticks* until 1704, by which time Hooke had died and so could no longer attack Newton's views. The laws of optics reported there were, however, virtually all established while Newton was less than 30 years of age." Dobbs and Jacob, *Culture of Newtonianism*, p. 20.

dynamics occupied two limited periods:[325] the 1660s and the two and a half years dedicated to the *Principia Mathematica* (1687).[326]

Newton and Boyle both believed there to be very few Adepts among those who legitimately practiced alchemy, distinct from the "vulgar Chymists"[327] Boyle called "Pretenders to that excellent science"[328] and "Cheats."[329] In his *Producibleness of Chymical Principles*, Boyle cautioned: "look not with the same eyes on the opinions or performances of vulgar Chymists, and Chymical Philosophers," and then associated the "vulgar chymists and Aristotelians."[330] Newton also took a critical approach to the alchemical writers, and crossed out one passage from the collected writings in the *Theatrum Chemicum* and wrote: "I believe that this author is in no way adept."[331] Starkey personally communicated to Dury in early 1651 that he had corresponded with the mysterious New England Adept while in London, and claimed that Philalethes had discovered in his investigations (only) sixty other Adepts who had possessed the Philosophers' Stone.[332]

---

[325] "In 1666, Newton put aside, unpublished, his work on gravitation, having found that the inverse square law 'answered pretty nearly'...because an essential link in his argument was missing...the proof was only hit on with some trouble in 1685, a year before finishing of the *Principia*...In 1684, Hooke had come to the conclusion that the motion of the planets could be explained on the basis on an inverse square law of attraction, but could not prove it. As a result, he discussed the problem with [Edmond Halley]...and Halley went to Cambridge to question Newton on it...The incident, however, revived his interest in science...he began to write the *Principia*, which came out in 1687 after prodigious labour...Again Hooke accused Newton of plagiarism, but received a devastating denial and failed to obtain any support for his claim; Newton's response to the argument was to eliminate almost every mention of Hooke from the manuscript." Milo Keynes, "The Personality of Isaac Newton," *Notes and Records of the Royal Society of London*, 49 (1995), pp. 23-25.

[326] Westfall, "Role of Alchemy," pp. 195–196; Westfall, *Never at Rest*, pp. 290-291.

[327] "Newton himself sharply distinguished between common or vulgar chemistry and 'vegetable' chemistry. The former comprised all ordinary reactions and took place by mechanical interactions of the corpuscles. The latter required the action of the vital spirit, God's universal agent for effecting the maturation of matter." Betty Jo Teeter Dobbs, "Newton's Copy of *Secrets Reveal'd* and the Regimens of the Work," *Ambix*, 26 (1979), p. 155.

[328] Boyle, "Essay of the holy Scriptures," fol. 55; in *Works of Boyle*, vol. XIII, p. 203.

[329] Boyle, *Sceptical Chymist*; in Boyle, *Works*, vol. I, p. 463.

[330] Boyle, *Producibleness of Chymical Principles*; in Ibid., vol. I, pp. 589-590, 661.

[331] Yahunda MS 259, no. 9; in Westfall, *Never at Rest*, p. 292.

[332] Hartlib, *Ephemerides*, [March? 1650/51], D-D4; in Wilkinson, "Hartlib Papers II," p. 90.

Newton relied heavily upon the works of George Starkey,[333] and his prodigious alchemical collection[334] included nine of Starkey's books,[335] this number equaled only by the books of the highly regarded German alchemist Michael Maier,[336] physician to the Holy Roman Emperor Rudolph II.[337] Newton included Starkey's alter ego Philalethes and Maier among fourteen of "the best Authors" on the subject: "This

[333] "The philosophical tradition of alchemy had always regarded its knowledge as a secret possession of a select few who were set off from the vulgar herd both by their wisdom and by the purity of their hearts...The concept of a secret for a select few aside, all the foregoing characteristics applied as well to the mechanical philosophy which Newton had recently embraced. In the nature of the truth they offered, however, the two philosophies differed profoundly. In the mechanical philosophy, Newton had found an approach to nature which radically separated body and spirit, eliminated spirit from the operations of nature, and explained those operations solely by the mechanical necessity of particles of matter in motion. Alchemy, in contrast, offered the quintessential embodiment of all the mechanical philosophy rejected...Among his 'Notable Opinions'...Newton included the argument of Effararius the Monk that the stone is composed of body, soul, and spirit, that is, imperfect body, ferment, and water...Newton also met in alchemy another idea that refused to be reconciled with the mechanical philosophy. Where that philosophy insisted on the inertness of matter, such that mechanical necessity alone determines its motion, alchemy asserted the existence of active principles in matter as the primary agents of natural phenomena. Especially it asserted the existence of one active agent, the philosophers' stone, the object of the Art. Images of every sort were applied to the stone, all expressing a concept of activity utterly at odds with the inertness of mechanical matter characterized by extension alone...In Sendivogius and Philalethes, the activity sometimes took on the specific form of an attraction, and they called it a magnet...To them, the magnet offered an image of the operation of nature." Richard S. Westfall, *The Life of Isaac Newton*, Cambridge, Cambridge University Press, 1993, pp. 116–118.

[334] "Magnetism was a power that Newton often cited, together with electricity, as an example of a force known to act in the world. He saw it, however, as a power that was limited in its operation, being confined, so he thought, to iron and some of its ores." R. W. Home, "Force, electricity, and the powers of living matter in Newton's mature philosophy of nature," *Religion, Science and Worldview: Essays in Honor of Richard S. Westfall*, M. J. Osler and P. L. Farber (eds.), Cambridge, Cambridge University Press, 1985, p. 106.

[335] "[I]t seems for most of his life Newton actually held, as did all but the Aristotelians among his contemporaries, that magnetic phenomena were caused by mechanical processes, namely circulating streams of subtle material effluvia which passed axially through a magnet, emerged from one pole, and returned to the other through the surrounding air before renewing their course...in one of the famous 'aether' Queries that Newton added to the second English edition (1717) of this work [*Opticks*], he uses the existence and activity of the magnetic effluvia, which he here again takes as beyond dispute, to justify certain assumptions he is making about the aether: 'If any one would ask how a Medium can be so rare', he suggests, 'let him tell me...how the Effluvia of a Magnet can be so rare and subtile, as to pass through a Plate of Glass without any Resistance or Diminution of their Force, and yet so potent as to turn a magnetick Needle beyond the Glass'...If Newton's faith that the magnetic force was caused mechanically ever wavered, it did so only once, and that very briefly. In a draft 'Conclusion' to his *Opticks* which he drew up during the early 1690s, Newton seems definitely to put the 'attractive vertue' of the particles of a magnet in the same category as those other primitive unexplained forces, gravity and electricity: 'The particles of bodies have certain spheres of activity', he here asserts, 'within which they attract or shun one another. For ye attractive vertue of the whole magnet is composed of ye attractive vertues of all its particles & the like is to be understood of the attractive vertues of electrical and gravitating bodies'." R. W. Home, "'Newtonianism' and the Theory of the Magnet," *History of Science*, 15 (1977), pp. 258-259.

[336] Newton appears to have been reading Michael Maier in 1669 or even earlier. *Correspondence of Newton*, vol. I, pp. 12–13 n. 5.

[337] Harrison, *Library of Newton*, pp. 188–189, 243; Karin Figala, "Newton's alchemy," *The Cambridge Companion to Newton*, I. B. Cohen and G. E. Smith (eds.), Cambridge, Cambridge University Press, 2002, pp. 370-386.

process J take to be ye work of the best Authors, Hermes, Turba, Morien, Artephius, Abraham ye Jew & Flammel, Scala, Ripley, Maier, the great Rosary, Charnock, Trevisan, Philaletha, Despagnet."[338] Newton also compiled an index of his alchemical readings that extends over 100 pages and contains citations to at least 100 different authors, in which there are nearly 300 references to Starkey and 140 references to Maier, providing a rough estimate of the importance of Starkey to Newton's alchemical quest.[339]

Newton maintained a continued interest in alchemy while residing in London,[340] and belonged to a shadowy circle of English alchemists that have been named the "Philalethes-school," the members of which secretly purchased and circulated alchemical manuscripts between themselves.[341] Among these figures was the mysterious William Y-worth or Yarworth, a close associate of Newton who adopted Starkey's title "Philosopher by Fire" along with Starkey's processes and recipes, publishing a popular book on distilling alcohol and the *Aqua vitae*, or "Water of life," which he entitled *Introitus Apertus*,[342] reprinted in 1705 as *The Compleat Distiller*.[343]

Isaac Newton was also heavily influenced by Robert Boyle; he began his diligent study of Boyle's writings almost from the time he matriculated in Cambridge in 1661.[344] Newton's earliest alchemical

[338] King's College, Keynes MS 49, fol. 1^r; in Dobbs, "Newton's Copy of Secret's Reveal'd," p. 157.

[339] King's College, Keynes MS 30; in Richard S. Westfall, "Isaac Newton's Index Chemicus," *Ambix*, 22 (1975), pp. 174–185; Westfall, "Role of Alchemy," pp. 203-206.

[340] "For nearly twenty-four years he reigns as President of the Royal Society...He has become the Sage and Monarch of the Age of Reason. The Sir Isaac Newton of orthodox tradition—the eighteenth-century Sir Isaac, so remote from the child magician born in the first half of the seventeenth century—was being built up. Voltaire returning from his trip to London was able to report of Sir Isaac—"twas his peculiar felicity, not only to be born in a country of liberty, but in an Age when all scholastic impertinences were banished from the World. Reason alone was cultivated and Mankind cou'd only be his Pupil, not his Enemy.' Newton, whose secret heresies and scholastic superstitions it had been the study of a lifetime to conceal! But he never concentrated, never recovered 'the former consistency of his mind'. 'He spoke very little in company.' 'He had something rather languid in his look and manner.'" John Maynard Keynes, "Newton, the Man," *Newton Tercentenary Celebrations 15–19 July 1946*, Cambridge, Cambridge University Press, 1947, pp. 33-34.

[341] Karin Figala, "Zwei Londoner Alchemisten um 1700: Sir Isaac Newton und Cleidophorus Mystagogus," *Physis*, 18 (1976), pp. 245-273; Karin Figala and Ulrich Petzold, "Alchemy in the Newtonian circle: personal acquaintances and the problem of the late phase of Isaac Newton's alchemy," *Renaissance and Revolution: Humanists, scholars, craftsmen, and natural philosophers in early modern Europe*, J. V. Field and F. A. J. L. James (eds.), Cambridge, Cambridge University Press, 1993, pp. 173–192.

[342] William Y-worth, *Introitus Apertus ad Artem Distillationis...*, London, Printed for *J. Taylor* at the *Ship* in St. *Paul's Church-yard*, 1692.

[343] William Y-worth, *The Compleat Distiller: Or The Whole Art of Distillation Practically Stated, And Adorned with all the New Modes of Working now in Use. In which is Contained, The way of making Spirits, Aquavitæ, Artificial Brandy...*, London, Printed for J. Taylor, at the *Ship* in St. *Paul's* Church-Yard, 1705.

[344] Dobbs, *Foundations*, p. 123.

experiments were based on Boyle,[345] and at the time of his death Newton possessed more books written by Boyle than any other author.[346] Shortly after his return to Cambridge following the Great Plague, in 1667 or 1668, Newton compiled a dictionary of chemical terminology in which the only book mentioned is Boyle's *Origine of Formes and Qualities* (1666). In this dictionary the words "Elixar" and "Alcahest" are left undefined,[347] and these definitions remained elusive to Newton who used his notes in ongoing investigations,[348] a strong indication of where his knowledge on the subject encountered its limits.[349] Newton meticulously studied Starkey's writings and annotated his copy of van Helmont's *Ortus Medicinæ*,[350] yet it was the establishment of Newtonianism that brought about the decline of Helmontian chemistry and medicine in Europe and America.[351]

Newton and Boyle met one another in early 1675,[352] and in February the following year the Royal Society published Boyle's famous essay, "Incalescence of Quicksilver with Gold."[353] Commencing on the subject of alchemy a correspondence between these Olympian scientists began late in 1676.[354] Boyle's publication captivated Newton's interest, as it referred to the transmutation formula that Boyle received from George Starkey, dated to April/May 1651,[355] which Newton also possessed as a partial translation.[356] Without crediting Starkey and transparently signing the work "B. R.," Boyle

[345] Westfall, *Never at Rest*, p. 293.

[346] Harrison, *Library of Newton*, pp. 107–109.

[347] Bodleian Library, MS Don. b. 15; in Dobbs, *Foundations*, pp. 121–122.

[348] Westfall, *Never at Rest*, p. 292.

[349] "To study Newton's alchemy and chemistry is to study failure, and a dead form of intellectual endeavour." Marie Boas Hall, "Newton's Voyage in the Strange Seas of Alchemy," *Mysticism in the Scientific Revolution*, p. 241.

[350] Harrison, *Library of Newton*, p. 158; Countway MS, eighth item, ff. 21, 24^v; in Westfall, *Never at Rest*, p. 292; King's College, Keynes MS 16; in Clericuzio, "Helmont to Boyle," p. 334; William R. Newman, "The background to Newton's chymistry," *Cambridge Companion to Newton*, pp. 358-369.

[351] "Newton, once fixed upon an idea, pursued it relentlessly until it surrendered its full secret or sheer exhaustion forced him to relinquish his hold for a time, when he turned to something else for what he called 'divertisement.' From the five or six inquiries into which Newton plunged, he returned each time with a heroic discovery; only in alchemy did he fail." Frank E. Manuel, *A Portrait of Isaac Newton*, Cambridge, Belknap Press of Harvard University Press, 1968, p. 135.

[352] "Pray present my humble service to Mr Boyle w[he]n you see him & thanks for ye favour of ye convers[ation] I had w[i]th him at spring." Newton to Oldenburg, 14 December 1675; in *Correspondence of Newton*, vol. I, p. 393.

[353] [Robert Boyle], "Of the Incalescence of *Quicksilver* with *Gold*, generously imparted by *B. R.*," *Philosophical Transactions*, 10 (1675–1675/76), pp. 515-533; in Boyle, *Works*, vol. IV, pp. 219-230.

[354] Westfall, *Never at Rest*, p. 371.

[355] Starkey to Boyle, [April/May 1651], Royal Society Library, Boyle Letters, vol. VI, fol. 100; in *Correspondence of Boyle*, vol. I, pp. 90–103; Newman, *Gehennical Fire*, p. 76; Principe, *Aspiring Adept*, pp. 160–162.

[356] Newman, "Starkey's *Key*," pp. 564-574.

claimed to have discovered a special "Mercury" that he considered a great breakthrough in the preparation of medicines, but revealed no details of the process and refused to answer any questions about it, citing the "political inconveniences which might ensue if it...fall into ill hands."[357] Newton immediately recognized Boyle as the author, and in a letter to Henry Oldenburg, Secretary of the Royal Society,[358] Newton dismissed Boyle's hypothesis and urged "high silence" over publicly divulging the secrets of alchemy,[359] citing the possibility of the "immense dammage to ye world if there should be any verity in ye Hermetick writers," because there were many things which "none but they [the Adepts] understand."[360]

With only one surviving letter of Boyle to Newton dated 1682,[361] their communications appear to have been infrequent[362] after Newton's

---

[357] [Boyle], "Incalescence of Quicksilver with Gold"; in Boyle, *Works*, vol. IV, p. 228.

[358] "The paper called forth the celebrated letter of Isaac Newton (then aged thirty-two) to Henry Oldenburg in which he heaped scorn on the credulity of the 'noble author'...At the end of the letter Newton, also under a thin cloak of anonymity, wrote thus: 'Because the author seems desirous of the sense of others in this point, I have been so free as to shoot my bolt; but pray keep this letter private to yourself.' Oldenburg, as chatterbox Secretary of the young Society, had always deemed it his responsibility to disseminate information and would almost certainly have seen to it that Boyle knew of Newton's letter. Boyle's reply came two years later, i.e. in 1678, in the brief [anonymous] tract...*Of a Degradation of Gold* [*Made by an Anti-Elixir*]." Fulton, *Bibliography of Boyle*, p. 93.

[359] In his *Ordinall of Alchimy*, Thomas Norton (*ca.* 1433–1514), a member of King Henry VII's household, calls "holi *Alkimy*" a "singular grace & gifte of th'almightie," and warns:

> For this *Science* must ever secret be,
> The Cause whereof is this as ye may see ;
> If one evill man had hereof all his will
> All Christian Pease he might hastilie spill,
> And with his Pride he might pull downe
> Rightfull *Kings* and *Princes* of renowne :
> Wherefore the sentence of perill and jeopardy,
> Upon the *Teacher* resteth dreadfully.

Commenting on this work, Elias Ashmole considered that violating alchemical secrecy "might render [one] *Criminall* before *God*, and a *presumptuous violator* of the *Calestiall Seales*." Thomas Norton, "The Ordinall of Alchimy," *Threatrum Chemicum Britannicum*, E. Ashmole (ed.), London, Printed by *J. Grismond* for Nath: Brooke, at the Angel in *Cornhill*, 1652, p. 14; Elias Ashmole, "Annotations and Discourses, Upon Some part of the preceding Worke," Ibid., p. 439.

[360] Newton to Oldenburg, 26 April 1676; in *Correspondence of Newton*, vol. II, pp. 1-2.

[361] Boyle to Newton, 19 August 1682; in Ibid., vol. II, p. 379.

[362] "By the end of 1676, as absorbed in theology and alchemy as he was and distracted by correspondence and criticism on optics and mathematics, Newton had virtually cut himself off from the scientific community. Oldenburg died in September 1677, not having heard from Newton for more than half a year. Newton terminated his exchange with Collins by the blunt expedient of not writing. It took him another year to conclude the correspondence on optics, but by the middle of 1678, he succeeded. As nearly as he could, he had reversed the policy of public communication that he began with his letter to Collins in 1670 and retreated to the quiet of his academic sanctuary. He did not emerge for nearly a decade...The Newton Humphrey found had immersed himself in unremitting study to the extent that he grudged even the time to eat and sleep. During five years, Humphrey saw him laugh only once, and John North, master of Trinity from 1677 to 1683, feared that Newton would kill himself with study...Theology was

famous letter to Boyle in February 1679.[363] This letter is the only surviving correspondence of Newton[364] to Boyle,[365] and in it Newton hypothesizes a physical ether[366] that pervades everything,[367] and he concludes a mechanical explanation of alchemy by remarking that he sometimes thought "that ye true permanent Air may be of a metallic

not Newton's only occupation during these years. The manuscript remains testify that alchemy vied with theology to command his attention." Westfall, *Life of Newton*, pp. 133–140.

363 Newton to Boyle, 28 February 1678/79; in *Correspondence of Newton*, vol. II, pp. 288-295.

364 "Newton's law of gravitation has bodies such as the moon and the earth attracting one another with a force proportional to the product of their masses and inversely proportional to the square of the distance between their centres. His ether theory of gravitation would explain this tendency as the effect of forces, not of attraction but of repulsion, exerted by the particles of a rarified medium dispersed unevenly throughout the vacuities in the gravitating bodies themselves and through the space that separates them. It is the medium itself that Newton calls an 'aether'." Cantor and Hodge, "Ether theories," p. 1.

365 "Under the influence of the Cambridge Platonists, especially Henry More and Ralph Cudworth, Newton believed that matter is passive. To attribute active powers or self-movement to matter would ineluctably lead to atheism, for what need would there be for God? Alchemy was one source that Newton hoped would lead to revelations about the origin of active powers in nature." Alan E. Shapiro, *Fits, Passions, and Paroxysms: Physics, method, and chemistry and Newton's theories of colored bodies and fits of easy reflection*, Cambridge, Cambridge University Press, 1993, p. 74.

366 "There has been in the past an immense amount of confusion regarding Newton's attitude toward a 'cause' of gravity, partly because of the misdating of one key manuscript (*De gravitatione*) but primarily because Newton's pursuit of alchemy, theology, and other forms of ancient wisdom have been arbitrarily and artificially separated out from his 'real' scientific interests by almost all post-Newtonian scholarship." Betty Jo Teeter Dobbs, "'The Unity of Truth': An Integrated View of Newton's Work," *Action and Reaction: Proceedings of a Symposium to Commemorate the Tercentenary of Newton's Principia*, P. Theerman and A. F. Seeff (eds.), Newark, University of Delaware Press, 1993, pp. 118–119.

367 "In attempting to account for gravitation motion...at the end of the *Opticks*, Newton suggests... that the ether is the cause. The ether he regards as made up of corporeal particles...which permeate all space and the pores between the particles of which bodies are made...the existence of this fluid substance is only an hypothesis, and 'I do not know what this Aether is'...In the *Principia*... [Newton writes:] 'for I am induced by many reasons to suspect that they [the phenomena] may depend upon certain forces by which the particles of bodies, by some causes hitherto unknown, are either mutually impelled towards one another, and cohere in regular figures, or are repelled and recede from one another.' Since, however, forces other than quantity of matter seemed to Newton not to be mathematically determinable, no adequate mathematical account of the other phenomena seemed possible, and 'These forces being unknown, philosophers have hitherto attempted the search of Nature in vain.' The other forces to which reference is made are those unknown ones which cause electrical and magnetic effects, fermentation, and other chemical reactions, all of which seem to require knowledge of more than simply the mass and motion of the particles of which bodies are made." Henry G. Van Leeuwen, *The Problem of Certainty in English Thought 1630–1690*, The Hague, Martinus Nijhoff, 1963, pp. 110–114.

original,"[368] which contains "Spirits"[369] that "flote in it."[370] Newton then concludes this correspondence with Boyle by conjecturing that ether is "ye cause of gravity."[371] In a manuscript probably related to

[368] "In Newton's early theory aether assumed the role of a divine creative *quinta essencia.*" Klaus Vondung, "Millenarianism, Hermeticism, and the Search for a Universal Science," *Science, Pseudo-Science, and Utopianism in Early Modern Thought*, S. A. McKnight (ed.), Columbia, University of Missouri Press, 1992, p. 139. "What could an immaterial aether be? To Newton, it was the infinite omnipotent God, who by His infinity constitutes absolute space and by His omnipotence is actively present throughout it." "Newton did not believe that the universal gravitational attraction of all bodies, the force by which his *Principia* explained the functioning of the heavenly system, is a power inherent in matter...the Creator was the 'agent acting constantly according to certain laws' (as he put it to Richard Bentley) that makes bodies move as though they attract each other." Richard S. Westfall, *Force in Newton's Physics: The Science of Dynamics in the Seventeenth Century*, New York, American Elsevier, 1971, p. 396; Westfall, "The Rise of Science and the Decline of Orthodox Christianity: A Study of Kepler, Descartes, and Newton," *God and Nature*, p. 233.

[369] "In his student notebook (mid to late 1660s), in his alchemical treatise 'Of Natures obvious laws & processes in vegetation' (early 1670s), in the 'Hypothesis of Light' he sent to the Royal Society in 1675, in his letter to Boyle of 1678/79, and in correspondence and unpublished papers of the early 1680s, there exists a continuous record of Newton's use of speculative aethereal mechanisms as an explanation for gravity." Betty Jo Teeter Dobbs, "Newton's Rejection of the Mechanical Aether: Empirical Difficulties and Guiding Assumptions," *Scrutinizing Science: Empirical Studies of Scientific Change*, A. Donovan, L. Laudan, and R. Laudan (eds.), Dordrecht, Kluwer Academic Publishers, 1988, p. 70.

[370] "The major innovation of Newton's scientific work, the concept of force, was derived from his beliefs in the occult powers of the natural magic tradition, or which, for him, the most important part was alchemy. Those 'mechanicall coalitions or separations of particles', that he referred to in his unpublished alchemical account, *Of Nature's Obvious Laws and Processes in Vegetation of Metals* (*ca.* 1674), were 'an exceeding subtile & inimaginably small portion of matter diffused through the masse w[ch] if it were separated there would remain but a dead & inactive earth.' In the paper, 'An Hypothesis Explaining the Properties of Light', that he gave to the Royal Society in 1675, Newton suggested that all bodies may be composed of 'certain aethereal spirits, or vapours' condensed in different degrees and 'wrought into various forms', adding that 'light and aether mutually act upon one another, generating fluids out of solids, and solids out of fluids'. In the alchemical account, *Of Nature's Obvious Laws*, he put: 'y[e] aether is but a vehicle to some more active sp[iri]t & y[e] bodys may bee concreted of both together...This sp[iri]t perhaps is y[e] body of light becaus[e] both have a prodigious active principle, both are perpetuall workers.' These scientific and magical-alchemical concepts obviously show a similarity." Keynes, "Personality of Newton," p. 34; Cf. Shapiro, *Fits, Passions, and Paroxysms*, pp. 72-89.

[371] "Newton's insistence that the natural world cannot be explained solely in terms of the arrangement and inertial movements of totally passive particles of matter is now regarded as one of the initial premises upon which his subsequent great achievements were built...It was Newton who transformed the mechanical philosophy by introducing into it his notion of *active principles*. These principles were to be called upon to account for those spontaneous workings of nature which could not be explained according to the principles of strict mechanism:

> Seeing therefore the variety of motion which we find in the world is always decreasing, there is a necessity of conserving and recruiting it by active Principles, such as are the cause of gravity, by which Planets and Comets keep their Motions in the Orbs, and Bodies acquire great Motion in falling; and the cause of Fermentation, by which the Heart and Blood of Animals are kept in perpetual Motion and Heat...For we meet with very little Motion in the World, besides what is owing to the Active Principles...Particles have not only a *vis inertiae* [force of inertia], accompanied with such passive Laws of Motion as naturally result from that Force, but also that they are moved by certain Active Principles...

[D]uring Newton's own lifetime, a number of leading natural philosophers were dismayed and appalled by Newton's notions of active principles, because they saw them as a betrayal of the principles of the mechanical philosophy. Newton's active principles seemed to represent a return to the lazy and slipshod occult qualities of Aristotelianism. As Leibniz, Newton's most

the "classical" scholia of the abortive 1690s edition of the *Principia*,[372] Newton associates the "active principle" with God:[373]

implacable rival, said: Gravity [or any of his active principles] must be a scholastic occult quality or the effect of a miracle. Newton was quick to rebut these charges...he did not deny the *occult* nature of his active principles. What Newton took exception to was the charge that they were *scholastic* or *Aristotelian*:

> These Principles I consider, not as occult Qualities...but as general Laws of Nature, by which the Things themselves are form'd; their Truth appearing to us by Phaenomena, though their Causes be not yet discover'd. For these are manifest Qualities, and their Causes only are occult. And the *Aristotelians* gave the name of occult Qualities, not to manifest Qualities, but to such Qualities only as they supposed to lie hid in Bodies, and to be the unknown Causes of manifest Effects: Such as would be the Causes of Gravity, and of magnetik and electrick Attractions, and of Fermentations...occult qualities are decried *not because their causes are unknown* but because the Schoolmen believed that those things w[hi]ch were unknown to their Master Aristotel, *could never be known*.

Newton...refused to make any capitulations to Aristotelianism, and yet he accepted the use of occult qualities and powers...Newton was drawing his notion of occult qualities not from scholastic traditions but from a completely separate tradition—the tradition of natural magic... had Newton *not* been steeped in alchemical and other magical learning, he would never have proposed forces of attraction and repulsion between bodies as the major feature of his physical system." John Henry, "Newton, Matter, and Magic," *Let Newton Be!*, J. Fauvel, R. Flood, M. Shortland, and R. Wilson (eds.), Oxford, Oxford University Press, 1988, pp. 134–144.

[372] "There are not only visible traces of aether in the *Principia*, but soon after the book was published, he wrote to Leibniz [1693] that 'some exceedingly subtle matter seems to fill the heavens'...In 1693 Newton even went so far as to write to Leibniz: 'But if, meanwhile, someone explains gravity along with all its laws by the action of some subtle matter, I shall be far from objecting'...there are a number of direct references to aether in the *Principia*. Toward the end of the scholium to the Laws of Motion, Newton writes...about 'the whole earth, floating in the free aether'...At the end of the scholium with which sect. 11, Bk 1 concludes, he says: 'I use the word *Attraction* here in a general sense for any endeavour whatever of bodies to approach one another, whether that endeavour...arises from the actions of aether or of air or of any medium whatsoever – whether corporeal or incorporeal – in any way impelling toward one another the bodies floating therein'...[In 1686 Newton] claimed that his aether hypothesis implied an inverse-square law." I. Bernard Cohen, "The *Principia*, Universal Gravitation, and the 'Newtonian Style', Relation to the Newtonian Revolution in Science: *Notes on the Occasion of the 250th Anniversary of Newton's Death*," *Contemporary Newtonian Research*, Z. Bechler (ed.), Dordrecht, D. Reidel Publishing Company, 1982, pp. 58, 73, 92 n. 10.

[373] "During his undergraduate years, probably about 1664, Newton accepted the Cartesian postulate of dense aether, acting mechanically by impact, as the cause of terrestrial gravitation... However, in 1684 or early in 1684/85, when he was engaged in writing the *Principia*, Newton rejected all forms of a dense mechanical aether as the cause of gravitation, apparently because he had recognized their incompatibility with celestial motions...The first tentative solution to the problem of a cause for gravity to emerge in Newton's later years, then, was that the omnipresent supreme Deity subsumed gravity directly. There can be no doubt that Newton thought God to be literally omnipresent...Between 1684 and 1710 Newton's conceptualization of the 'cause' of gravity was probably more directly mediated by ancient cosmic thought and especially by the Platonizing version of ancient Stoicism than by alchemical doctrines of an active principle at work in the realm of micromatter. The analogies at work in the evolution of Newton's thought on the cause of gravity would seem to have been cosmic ones based on the musical harmonies of ancient Pythagorean-Platonic metaphysics and on the Divine Mind of Presocratic and Stoic thought." Betty Jo Teeter Dobbs, "Newton's Alchemy and his 'Active Principle' of Gravitation," *Newton's Scientific and Philosophical Legacy*, P. B. Scheurer and G. Debrock (eds.), Dordrecht, Kluwer Academic Publishers, 1988, pp. 55-80; See also Dobbs, "Newton's Alchemy and His Theory of Matter," *Isis*, 73 (1982), pp. 511-528; Dobbs, "Stoic and Epicurean doctrines in Newton's system of the world," *Atoms, pneuma, and tranquillity: Epicurean and Stoic Themes in European Thought*, M. J. Osler (ed.), Cambridge, Cambridge University Press, 1991, pp. 232-238; Dobbs, "Gravity and Alchemy,"

> The Epicureans making a distinction of the whole of nature into body and void, denied the existence of God, but very absurdly. For two planets distant from one another by a long and empty interval will not approach one another by any force of gravity, nor will they act upon one another, except by the mediation of some *active principle* which comes between them both, and by means of which force is propagated from each into the other...consequently those ancients who more rightly held unimpaired the mystical Philosophy as Thales and the Stoics taught that a certain infinite spirit pervades all space *into infinity*, and contains the vivifies the entire world: and this spirit was their supreme divinity, according to the Poet cited by the Apostle, in him we live and move and have our being.[374]

In his "Observation I,"[375] probably written in 1716 and which relates to a revision of the 1717 edition of the *Opticks*,[376] Newton returned to the ether theory:[377]

The Scientific Enterprise; The Bar-Hillel Colloquium: Studies in History, Philosophy, and Sociology of Science, E. Ullmann-Margalit (ed.), Dordrecht, Kluwer Academic Publishers, 1992, pp. 205-222.

[374] U.L.C. Add. 3965.12, fol. 269r; in J. E. McGuire, "Force, Active Principles, and Newton's Invisible Realm," *Ambix*, 15 (1968), p. 196.

[375] "In the decade following 1706, Newton changed his mind about God's rôle, and resolving his ambiguity, began to commit himself to seemingly mechanical means. At the beginning of the 1717 edition of the *Opticks*, Newton wrote that he had added some indications of how gravity might be the result of an aethereal mechanism '...to shew that I do not take Gravity for an essential Property of Bodies...' Similarly, he now attributed motion only to the effects of an active principle, now in effect materialized, instead of to a metaphysical active principle, or dictates of a will, as he had in 1706." David Kubrin, "Newton and the Cyclical Cosmos: Providence and the Mechanical Philosophy," *Journal of the History of Ideas*, 28 (1967), p. 339.

[376] "This late return—Newton was now in his seventies—to an aether theory at first sight appears inconsistent. After all, he had long ago shown that an aether, involving contact action through a *plenum*, would be necessarily 'in exile from the nature of things,' because of the resistance it would oppose to planetary motions. How could he now propose a medium that would by 'impulse' explain gravitational action? If it could exert an impulse, would it not also offer resistance to motion? In a draft-letter reacting to Leibniz' views (c. 1714) he suggested one way out of the dilemma: 'a substance in which bodies move and float without resistance, and which has therefore no *vis inertiae* [force of inertia] but acts by other laws than those that are mechanical'. And he went on to defend the propriety of non-mechanical modes of explanation in general, for, after all even 'Mr. Leibniz himself will scarce say that thinking is mechanical'... This may have been why the aether Newton did propose in 1717 was of very small, but still finite, density...[I]n *Principia* [1713, second edition] he inserted a passage which refers to the earth as 'floating in the non-resisting aether.'" Ernan McMullin, *Newton on Matter and Activity*, Notre Dame, University of Notre Dame Press, 1978, pp. 96-97, 149 n. 193.

[377] "In the 1670s Newton's cosmogony was certainly alchemical in character... [Newton] describes the operations of this aether in unmistakable alchemical terms...There is little doubt that the aether is conceived as a source of activity for the maintenance of the cosmos. And as it perpetually circulates, it changes into countless forms of things, and that the possibilities for transmutation are limitless...Newton cites Descartes, the Peripatetics, Moses, Thales, Epicurus, Phythagoras and Democritus as thinkers who held to the doctrine of transmutation. It is clear that the theories of Epicurus and Democritus, and those of Moses, Thales and the Phythagoreans—all of whom were interpreted by Newton and the Cambridge Platonists to be atomists—are closer to Newton's theory of matter than to that of the Cartesians and Peripatetics." J. E. McGuire, "Transmutation and Immutability: Newton's Doctrine on Physical Qualities," *Ambix*, 14 (1967), pp. 84-91. See also R. W. Home, "Newton's subtle matter: the *Opticks* queries and the mechanical philosophy," *Renaissance and Revolution*, pp. 193-202.

> There are therefore Agents in Nature able to make the particles of bodies attract one another very strongly & to stick to-gether strongly by those attractions. One of these Agents may be the Aether above mentioned whereby light is refracted. Another may be the Agent or Spirit which causes electrical attraction...And as there are still other mediums which may cause attractions, such as are the Magnetick effluvia ; it is the business of experimental Philosophy to find out all these Mediums with their properties...To distinguish this Medium from the bodies which flote in it, & from their effluvia & emanations & from the Air, I will hence forward call it Aether & by the word bodies I will understand the bodies which flote in it, taking this name not in the sense of the modern metaphysicians, but in the sense of the common people & leaving it to the Metaphysicians to dispute whether the Aether & bodies can be changed into one another.[378]

In an unfinished and unpublished essay entitled "De Aere et Aethere," Newton connects the ether[379] with magnetism:[380]

> "Newton distinguished between gravity and those physical properties—hardness, extension, inertia—that he held to be intrinsic to the nature of matter. As he emphasized in a famous statement to Bentley...gravity was not to be considered as 'innate, inherent and essential to Matter'. In query 31 of the 1717 *Opticks*, Newton contrasted passive principles such as *vis inertiae* (force of inertia) and the 'passive Laws of Motion as naturally result from that Force' with 'active Principles, such as that of Gravity'...Newton conceived the ether of

[378] U.L.C. Add. 3970.9, ff. 622^r, 623^r; in McGuire, "Newton's Invisible Realm," pp. 179–181.

[379] "[N]atural philosophers frequently speculated about the structural relationship between ether and gross matter, and the role of God in establishing this relationship. Ether was usually considered a more basic substance than gross matter, and thus it was suggested that ether was the protoplast out of which God had formed matter...Newton's concern with the ether as 'protoplast' was related to the alchemical tradition...An important theme in Western philosophy has been the doctrine of dualism in which two distinct substances, mind and matter, are postulated in order to distinguish God from the material universe and our souls (or minds) from our bodies. Although these two substances are considered to be incommensurable they must be able to interact with each other, for example, in the acts of perception and volition. However, this dualistic ontology poses the problem of how these two radically different substances—the one incorporeal, penetrable, and intelligent and the other extended, impenetrable, and passive—are able to affect one another. Dualists...have sometimes adopted an ethereal fluid as the intermediary between mind and matter...Probably the most frequently cited and most influential example of ether as intermediary appeared in queries 23 and 24 to Newton's *Opticks* (1717). Here, Newton proposed the scarcely original theory that an ether fills the nervous system...in explaining how we move parts of our bodies, Newton considered that the 'power of the Will' causes the ether in the brain to vibrate." G. N. Cantor, "The theological significance of ethers," *Conceptions of ether*, pp. 139–140, 145.

[380] "The connections in Newton's mind between gravity and alchemy, however dramatically they may have varied from time to time, constitute yet another argument for the unity of Newton's thought. He simply did not ultimately divide his studies into the logic-tight compartments that most recent scholars have deemed appropriate. Not only was there a connection between gravity and alchemy, there was also one between gravity and God and one between alchemy and God." Betty Jo Teeter Dobbs, *The Janus Faces of Genius: The role of alchemy in Newton's thought*, Cambridge, Cambridge University Press, 1991, p. 253.

the 1717 Opticks as an active principle communicating God's causal agency and as a physical mode (though not a contact-action model) establishing the intelligibility of the distance force of gravity...In query 31 of the 1717 Opticks, Newton linked active principles with the 'great and violent' processes of chemistry, questioning the reducibility of chemistry to the 'passive Laws of Motion'...The association between ether and chemical active principles was heightened by the publication of Newton's 'Hypothesis on light' and his letter to Oldenburg, where the operations of ether were linked to chemical processes, and where ether served the same function as active principles in maintaining the activity of nature. Newton suggested that 'the whole frame of nature may be nothing but aether condensed by a fermental principle'. He supposed that nature may be nothing but various Contextures of some certaine aethereall Spirits of vapours condens'd as it were by precipitation, much after the manner that vapours are condensed into water or exhalations into grosser Substances and after condensation wrought into various formes, at first by the immediate hand of the Creator, and ever since by the power of Nature...Thus perhaps may all things be originated from aether. Newton added that 'nature is a perpetuall circulatory...worker'; by the chemical transformation of ethereal spirits and by 'nature making a circulation', the activity of the cosmos was conserved. In stressing that ether was an underlying first principle from which all things originated, in supposing that the activity of the cosmos was conserved by the circulation of ethereal spirits, in arguing for the generation of all things from ether, and in relating the operations of ether to chemical processes, Newton echoed seventeenth-century alchemical and neo-Platonist writers. In a manner analogous to his disjunction between active principles and the passive principles of matter, this chemical, active ethereal cosmology lay outside the framework of the 'passive Laws of Motion'." [381]

And just as bodies of this Earth by breaking into small particles are converted into air, so these particles can be broken into lesser ones by some violent action and converted into yet more subtle air which, if it is subtle enough to penetrate the pores of glass, crystal and other terrestrial bodies, we may call the spirit of air, or the aether. That such spirits exist is shown by the experiments of Boyle in which metals, fused in a hermetically sealed glass for such a time that part is converted into calx, become heavier...I believe everyone who sees iron filings arranged into curved lines like meridians by effluvia circulating from pole to pole of the [lode-]stone will acknowledge that these magnetic effluvia are of this kind. So also the attraction of glass, amber,

[381] P. M. Heimann, "Ether and imponderables," *Conceptions of ether*, pp. 65-67.

jet, wax and resin and similar substances seems to be caused in the same way by a most tenuous matter of this kind...[382]

In his *History of the Air*, Boyle also assumes the existence of "the æther [or vacuum] in the intermundane or interplanetary spaces."[383] According to Boyle, there exist "aerious, etherial, luminous" spirits in "all mixed bodies," and these ethereal spirits are material, and "the only principles of energy, power, force and life, in all bodies wherein they are, and the immediate causes through which all alteration comes to the bodies themselves."[384] In his *Hidden Qualities of the Air*,[385] Boyle hypothesizes that "there may be in the air some secret powerful substance," which may be of "a solar, or astral, or some other exotic nature,"[386] and "Some of the mysterious writers about the

[382] A. Rupert Hall and Marie Boas Hall (eds. and trans.), *Unpublished Scientific Papers of Isaac Newton*, Cambridge, Cambridge University Press, 1962, pp. 227-228.

[383] "Newton began to prepare a second part of book 3 of the *Opticks* devoted to the role of ether... he suppressed publication of this addition to the *Opticks* and instead cautiously tailored some of his material into the seven ether queries added to the 1717 edition. In that edition he published further experimental evidence favouring ether's existence. This is the famous two-thermometer experiment showing that a vacuum posed no resistance to the propagation of heat. Heat thus appeared to be 'convey'd through the *Vacuum* by the vibrations of a much subtiler Medium than Air, which after the Air was drawn out remained in the *Vacuum*'. This experiment was performed at the Royal Society under Newton's direction but this time by J. T. Desaguliers in the autumn of 1717...his ethereal spirits...are active and material—but not mechanical in the sense of the mechanical philosophy." Cantor and Hodge, "Ether theories," pp. 23-24.

[384] Robert Boyle, *The General History of the Air*, London, Printed for *Awnsham* and *John Churchill...*, 1692; in Boyle, *Works*, vol. V, pp. 613, 641.

[385] "*The Spring and Weight of the Air*', the first scientific work of Robert Boyle, quickly established a wide reputation for its author. An air-pump...led him almost at once to devise experiments and to make from them deductions of the highest scientific importance. He first proved the air had weight, and that by virtue of this it became compressed near the earth to an extent that enabled it to support a column of 29 inches of mercury. He observed, however, that there were variations from day to day in the height of the mercury column which the air would support (p. 130), and concluded from this 'that there may be strange Ebbings and Flowings, as it were, in the Atmosphere; or at least, that it may admit great and sudden Mutations, either as to its Altitude or its Density, from causes, as well unknown to us, as the effects are unheeded by us' (p. 136)... The generalization which carried Boyle's name to posterity, i.e. that the volume occupied by a gas is reciprocal of its pressure [Boyle's Law], did not appear in the first edition (1660), but was brought forth later (1662) as a result of Linus's attack upon the deductions which Boyle had made concerning the air's weight. In his 'Defence' against Linus...he describes, in Chapter V, 'Two new Experiments...' With a U-shaped tube, sealed at one end and containing mercury at the bottom of the U so shaken as to be at equal pressure on the two sides, it required a column of 29 inches of mercury in the open limb of the tube to reduce to one-half the volume of air in the sealed end. Boyle records this observation in the following words (p. 58): 'and continuing this pouring in of Quicksilver till the Air in the shorter leg was by condensation reduced to take up but half the space it possess'd (I say, possess'd, not fill'd) before; we cast our eyes upon the longer leg of the Glass...and we observed...that the Quicksilver in the longer part of the Tube was 29. Inches higher than the other'...In other words, when the pressure on the gas was two atmospheres, the volume was reduced to one half, and so on for three and four atmospheres, until, in the original experiment, the tube broke and Boyle lost most of his quicksilver." Fulton, *Bibliography of Boyle*, pp. 9-10.

[386] "Newton's most fundamental contribution to the development of matter-theory was his replacement of the sort of 'corpuscular philosophy' favored by, say, Boyle, with a view of Nature

philosophers-stone speak great things of the excellency of what they call their philosophical magnet,[387] which, they seem to say, attracts and (in their phrase) corporifies the universal spirit, or (as some speak) the spirit of the world."[388]

In his letter to Boyle, Newton also apologizes for the unscientific speculations that he made about Boyle's essay "Incalescence of Quicksilver with Gold," but then calls the theory "fansying."[389] Yet within a month of Boyle's death on New Year's Eve 1691, Newton began to resolutely acquire through Locke, who was among those entrusted with Boyle's manuscripts, the papers detailing the procedure.[390] There is curiously no evidence that Newton believed the formula in his possession was incomplete, and pursued Boyle's papers because of the possibility that Starkey had communicated something to Boyle that was missing from his manuscript.[391] In his letter to Locke, Newton writes:[392]

based on particles and the forces between them...Newton's favorite simile for the ether was how it is 'much like air in all respects, but far more subtile.' Indeed his early acquaintance with 'Mr. Boyles receiver' and the 1660 *Spring of the Air* no doubt help to explain his own insistence on the explanatory possibilities inherent in the ether's 'elasticity.'" Arnold Thackray, *Atoms and Powers: An Essay on Newtonian Matter-Theory and the Development of Chemistry*, Cambridge, Harvard University Press, 1970, pp. 26, 33.

[387] "[Boyle] explicitly included 'the Cartesians' among the 'mechanical philosophers' in his instructive essay *Excellence and grounds of the corpuscular or mechanical philosophy* (1674)...the Cartesian ether is, he argues consistently enough, in principle detectable through its mechanical effects; although his 'attempt to examine the motions and sensibility of the Cartesian Materia Subtilis, or the Aether'—with a pair of bellows in a receiver exhausted of air—gave only negative results. More generally, Boyle takes the Cartesian ether to show that any apparently unmechanical agent may well be given an underlying mechanical interpretation. In its constitution and operation the Cartesian subtle matter is explicitly corporeal and mechanical, of course. But is it hardly, Boyle urges, any less ubiquitous and active than the 'universal spirit of some spagyrists [i.e., Paraclesans], not to say, the *Anima Mundi* of the Platonists'. Like these agents, it is, in Boyles words, an 'active principle'...Boyle is right: There are manifest overall functional analogies between Descartes' ether and the Platonists' *anima mundi*—and all the other entities in that explanatory family, too, from the Stoic *pneuma* and the Epicurean *aether* to the Paracelsan *spiritus*...This clash—between the mechanical philosophies of men like Descartes and Boyle and the neo-Platonic alternatives deliberately opposed to them, especially by More and Cudworth during Newton's early years at Cambridge—provides the indispensable background for Newton's physics and metaphysics, and so for his ether theorising." Cantor and Hodge, "Ether theories," pp. 15-16, 19.

[388] Robert Boyle, *Suspicions about some Hidden Qualities of the Air...*, London, Printed by *W. G.* and are to be Sold by *M. Pitt*...,1674; in Boyle, *Works*, vol. IV, pp. 90, 92, 96; Cf. John Henry, "Occult Qualities and the Experimental Philosophy: Active Principles in Pre-Newtonian Matter Theory," *History of Science*, 24 (1986), pp. 335-381.

[389] Newton to Boyle, 28 February 1678/79; in *Correspondence of Newton*, vol. II, p. 288.

[390] Newton to Locke, 26 January, 16 February and 2 August 1692; in Ibid., vol. III, pp. 193, 195, 215-219.

[391] Karin Figala, "Newton as Alchemist," *History of Science*, 15 (1977), p. 107; Principe, *Aspiring Adept*, p. 178.

[392] "[Locke] sent him a transcript of the first part of Boyle's recipe...and offered to forward to him the other parts...Newton replied on August 2, 1692, to dissuade Locke from wasting his time and money on Boyle's recipe...Newton's vacillation...is at its pitch as he changes his mind three times

> In ye margin...was ye foundation of what he [Boyle] published many years ago...And yet in all this time I cannot find that he has either tried it himself or got it tried w[i]th success by any body els[e]...And I doubt it ye more because I heard some years ago of a company who were upon this work in London...I enquired after them & learnt that two of them were since forced to other means of living & a third who was the chief Artist was still at work but was run so far into debt yt he had much ado to live...[Mr B.] acknowledged yt ye Rx was gone about among several Chymists, & therefore I intend to stay till I hear yt it succeeds w[i]th some of them...For Mr B. has reserved a part of it from my knowledge. I know more of it then he was told me, & by that & an expression or two w[hi]ch dropt from him I know that what he has told me is imperfect & useless w[i]thout knowing more then I do... For Mr B. to offer his secret upon conditions & after I had consented, not to perform his part looks od[d]ly...I do not desire to know what he has communicated but rather that you would keep ye particulars from me...For when I communicated a certain experim[en]t to him he presently by way of requital subjoined two others, but cumbered them w[i]th such circumstances as startled me & made me afraid of any more. For he expected yt I should presently go to work upon them & desired I would publish them after his death. I have not tried either of them nor intend to try them but since you have inspection of his papers, if you designe to publish any of his remains, You will do me a great favour to let these two be published among ye rest. But then I desire that it may not be known that they come through my hands. One of them seems to be a considerable Expt. & may prove of good use in medicine for analyzing bodies, the other is only a knack.[393]

In the subsequent year, following his successful acquisition of Boyle's papers and at the culmination of his alchemical investigations, Newton suffered his well-known mental breakdown.[394] On 13

in one note...His suspicions of Boyle's candor break through time and time again. At moments in the devious interrogation one hears the rumble of the approaching psychic crisis. There is hardly a sentence in this letter that is not qualified or negated by what follows...Newton also concluded from the margin of the paper that this was the same recipe for the same mercury...about which Boyle had published...in the *Philosophical Transactions*. And yet Newton doubted whether Boyle, despite his published report, had at any time during these two decades ever tried the recipe himself or gotten it performed by anyone else...Moreover, he had private evidence that the recipe was ineffective because he knew of a 'company' that had use it and had failed, its chief artist having run so far into debt that he had 'much ado to live.' When taxed for this information, Boyle had admitted that the recipe had been imparted to several chemists. Following hard on upon the opinion that the recipe was worthless was Newton's announcement to Locke that he could not try it in any case because something had been withheld from him by Boyle...The ending of the letter to Locke is as obfuscatory about his true desires and intentions as Newton can become." Manuel, *Portrait of Newton*, pp. 185–186.

[393] Newton to Locke, 2 August 1692; in *Correspondence of Newton*, vol. III, pp. 217-219.

[394] "Newton introduced Fatio de Duillier, the young Swiss mathematician who entered intensely into his life at this time, to the Art, and both Fatio and alchemy figured in the mounting emotional tension that wracked Newton's life in 1693. In the summer of that year, sometime

September 1693, Newton abruptly terminated his relationship with diarist Samuel Pepys, during whose tenure of office as President of the Royal Society the *Principia* had been published,[395] accusing him of conspiracy,[396] and three days later Newton accused Locke of being an atheist and wished him dead.[397]

Newton quickly recovered from his ailment and extended his apologies to Pepys in a letter of John Millington dated 30 September,[398] and apologized to Locke in a letter of 5 October.[399] In a letter of

after his reception of Fatio's letter of 13 May which he cited, Newton composed the essay *Praxis*, which I have called possibly his most important alchemical paper. At the climax of *Praxis*, Newton claimed to have achieved multiplication...I do not think that we should take this passage seriously, certainly not as evidence that Newton attained the alchemist's goal, but no more as evidence that he believed he had. Newton was overwrought at the time. Within the following three months he wrote the famous letters to Locke and Pepys, which both of them took as evidence of some derangement, as everyone who reads them must...twenty-five years of deep involvement in the Art reached their culmination in what was also, for a number of reasons the tragic climax of Newton's life." Richard S. Westfall, "The Influence of Alchemy on Newton," *Science, Pseudo-Science and Society*, M. P. Hanen, M. J. Osler, and R. G. Weyant (eds.), Waterloo, Ont., Wilfrid Laurier University Press, 1980, p. 164. Newton's *Praxis* gives very little indication of any practical operations, but is a series of quotations collected from numerous alchemical authors, especially Starkey's Philalethes and Michael Maier. The *Praxis* manuscript is reproduced in Dobbs, *Janus*, pp. 293-305.

[395] "Newton may have experienced a number of earlier psychic disturbances—his references to illnesses are vague—but to our knowledge he sank into delusion only once, during the grave events of September 1693, when he had passed his fiftieth birthday. Then the great man behaved as might any mortal who is struck down. He broke with his friends, crawled into a corner, accused his intimates of plotting against him, and reported conversations that never took place." Manuel, *Portrait of Newton*, p. 214.

[396] "I am extremely troubled at the embroilment I am in, and have neither ate nor slept well this twelve month[s], nor have my former consistency of mind. I never designed to get anything by your interest, nor by King James's favour, but am now sensible that I must withdraw from your acquaintance, and see neither you nor the rest of my friends any more, if I may but leave them quietly. I beg your pardon for saying I would see you again..." Newton to Pepys, 13 September 1693; in *Correspondence of Newton*, vol. III, p. 279.

[397] "Being of opinion that you endeavoured to embroil me w[i]th woemen & by other means I was so much affected with it as that when one told me you were sickly & would not live I answered twere better if you were dead. I desire you to forgive me this uncharitableness. For I am satisfied that what you have done is just & I beg your pardon for having hard thoughts of you for it & for representing that you struck at ye root of morality in a principle you laid down in your book of Ideas & designed to pursue in another book & that I took you for a Hobbist [atheist]." Newton to Locke, 16 September 1693; in Ibid., vol. III, p. 280.

[398] "[Newton] told me that he had writ[ten] to you a very odd letter, at which he was much concerned ; added, that it was in a distemper that much seized his head, and that kept him awake for above five nights together, which upon occasion he desired I would represent to you, and beg your pardon, he being very much ashamed he should be so rude to a person from whom he hath so great an honour. He is now very well, and though I fear he is under some small degree of melancholy, yet I think there is no reason to suspect it hath at all touched his understanding..." Millington to Pepys, 30 September 1693; in Ibid., vol. III, pp. 281-282.

[399] "I have b[e]en ever since I first knew you so intirely & sincer[e]ly your friend & thought you so much mine yt I could not have beleived what you tell me of your self had I had it from anybody else. And though I cannot but be mightily troubled that you should have had so many wrong & unjust thoughts of me yet next to the returne of good offices such as from a sincere good will I have ever done you I receive your acknowledgm[en]t of the contrary as ye kindest thing you could have done me since it gives me hopes I have not lost a freind I soe much valued...I wish for noe thing more than the opportunities to convince you yt I truly love and esteem you & yt I have

Newton to Locke on 15 October he describes his affliction as "a distemper w[hi]ch this summer has been epidemical put me further out of order, so that when I wrote to you I had not slept an hour a night for a fortnight together & for 5 nights not a wink. I remember I wrote to you but what I said of your book I remember not."[400] In another undated letter thought to be to Fatio in May of 1693, Newton writes: "About a fortnight since I was taken ill of a distemper w[hi]ch has been here very common, but am now pretty well again."[401]

The theories proposed by scholars for Newton's "distemper"[402] are overwork,[403] frustration, depression,[404] the destruction of his manuscripts by fire, and the collapse of his friendship with Fatio de Duillier.[405] It has also been suggested that mercury poisoning contributed to Newton's mental and physical illnesses,[406] and high levels of toxic substances in Newton's hair samples appear to lend support to this position, although this does *not* constitute conclusive evidence that his illness in 1693 was caused by mercury poisoning.[407]

still the same goodwill for you as if noe thing of this had happened." Newton to Locke, 5 October 1693; in Ibid., vol. III, pp. 283-284.

[400] Newton to Locke, 15 October 1693; in Ibid., vol. III, p. 284.

[401] Newton to ——, n.d.; in Ibid., vol. VII, p. 367.

[402] "I do not know if you are acquainted with the accident which has happened to the good Mr. Newton, namely, that he has had an attack of 'frenesie'..." Huygens to Leibniz, 8 June 1694, Christiaan Huygens, *Oeuvres Complètes*, The Hague, Martinus Nijhoff, 1905, vol. X, p. 618; in Manuel, *Portrait of Newton*, p. 433 n. 20.

[403] "Intellectual excitement always stretched him to the very limit and, on occasion, beyond it. His breakdown in 1693 was not altogether different from his behavior in 1677-8." Westfall, *Life of Newton*, p. 216.

[404] "Newton first suffered an emotional disturbance in 1662, and was mentally ill in the 1690s... Newton was a humourless, solitary, anxious, insecure and private man with obsessional traits. He was poor at human relationship, such as the expression of gratitude, and held unorthodox and heretical religious beliefs. He was clearly puritanical, with feelings of guilt, and had little capacity for enjoyment...He never used the word 'love', and expressions of gladness and of desire were rare...I do not think he had a depressive personality: I believe he was a suspicious and slightly, or even, very paranoid man with *sensitiver Beziehungswahn*. He had a sensitive, prickly personality at the edge of normality, which could tip over into psychosis, if the conditions were right and sufficiently emotionally upsetting; that is, he occasionally lost the capacity for reality testing...Newton's psychological symptoms of the 1690s were, I believe, due to erethism [mercury poisoning] in an already eccentric, paranoid man, and it is known that toxic substances may exacerbate quirks of personality." Keynes, "Personality of Newton," pp. 2-43.

[405] Manuel, *Portrait of Newton*, pp. 213-225; Westfall, *Never at Rest*, pp. 537-539; A. Rupert Hall, *Isaac Newton: Adventurer in Thought*, Oxford, Blackwell Publishers, 1992, pp. 244-245.

[406] Newton was rarely physically ill until his last years, although in 1689 Newton writes of "Being confined to my Chamber by a cold & bastard Pleurisy..." Newton to Covel, 10 May 1689, in *Correspondence of Newton*, vol. III, p. 22.

[407] L. W. Johnson and M. L. Wolbarsht, "Mercury Poisoning: A Probable Cause of Isaac Newton's Physical and Mental Ills," *Notes and Records of the Royal Society of London*, 34 (1979), pp. 1-9; P. E. Spargo and C. A. Pounds, "Newton's 'Derangement of the Intellect': New Light on an Old Problem," *Notes and Records of the Royal Society of London*, 34 (1979), pp. 11-32.

Beyond the failures of Newton and Boyle,[408] there were a number of other unsuccessful attempts made with Starkey's transmutation formulas,[409] which likely contributed to Starkey's falling out with the Hartlib circle.[410] Starkey repeatedly deceived Boyle and other members of Hartlib's circle with fables of Philalethes, and to protect his secrets Starkey also appears to have, on more than one occasion, simply told Boyle and others what they wanted to hear.[411]

It was in a letter to Fatio in 1689 that Newton claimed that he had earlier declined to take up a correspondence with Boyle on alchemical matters because he was "in my opinion too open & too desirous of fame,"[412] but in his correspondence with Locke in August 1692 Newton complained that Boyle had offered him information only under the most strict conditions, and even then concealed his full knowledge of the subject.[413] Yet Newton was rather standoffish with Boyle and less than forthright when it came to the subject of alchemy,[414] and referring to his communications with Boyle, Newton told Locke: "I suspect his reservedness might proceed from mine."[415] Robert Boyle struggled with his anxieties concerning the potential social and political consequences of openly revealing the Philosophers' Stone,[416]

[408] "One might question as well whether Boyle's own chronic sickness was caused or aggravated by similar mercury poisoning." Principe, *Aspiring Adept*, p. 179. "One might even speculate that part of Starkey's inconsistent behavior was due to heavy metal poisoning." Newman and Principe, *Alchemy*, p. 225.

[409] Hartlib, *Ephemerides*, [February 1655], Hartlib Papers 29/5/12A; in Newman and Principe, *Alchemy*, pp. 264-265; Hartlib, *Ephemerides*, [1655], 27-278; in Wilkinson, "Hartlib Papers II," pp. 104–105; Johann Hiskias Cardilucius, *Magnalia Medico-Chymica Continuata*, Nuremberg, Wolffgang Moritz Endters, 1680, [preface]; in Newman, *Gehennical Fire*, p. 82; Newton to Locke, 2 August 1692; in *Correspondence of Newton*, vol. III, pp. 217-218; Johannis Ferdinandi Hertodt, "Epistolâ *contra* Philalethæ Processus," *Bibliotheca Chemica Curiosa*, vol. II, pp. 697-699.

[410] "Factionalism, manipulation, and appropriation seem notable attributes of many of the Hartlibians. The Clodius-Digby alliance may well have been in part responsible for turning Hartlib against Starkey. Various members attempted to appropriate Starkey's work: Worsley claimed Starkey's Philosophical Mercury and *luna fixa* as his own to Moriaen, Clodius may have misrepresented Starkey's early alkahest as his own to Hartlib, and Boyle not only divulged secrets entrusted to him by Starkey, but often gave the impression—sometimes implicitly and sometimes explicitly—that they were his own discoveries." Newman and Principe, *Alchemy*, p. 268.

[411] William R. Newman, "George Starkey and the selling of secrets," *Samuel Hartlib and Universal Reformation: Studies in intellectual communication*, M. Greengrass, M. Leslie, and T. Raylor (eds.), Cambridge, Cambridge University Press, 1994, pp. 201-202.

[412] Newton to Fatio, 10 October 1689; in *Correspondence of Newton*, vol. III, p. 45.

[413] Newton to Locke, 2 August 1692; in Ibid., vol. III, p. 218.

[414] Manuel, *Portrait of Newton*, pp. 178–187; Principe, *Aspiring Adept*, p. 178.

[415] Newton to Locke, 2 August 1692; in *Correspondence of Newton*, vol. III, p. 218.

[416] "It is objected by som[e] Chymists that had they the Elixir it selfe (the Universall Antagonist of all Diseases) they shud think it unlawfull to impart it : partly because such extraordinary Discovery ; being rather inspir'd then acquir'd ; ought not to be profan'd by bei[n]g divulg'd ; & partly because in these Debauched Times ; it wud be an invitation to all kind of Riot and intemperance by securing us from the Danger of it's Effects. Tis strange that Conscience shud

and equally so with his Puritan beliefs regarding the morality and dangers of communicating with the supernatural.[417] On the subject of alchemy and communicating with angels and spirits,[418] Boyle writes in his "Dialogue" on good and evil spirits:[419] "the acquisition of the Philosophers' stone may be an inlett into another sort of knowledge and a step to the attainment of some intercourse with good spirits."[420] These factors help to explain Boyle's defensiveness and secrecy,[421] which failed to protect him from the loss of his writings to thieves.[422]

---

be fallen out with Charity. but the Objection itselfe furnishes us with an Answer ; for if the Elixir be a thing that we ow[e] wholly to God's Mercy not our own Industry, me thinks we shud the less grudge to impart what we did not labor'd to Acquire ; since our Savior's Rule in the like Case is Freely ye have received, Freely give. Shud God to one of our Divines reveale som[e] newer Truths, wud we not in him Condemn their Concealment, & in Effect ; tho God shud ad[d]ress those Speciall Favors but to one man, yet he intends um for the Good of all Mankind. As for the other side of the objection shall we think it unlawfull to do Evill that Good may com[e] of it think it just to forbeare/decl[ine]/Dutys lest Evil might ensue ? & shall we let Good men languish/perish/for want of Releefe ; lest others shud be encourag'd to expect it ? Besides that the same Reason wud justify the Concealment of all other Secrets...To omit that to Think to restraine Vices by refusing men the Remedys of their Excesses ; is an Expectation as Vaine as the Dessein [Design] is uncharitable...Sure many of these envious salamanders will one day find their kno[w]ledge to be their Crime, when those Secrets that were bestow'd as fewell for their Charity shall serve but to aggravate their Guilt. Then they will be forc't to Endure far greater torments then those that they declin'd to Cure : & with as little Pitty as they heer made use of : it being unjust for um to Expect that Ease/Mercy/that they refuse to give/Share/grant." Boyle Letters, vol. I, fol. 146, [in margin]; in Boyle, *Letters and Papers* [microform]. The full text is reproduced in Maddison, "Earliest Published Writing of Boyle," pp. 168–172.

[417] Hunter, "Alchemy of Boyle," pp. 396-398; Lawrence M. Principe, "Robert Boyle's Alchemical Secrecy: Codes, Ciphers and Concealments," *Ambix*, 39 (1992), pp. 70-71; Principe, *Aspiring Adept*, pp. 187, 310-317; Wojcik, *Boyle and Reason*, 1997, pp. 135, 141–144.

[418] In Hartlib's *Chymical Addresses* containing Boyle's first printed work "An Invitation to a free and generous Communication of Secrets and Receits in Physick," which like the *Sceptical Chymist* is an attack upon alchemical secrecy, another contributor to the collection states "that (at least, by the extraordinary help of good or evil Angels) men might come to have some knowledge of it [the Philosophers' Stone]." [Theophraste] Renaudot, "A Conference Concerning the Philosophers' Stone," *Chymical, Medicinal, and Chyrurgical Addresses*, p. 112; Margaret Rowbottom, "The Earliest Published Writing of Robert Boyle," *Annals of Science*, 6 (1950), pp. 376-387; Fulton, *Bibliography of Boyle*, pp. 1-3.

[419] "[Boyle] expected to obtain...the secret of the preparation of the Philosophers' Stone...widely regarded as the panacea...Both men saw alchemy as a corrective to an overtly mechanized and potentially atheistic world view...For Boyle, traditional alchemy provided an interface between the natural, mechanical realm and the supernatural, miraculous realm. The Philosophers' Stone which Boyle sought to prepare was a physical substance that caused physical changes, but which could also, as Boyle came to believe, attract and manifest spirits and angels by some unknown 'congruities or magnatisms.' Indeed in a remarkable document on the Stone's ability to attract rational spirits, Boyle explicitly relates such manifestations of spirits to the confounding of atheism...the Stone represented a 'boundary' where corporeal and incorporeal met." Lawrence M. Principe, "The Alchemies of Robert Boyle and Isaac Newton: Alternate Approaches and Divergent Deployments," *Rethinking the Scientific Revolution*, pp. 214-216. Boyle's "Dialogue on Spirits" is reproduced in Principe, *Aspiring Adept*, pp. 310-317.

[420] Royal Society Library, Boyle Papers, vol. VII, fol. 138^v; in Boyle, *Letters and Papers* [microform].

[421] Boyle, *Works*, vol. I, p. cxxxi.

[422] Royal Society Library, Boyle Papers, vol. XXXVI; in Boyle, *Works*, vol. I, pp. cxxv-cxxviii.

Newton seems to have held similar fears, and these activities also remained unlawful until 1689, legally punishable by hanging.[423]

The wariness in the interactions between Newton and Boyle fails to be explained by these various factors, which suggests that both scientists were carefully guarding their secrets from one another,[424] both intending to publicly announce the success of their efforts.[425] There seems to be little reason to believe that had Newton or Boyle been successful in their alchemical pursuits,[426] to which both men dedicated the best part of their lives,[427] that either man would have concealed their discovery of the "Universal Medicine"[428] from the

[423] Boyle initiated the repeal in 1689 of the 1404 Act of Henry IV prohibiting "the craft of multiplication," which rendered alchemy illegal. Steele, "Alchemy in England," p. 100; Maddison, *Life of Boyle*, p. 176. Newton believed that Starkey's recipe, which Locke received from Boyle, was his motivation for initiating the repeal. Newton to Locke, 2 August 1692; in *Correspondence of Newton*, vol. III, p. 217.

[424] "Newton was not the first of the age of reason. He was the last of the magicians, the last of the Babylonians and Sumerians...a posthumous child born with no father on Christmas Day, 1642, was the last wonder-child to whom the Magi could do sincere and appropriate homage... this strange spirit, who was tempted by the Devil to believe...that he could reach *all* the secrets of God and Nature by the pure power of mind—Copernicus and Faustus in one." Keynes, "Newton, the Man," pp. 27, 34.

[425] "That Newton might discover an elixir of life which would bestow immortality upon him—if not upon his fellowmen, about whom he had less concern—may be a remote motivation for his search, but one not to be wholly excluded in the light of his hypochondria and his omnipotence fantasies...To Newton Boyle was a rival as well as an inspiration...A fear that Boyle might conceivable have made or known of the discovery of the philosopher's stone, as his writings intimated, is apparent and the idea that Newton might be outstripped is intolerable." Manuel, *Portrait of Newton*, pp. 169–178.

[426] "Newton's work in alchemy had a religious motivation...Newton began work on the prophecies in the 1670s if not earlier, and he is thought to have still been working on his last version of their interpretation the night before he died in 1727." Dobbs and Jacob, *Culture of Newtonianism*, p. 32; Cf. Frank E. Manuel, *Isaac Newton: Historian*, Cambridge, Belknap Press of Harvard University Press, 1963; Stephen D. Snobelen, "Isaac Newton, heretic: the strategies of a Nicodemite," *British Journal for the History of Science*, 32 (1999), pp. 381-419; James E. Force, "The Virgin, the Dynamo, and Newton's Prophetic History," *Millenarianism and Messianism in Early Modern European Culture*, J. E. Force and R. H. Popkin (eds.), Dordrecht, Kluwer Academic Publishers, 2001, vol. III, pp. 67-94, Stephen D. Snobelen, "'The Mystery of this Restitution of All Things': Isaac Newton and the Return of the Jews," Ibid., pp. 95–118.

[427] "[T]he hatred of Catholicism permeated nearly every aspect of Newton's life. An adept in the history of the Church, he spent the bulk of his time at Trinity with the bunker mentality proper for one of the Elect, and engaged in theological study and not in scientific pursuits." Rob Iliffe, "Those 'Whose Business it is to Cavill': Newton's Anti-Catholicism," *Newton and Religion: Context, Nature, and Influence*, J. E. Force and R. H. Popkin (eds.), Dordrecht, Kluwer Academic Publishers, 1999, p. 119.

[428] "The Newtonian papers contain a very large number...that are devoted to alchemy. They date from about 1669 until the middle 1690s; something approaching half of them come from the years immediately following the *Principia*...[Newton's] papers reveal that he turned to serious reading of theology some time late in the 1660s...Theology quickly drove physics and mathematics out of Newton's active concern, and together with...alchemy, it largely dominated his consciousness for well over a decade. With the mid-1680s, the *Principia* broke the ascendance of theology over Newton's mind, and during the following two decades, he devoted much less attention to it. He returned to theology in the early eighteenth century...The religious studies that began shortly after Newton's student days, then, continued unbroken, though with a period of sharply reduced intensity, for the remaining sixty years of his life." I. Bernard Cohen and Richard S.

public.[429] After all, Boyle *did* make a public announcement, and Newton[430] and Boyle were both aware of the many prophesies concerning this event,[431] including the well-known prediction years

Westfall (eds.), *Newton: Texts Backgrounds Commentaries*, New York, W. W. Norton & Company, Inc., 1995, pp. 299-300, 327.

[429] "The more Newton's theological and alchemical, chronological and mythological work is examined as a whole corpus, set by the side of his science, the more apparent it becomes that in his moments of grandeur he saw himself as the last of the interpreters of God's will in actions, living on the eve of the fulfilment of the times. In his generation he was the vehicle of God's eternal truth, for by using new mathematical notations and an experimental method he combined the knowledge of the priest-scientists of the earliest nations, of Israel's prophets, of the Greek mathematicians, and of the medieval alchemists. From him nothing had been withheld. Newton's frequent insistence that he was part of an ancient tradition, a rediscoverer rather than an innovator, is susceptible to a variety of interpretations. In manuscript scholia to the *Principia* that date from the end of the seventeenth century he expounded his belief that a whole line of ancient philosophers had held to the atomic theory of matter, a conception of the void, the universality of gravitational force, and even the inverse square law...Newton's conviction that he was a chosen one of God, miraculously preserved, was accompanied by the terror that he would be found unworthy and would provoke the wrath of God his Father. This made one of the great geniuses of the world also one of its great sufferers." Frank E. Manuel, *The Religion of Isaac Newton: The Fremantle Lectures 1973*, Oxford, Oxford University Press, 1974, pp. 23-24.

[430] "[Newton's] interpretation of Daniel, his decoding of the book's cryptic language, and his discussion of theological questions, among them the Second Coming, all emanated from the sense of a special mission. This feeling grew from a belief that the 'wisdom' to understand the prophecy was transmitted from God to a chosen person—himself. The sense of chosenness grew in Newton, not as a result of his study of prophecies, but following his unique achievements in natural science, which in his opinion, were conveyed to him by God alone...The Second Coming of Christ had sacred and hidden dimensions, most of which related to the time of this event. This was the most central, sacred question that occupied the millenarians...Newton made such calculations...for Christ's Second Coming. The addition of 1,260 years to the year 381 shows the process of redemption started in 1641—one year before Newton was born...When Newton recognized his authority to interpret the Book of Daniel, he was convinced that (a) he had been chosen by God; (b) that the time for the end had arrived and the holy process begun (with the birth of Newton), and (c) that everything was about to be realized in the very near future...Newton believed that scientific knowledge constituted not only a source of authority in interpreting prophecies, but also serving as a practical means of gaining a better theological understanding. This can be learned from what Newton wrote about the purpose of his alchemical research:

> Alchemy tradeth not with metals as ignorant vulgar think...this philosophy is not that kind which tendeth to vanity and depict but rather to profit and to edification inducing first the knowledge of God...so that the scope is to glorify God in his wonderful work, to teach Man how to live well and to be charitably affected helping our neighbour...

Newton developed a sense of chosenness owing to his scientific understanding and employed his knowledge in the process of interpreting the Book of Daniel...Newton believed that the decoding of Daniel, its interpretation, and the medium through which the exegesis was carried out comprised one inseparable unit—and that it all had come together in his personality. Newton found proof of his unique status in his ability to understand the structure of the universe, which was the first message that God transmitted to his messenger, as he had earlier done with Moses." Matania Z. Kochavi, "One Prophet Interprets Another: Sir Isaac Newton and Daniel," *The Books of Nature and Scripture: Recent Essays on Natural Philosophy, Theology, and Biblical Criticism in the Netherlands of Spinoza's Time and the British Isles of Newton's Time*, J. E. Force and R. H. Popkin (eds.), Dordrecht, Kluwer Academic Publishers, 1994, pp. 105–120.

[431] "Newton *was* a very strange man. His reluctance to publish, his sensitivity to criticism, his repeated camouflaging of his true intentions, and perhaps too his preoccupation with the millenarian and the occult—all these point to a neurotic behavior pattern that makes the nervous breakdown of 1693 far from surprising." Arthur Quinn, "On Reading Newton Apocalyptically,"

earlier by the Fifth Monarchist Mary Rand,[432] who had prophesied the impending disclosure of the Philosophers' Stone[433] to the masses.[434]

This prophecy of Mary Rand is repeated by John Langius in the preface of Philalethes's *Secrets Reveal'd*,[435] the English version of the *Introitus Apertus*.[436] Newton heavily annotated his copy of *Secrets Reveal'd*,[437] in which Langius states:[438] "if those things which *Mary Rant*

Millenarianism and Messianism in English Literature and Thought 1650–1800; Clark Library Lectures 1981–1982, R. H. Popkin (ed.), Leiden, E. J. Brill, 1988, p. 178.

[432] "From the event-filled year of 1666...twenty-three year old Newton was already dabbling in divinely appointed dates...Millennialist diatribe concerning the four world empires had not died down with the collapse of the Protectorate (in 1659) or with the vociferations of the 'Fifth Monarchy Men' (and their uprisings of 1657 and 1661). There were Anglican divines keeping anti-Catholic interpretations of Daniel and Revelation alive during Restoration years, and nowhere more actively than at Cambridge. During his Cantabrian undergraduate days (1660-4) Newton had procured van Sleidan's *Key to Historie* (or 'Four Monarchies' as he wrote of it), and a Fellow at Trinity could hardly escape the discussions centered on Joseph Mede's extraordinarily influential *Key of the Revelation* (Lat. 1627, Eng. 1643) which had been republished and enlarged in 1664, and which through identifying the Papacy with Anti-Christ in the Apocalypse, re-acquired its earlier political significance when Prince James, brother to the heirless Charles II, publicly disclosed his Catholicism in 1673." Garry W. Trompf, "On Newtonian History," *The Uses of Antiquity: The Scientific Revolution and the Classical Tradition*, S. Gaukroger (ed.), Dordrecht, Kluwer Academic Publishers, 1991, p. 213. For Mary Rand, see B. S. Capp, *The Fifth Monarchy Men: A Study in Seventeenth-century English Millenarianism*, London, Faber and Faber, 1972, p. 244.

[433] "Boyle's apocalyptic eschatology belonged to a perception closer to the Protestant mainstream than was typical of most natural philosophers and virtuosi he associated with, though natural philosophy took on a particular significance within that eschatology. Boyle could perceive himself as one of God's instruments in bringing about a new 'Revolution' in both science and divinity before the final transformation of a new heaven and a new earth. He had cause to think that this might be close at hand, a sentiment shared by others in the Hartlib circle...It is, of course, perfectly legitimate to conjecture that millennial beliefs were secretly held by Boyle, much as his alchemical beliefs largely were, perhaps expressed in private correspondence which was subsequently destroyed. After all, one only needs to be reminded of Boyle's puritan friend, Richard Baxter, studying Revelation secretly in prison in 1686 while a Roman Catholic ruled England or Newton's secretive studies on the same book, to recognize how aware Boyle must have been 'of the damaging effect upon less tutored minds of such speculations'. Among the Boyle Papers, for example, can be found notes almost certainly written in the later part of Boyle's life by one of his assistants, Hugh Greg, on the meaning of various numbers in the book of Revelation as predicting the overthrow of Rome. It does not, with its elaborate number calculations, reflect the personal commitment of Boyle, but it may at least indicate that Boyle continued to be alert to millennial speculation." Malcolm Oster, "Millenarianism and the new science: the case of Robert Boyle," *Hartlib and Universal Reformation*, pp. 135–136.

[434] Oldenburg to Boyle, 10 September 1658, Beale to Hartlib, 22 March 1658/59, Hartlib to Boyle, 5 April 1659; in Webster, *Great Instauration*, pp. 392-393.

[435] Eirenaeus Philalethes (pseud.), *Secrets Reveal'd: Or, An Open Entrance To The Shut-Palace of the King. Containing, The greatest Treasure in Chymistry, Never yet so plainly Discovered. Composed By a most famous English-man, Styling himself Anonymus, or Eyrœneus Philaletha Cosmopolita: Who, by Inspiration and Reading, attained to the Philosophers Stone at his Age of Twenty three Years, Anno Domini, 1645...*, London, Printed by *W. Godbid* for *William Cooper* in Little St. *Bartholomews*, near *Little-Britain*, 1669, p. a5r.

[436] Ronald Sterne Wilkinson, "The Problem of the Identity of Eirenaeus Philalethes," *Ambix*, 12 (1964), pp. 33-51.

[437] Dobbs, *Foundations*, pp. 67, 88 n. 153; Dobbs, "Newton's Copy of Secret's Reveal'd," pp. 145–169.

[438] "Much of Newton's effort on the regimens centred on *Secrets Reveal'd*...Newton had only recently begun his study of alchemy when the English version of the book was published in 1669. At the time, Eirenaeus Philalethes was attracting much interest. His commentary on George Ripley's 'Epistle to King Edward' had been published in 1655, and that and other works were circulating in manuscript. Newton himself had laboriously copied out four of these circulating

(an *English* woman) by inward Revelation promised concerning the making of Gold (that it would become vulgar or common in the year 1661) come to pass within an hundred years after, then I doubt not at all but it hath some beginning from this."[439] Repeating a prophecy that was made by others before him, Paracelsus had predicted that some time following his death the Adept known as "Elias the Artist" would come and reveal the secrets of alchemy, a prophecy that van Helmont and Starkey also perpetuated.[440] In his *Secrets Reveal'd*, the "American philosopher" even declares that "Elias Artista" is already present in the world:[441] "*Helias* the Artist is already born, and now glorious things are declared of the City of God...These things I send before into the world, like a Preacher, that I may not be buried unprofitably in the World : Let my Book therefore be the fore-runner of *Elias*, which may prepare the Kingly way of the Lord."[442]

manuscripts, probably in 1668...A Latin version of *Secrets Reveal'd* had appeared in Amsterdam in 1667 as *Introitus apertus*...but so far as is known Newton never owned a copy of it. Actually, there is no reason to suppose that he had begun his alchemical studies as early as 1667, and he may not have known of the existence of that edition until the English one appeared. He did, however, buy a copy of the three Philalethes tracts that were published by the same Amsterdam firm in 1668, and he then proceeded to make detailed summaries of them. That meant that by 1669 he had already studied seven of the Philalethes tracts, and given Newton's propensity for scholarly thoroughness, it would indeed have been out of character for him *not* to have purchased a copy of *Secrets Reveal'd* when it appeared." Dobbs, "Newton's Copy of Secret's Reveal'd," p. 146.

[439] John Langius, "The Preface of John Langius To The Reader," *Secrets Reveal'd*, p. a5r.

[440] Helmont, "Pharmacapolium," *Oriatrike*, p. 459; Kittredge, "Robert Child," pp. 129-132; Walter Pagel, "The Paracelsian Elias Artista and the Alchemical Tradition," *Kreatur und Kosmos: Internationale Beiträge zur Paracelsusforschung*, Stuttgart, Gustav Fischer Verlag, 1981, pp. 6-19; Breger, "*Elias Artista*," pp. 49-72.

[441] "During the Reformation, in Protestant areas, alchemy thus assumed unparalleled significance in alignment with contemporary convictions that the End of Days was at hand, that the Reformation of the Church had itself initiated that great prophesied *renovatio* that was to usher in the Millennium...Of earliest historical significance is the attention bestowed on the apocalyptic figure of Elias (or Elijah)...Elias was traditionally recognized as a worker of miracles who did not die but ascended into heaven in a fiery chariot, from whence he would return to herald the last days of judgment or the advent of the Messiah. Joachim of Fiore in the twelfth century expected Elias to return to usher in a third great era of history (the first two corresponding to those of the Old and New Testaments), and, when Luther repeatedly stated in the sixteenth century that the day of judgment could not be much longer in coming, his disciples began to call him the reborn or third Elias (the second having been John the Baptist). Protestants of every ilk, from the Münster Anabaptists to the distinguished Calvinist theologian Alsted, expected the return of Elias as an established part of their eschatological hopes. In the hands of Paracelsus, however, the biblical precursor was transformed into the figure of 'Elias Artista,' an Elias of the arts, and hopes for full knowledge in natural science were merged with expectations of social and religious Utopia. God allows us some knowledge of nature now, Paracelsus suggested, but then much more will be revealed to us: 'For the arts have just as much their Elias as religion also has.'" Betty Jo Teeter Dobbs, "From the Secrecy of Alchemy to the Openness of Chemistry," *Solomon's House Revisited: The Organization and Institutionalization of Science*, T. Frängsmyr (ed.), USA, Science History Publications, 1990, pp. 83-84.

[442] Philalethes, *Secrets Reveal'd*, pp. 47-49.

Chapter 6. George Starkey's Philosophers' Stone

George Starkey often referred to his individual drugs as *Arcana*, or "Secrets."[443] Referring to these "Secrets," Starkey writes in *Natures Explication*: "First then let no man expect from me linear receipts, for that would be foolish in me to perform, and therefore fond of them to expect ; for I shall not write of trifles, but of commanding *Arcanaes*, which require to be discovered in the language of the *Magi*, lest fools and Mechanists, bring these so noble secrets into common Shops, to be adulterated."[444] Starkey also told Hartlib: "a Receipt is every mans meat...I could not possibly throw the Receipts into the mouth of every one that could but gape."[445]

Invoking the "silence of *Pythagoras*" as his defense,[446] Starkey writes in *Ripley Reviv'd*: "our Books are full of obscurity, and Philosophers write horrid Metaphors and Riddles to them who are not upon a sure bottom, which like to a running Stream will carry them down head-long into despair and errors."[447] Starkey adds: "But he who thinks, because Philosophers say it is such a deadly poison, that it is to be bought at the Apothecaries, or Druggists, he is mistaken ; for as it is first bought, I confess it is very venomous, but this malignity I conceive and know is fully taken away, before it become[s] the Philosophers *Mercury*."[448] In his unpublished manuscript entitled "Diana denudata,"

[443] Starkey, *Natures Explication*, p. 216; Sloane Collection, British Library, MS 3750; in Newman, *Gehennical Fire*, p. 300 n. 88.
[444] Starkey, *Natures Explication*, p. 239.
[445] Starkey to Hartlib, n.d.; in [Hartlib], *Common-Wealth of Bees*, p. 34.
[446] Philalethes, "Sir George Ripley's Preface," *Ripley Reviv'd*, p. 30.
[447] Philalethes, "First Six Gates of Sir George Ripley's Compound of Alchymie," Ibid., p. 135.
[448] Ibid., p. 244.

Starkey calls the Philosophers' Stone the "gold and ☿ [Mercury] of the Magi,"[449] and writes:

> Now how, & why it is called our ☿ [Mercury], our moone, & our unripe gold, our fire, & our furnace, it is very worth the Consideration, which trust me is the very key of all our secrets, & the threed of Ariadnes to conduct us in the windings of the Dedalean labyrinth, For it is only this one thing which is secret, & hidden, which if it should be told plainly, with the pondus and regimen, even fooles would deride the Art, it is soe easy, being indeed but the worke of women, & the play of Children, obscured and vayled by the Antient Sages, by al[l] meanes possible, being indeed an Arcanum not fit to be divulged.[450]

To this Starkey adds: "Therefore to amuse & to amaze the vulgar erring Artists, wee speake of our ☉ [Gold], our ☽ [Silver], our ☿ [Mercury], our fire, our water, our furnace &c, which is al[l] but one thing...it is called our fire, our furnace, our vinegre, &c, with many other infinite appellations which it hath on one account or other."[451] Starkey ends this manuscript with the following:

> Such a meane the wise Philosophers with all their might have sought & found, & left the record of their search in writing, withall so veyling the maine secret that only an immediate hand of god must direct an Artist who by study shal[l] seeke to atteyne the same. This meane Substance is the Key of the whole worke, it is the only hidden secret which they in their Bookes have Concealed, concerning which all theyr allegoryes, Metaphors & darke sentences doe treat, learne this, & al[l] the hard, darke sentences of the wise wil[l] appeare plaine & easy to thee.[452]

Referring to the Philosophers' Stone and the production of the alchemical Elixir, Starkey writes in *Ripley Reviv'd*: "it is found in a Dunghil[l], according to *Morien...Artephius* and *Flammel* say it is but the play of Children and work of Women...this Art is easie to him that understands it, as *Artephius* plainly expresseth ; but to him that is ignorant of it, there is nothing can appear so hard."[453] In *Secrets Reveal'd* Starkey writes: "if this one Secret were but openly discovered, Fools themselves would deride the Art ; for that being known, nothing remains, but the Work of Women and the Play of Children, and that

449 [George Starkey], "Diana denudata," Royal Society Manuscript 179, fol. 22r; in *Starkey Notebooks and Correspondence*, p. 236.
450 Ibid., fol. 13r; in Ibid., p. 232.
451 Ibid., fol. 15r; in Ibid., p. 233.
452 Ibid., ff. 24v-25r; in Ibid., p. 238.
453 Philalethes, "Sir George Ripley's Preface," *Ripley Reviv'd*, pp. 13–14.

is Decoction."[454] Starkey also writes in *Ripley Reviv'd*: "Yet trust me, for I speak knowingly, the Art is both true and easie ; yea so easie, that if you did see the Experiment, you could not believe it. I made not five wrong Experiments in it, before I found the truth...and in less then full two years and a half, of a vulgar *Ignoramus*, I became a true *Adept*, and have the Secret through the goodness of God."[455]

In his unpublished manuscript entitled "Aphorismi hermetici," Starkey writes, "Ascend sayth Flammel into the mountaine, & there shalt thou see the vegetable Saturnia, the Royal & triumphant Herbe. Let its juice be taken pure, rejecting its faeces, for with it mayst thou performe the greatest part of thy worke."[456] Starkey also writes in this manuscript:

> First from [the] testimony [of] Bernard Trevisan...hath this expression. The thing is very easy, yea soe easy that if I should tell it in plaine words you would scarce beleeve it, Al[l] the difficultly consists in our words & our meaning in them. Also another Philosopher sayth that if the Art were told in plaine words, even very fooles would laugh at it. Hence it is called the worke of women & the play of Children. The Author likewise of the Dialouge betweene Senior & Adolphus... sayth of the Art that it is so easy that it may be learned in the space of 12 houres & be brought to action in eight dayes. Artephus sayth expressely, the worke is no hard labour to him that knowes it, in his secret Booke, & Flammel speaking of Perenelle his wife sayth of her that had she attempted it alone, doubtlesse shee had performed it.[457]

Starkey made numerous references to the Magi in his writings, which clearly refer to the ancient Persian priests rather than the Magus of the seventeenth century,[458] the latter of whom Starkey largely ignores.[459] In Philalethes's poem *The Marrow of Alchemy*, a two-part work devoted to the "Philosophers Elixer,"[460] Starkey refers to

---

454 Philalethes, *Secrets Reveal'd*, p. 89.

455 Philalethes, "Sir George Ripley's Preface," *Ripley Reviv'd*, p. 87.

456 [George Starkey], "Aphorismi hermetici," Royal Society Manuscript 179, fol. 40r; in *Starkey Notebooks and Correspondence*, p. 250.

457 Ibid., ff. 35v-36r; in Ibid., p. 248.

458 Child related to Hartlib that Starkey was "an excellent Schollar both for Hebr[ew] and Greeke and of a vast Stupendious memory, knowing almost all Helmont by heart." Hartlib, *Ephemerides*, [December 1650], L4; in Wilkinson, "Hartlib Papers II," p. 87.

459 Boyle informed Hartlib that Starkey was "about to refute [Thomas] Vaughan," the famous alchemist and mystic who was also closely connected with the Hartlib circle at the time of Starkey's arrival in London. Ibid., [1650/51], A-B6; in Ibid., p. 88.

460 In an unpublished preface to *Sir George Riplye's Epistle to King Edward unfolded*, Starkey refers to this work as "my small chemicall Ballade intituled The Marrowe of Alchemy." [George Starkey], Preface to *Sir George Riplye's Epistle to King Edward unfolded*, Sloane Collection, British Library, MS 633, fol. 3v; in *Starkey Notebooks and Correspondence*, p. 314.

the "ancient *Magi*"[461] and declares: "the Magi all have it decreed To be the only matter."[462] As Philalethes, Starkey also wrote an unpublished manuscript entitled "Cabala Sapientum, or an Exposition upon the Hieroglyphicks of the Magi," which unfortunately is not extant.[463] In Philalethes's *Ripley Reviv'd*, Starkey also refers to the ancient Egyptians: "The Ancient *Egyptians* taught much by Hieroglyphicks, which way many Fathers of this Science have followed ; but most especially they have made use of Mystical or Cabalistical descriptions."[464] In an unpublished preface to *Sir George Riplye's Epistle to King Edward unfolded*, Starkey refers to this as a "Treatise comming abroade into the worlde so naked, as scarce havinge enough of the cloathinge of Language to hide & keepe secret the whole, & most hidden mystery of the Magi."[465]

Employing the "language of the *Magi*," the "American philosopher" writes in *Secrets Reveal'd*: "Whosoever desires to enjoy the secret *Golden-Fleece*, let him know, That our Gold-making POWDER (which we call our *Stone*)...is called *Our Gold*."[466] Starkey adds: "our Fiery *Dragon*...the Vegetable *Saturnia*...is therefore in very deed a *Chaos*...This *Chaos* is called, our *Arsenick*, our *Air*, our ☽ [Silver], our *Magnet*, our *Chalybs* or *Steel*...The Wise Magi have delivered many things of their *Chalybs* to Posterity."[467] In Philalethes's *George Ripley's Vision*, Starkey declares that the "Juice of Grapes then, is our *Mercury*, drawn from the Chameleon or Air of our Physical *Magnesia*, and *Chalybs* Magical," which "is the same which the fair *Medea* did prepare, and pour upon the two Serpents which did keep the Golden Ap[p]les, which grew in the hidden Garden of the Virgins *Hesperides*."[468] Starkey also writes in Philalethes's *Transmutation of Metals*:

> [T]his is the Keeper of our Gates, our Balsam, Oyle, Honey ; our Urine, May-Dew, our female, Mother, Egg, Secret Furnace, Stove, Sieve, Marble, true Fire, venomous Dragon, Treacle, burning Wine, Green Lyon, Bird of Hermes, Goose of Hermogenes, Double Sword in the hand of the Cherub, that defends the way of the Tree of Life, and

[461] Eirenaeus Philalethes (pseud.), *The Marrow Of Alchemy: Being an Experimental Treatise, Discovering the secret of the most hidden Mystery of the Philosophers Elixer. The Second Part*, London, Printed by *R. I.* for *Edw. Brewster* at the Sign of the Crane in *Pauls* Church-yard, 1655, p. 59.
[462] Philalethes, *Marrow of Alchemy* [Part I], p. 56.
[463] Philalethes, "Sir George Ripley's Epistle to King Edward IV," *Ripley Reviv'd*, p. 48.
[464] Philalethes, "An Exposition upon Sir George Ripley's Vision," Ibid., p. 1.
[465] [George Starkey], Preface to *Sir George Riplye's Epistle to King Edward unfolded*, Sloane Collection, British Library, MS 633, fol. 2r; in *Starkey Notebooks and Correspondence*, p. 311.
[466] Philalethes, *Secrets Reveal'd*, pp. 1-2.
[467] Ibid., pp. 5-6.
[468] Philalethes, "Sir George Ripley's Vision," *Ripley Reviv'd*, pp. 6-7.

> is famous for Infinite other Names ; and it is our Vessel, true, hidden, also it is our Philosophical Garden in which our Sun rises and sets, it is our Royal Mineral, and Triumphant Vegetable Saturnia, also the Caduce[us] of Mercury...[469]

Starkey writes in his *Pyrotechny*: "And wonder not, *Reader*, that I allude to *Eden* the Garden of God, whose guardian angel stands sentinel with a naked flaming sword, and no man returns but he that passes thorow untoucht of that flaming blade, and he that hath passed thorough this fiery trial, hath freedom of accesse to the midst of the Garden, where (thou knowest) God planted his *Arborem Vitæ*... to him that overcomes is onely given to eat of the hidden Manna."[470] In the same work Starkey goes on to compare "the Masterie of *Hermes*, or this peerless Key, the Liquour *Alchahest*" with "the Medicine of the *Magi*, their *Aurum potabile* ; attained by means of their Stone."[471] Starkey then identifies the Alkahest as an herbal medicine: "*Paracelsus* glories, and not in vain of most excellent and several cures performable by one Herb duely prepared, as to instance in wormwood...Of this preparation of Herbs the noble *Helmont* speaketh in his *Pharmacapolium*, and *dispensatorium modernum*, where he gives counsel by way of Legacie to such who have not tasted the Virtue, of the *Circulatum majus*, that is the *Alchahest*."[472]

In *Secrets Reveal'd*, the "American philosopher" discloses that the "Philosophers Elixer" called "our Water" is composed of "Fire," "the Liquor of the Vegetable *Saturnia*,"[473] and the "bond of ☿ [Mercury]," telling the reader:

> Therefore learn to know, who the Companions of *Cadmus* are, and what that *Serpent* is which devoured them, what the hollow *Oak* is which *Cadmus* fastened the *Serpent* through and through unto ; Learn what *Diana's Doves* are, which do vanquish the *Lion* by asswaging him : I say the Green *Lion*, which is in very deed the *Babylonian Dragon*, killing all things with his Poyson : Then at length learn to know the *Caducean Rod* of *Mercury*, with which he worketh Wonders, and what the *Nymphs*

[469] Eirenaeus Philalethes (pseud.), "Ars Metallorum Metamorphoseωs," *Three Tracts of the Great Medicine of Philosophers for Humane and Metalline Bodies*, London, Printed and sold by *T. Sowle*, at the *Crooked-Billet* in *Holy-well-Lane Shoreditch*, 1694, p. 82.

[470] Starkey, *Pyrotechny*, pp. A3v-A4r.

[471] Ibid., p. 49.

[472] Ibid., p. 94.

[473] In two letters addressed to Hartlib, Starkey provides a hint, as he refers to it, that the *Aqua vitæ*, or "Water of life," is composed of fermented cereal. Starkey to Hartlib, n.d.; in [Hartlib], *Common-Wealth of Bees*, pp. 30-35.

> are, which he infects by Incantation, if thou desirest to enjoy thy wish.[474]

Starkey concludes his *Secrets Reveal'd* with a noble promise: "He who hath once, by the Blessing of God, perfectly attained this Art...he hath a Medicine Universal, both for prolonging Life, and Curing of all Diseases, so that one true *Adeptist* can easily Cure all the sick People in the World, I mean his Medicine is sufficient."[475] While annotating the Paracelsian Michael Maier's[476] *Symbola Aureæ Mensæ*[477] in his search for Starkey's "Medicine Universal," Newton writes:[478]

> The Dragon kil[le]d by Cadmus is ye subject of our work, & his teeth are the matter purified.
>
> Democritus (a Graecian Adeptist) said there were certain birds (volatile substances) from whose blood mixt together a certain kind of Serpent ↗(☿)↙ [Mercury] was generated w[hi]ch being eaten (by digestion) would make a man understand ye voyce of birds (ye nature of volatiles how they may bee fixed)[.]
>
> St John ye Apostle & Homer were Adeptists.

[474] Philalethes, *Secrets Reveal'd*, pp. 5-6.

[475] Ibid., pp. 118–119.

[476] Thorndike, *History of Magic*, vol. VII, p. 167; R. J. W. Evans, *Rudolf II and His World: A Study in Intellectual History 1576–1612*, Oxford, Clarendon Press, 1973, pp. 200, 205; Wlodzimierz Hubicki, "Maier, Michael," *Dictionary of Scientific Biography*, vol. IX, p. 23; Bruce T. Moran, "Privilege, Communication, and Chemiatry: The Hermetic-Alchemical Circle of Moritz of Hessen-Kassel," *Ambix*, 32 (1985), p. 118.

[477] Michael Maier, *Symbola Aureæ Mensæ Duodecim Nationum...*, Francofurti, Typis Antonij Hummij, impensis Lucæ Iennis, 1617; Harrison, *Library of Newton*, p. 189; *Correspondence of Newton*, vol. I, pp. 12–13 n. 5.

[478] Newman and Principe have concluded that the *Introitus Apertus* (*Secret's Reveal'd*) contains evidence of Starkey's vain attempt to produce the Philosophers' Stone (and his medicines) through the transmutation of metals. From the standpoint of Aristotelian science, the transmutation of base metals into gold remains an untenable theory, and was rightfully and completely rejected by the Aristotelians throughout academia by the late eighteenth and early nineteenth century. There is no doubt that Starkey possessed metallurgical skills among his multitude of talents, and in fact, as we recall, Starkey gave a transmutation recipe (the *Key*) to Boyle in 1651, which is clearly a medically useless formula that possesses a scientific validity equal to the *persona* of Philalethes. Furthermore, Newman has determined that the transmutation formulas of Alexander von Suchten are found throughout Starkey's writings, and Starkey began his plagiarizing of Suchten with the fabricated transmutation recipe that Boyle received in 1651, parts of which were "borrowed verbatim" from Suchten's *Mysteria gemina*, and Newman concludes that "Not only does Starkey agree with Suchten in his manual practice, but even the elaborate hylozoic theory of the *Key* is derived from the German iatrochemist." According to Walter Pagel, "Alexander of Suchten of Danzig, vagrant Paracelsist, was the first to prove by means of the balance the impossibility of metal transmutation (1570–1604)." Pagel, *Smiling Spleen*, p. 13; Newman, "Prophecy and Alchemy," pp. 104–106; Newman, *Gehennical Fire*, pp. 135–143; William R. Newman and Lawrence M. Principe, "The Chymical Laboratory Notebooks of George Starkey," *Reworking the Bench: Research Notebooks in the History of Science*, F. L. Holmes, J. Renn, and H. Rheinberger (eds.), Dordrecht, Kluwer Academic Publishers, 2003, pp. 25-41.

Sacra Bacchi (vel Dionysiaca) instituted by Orpheus were of a Chymicall meaning.[479]

[479] The rites of Bacchus (Dionysus). King's College, Keynes MS 29, fol. 1^{v}; in Dobbs, *Foundations*, pp. 67, 88 n. 153, 90.

Chapter 7. Starkey, Boyle, and Newton's Divine Quest

George Starkey and his predecessors Jean Baptiste van Helmont and Paracelsus von Hohenheim were devoted followers of the Christian faith, unwavering in their belief that knowledge of the "immortal liquor" was a divine gift granted only by God. These Adepts of the "Hermetic Science" were Hippocratic physicians and Neoplatonic philosophers reviving the pre-Socratic, Orphic-Pythagorean tradition of the Classical world, characterized by the religious use of secret decoctions of ergot, opium, and the extremely potent Dionysian wine, the most famous of these institutions functioning for nearly two thousand years at Eleusis in Greece.[480] It would seem that the error of Newton and Boyle[481] was not their unconditional belief in

[480] P. G. Kritikos and S. P. Papadaki, "The History of the Poppy and of Opium and their Expansion in Antiquity in the Eastern Mediterranean Area," G. Michalopoulos (trans.), *Bulletin on Narcotics*, 19 (1967), pp. 5–10, 17-38; Carl Kerényi, *Eleusis: Archetypal Image of Mother and Daughter*, R. Manheim (trans.), New York, Bollingen Foundation, 1967, pp. 55-57, 120–144, 158–168; Calvert Watkins, "Let Us Now Praise Famous Grains," *Proceedings of the American Philosophical Society*, 122 (1978), pp. 9–17; R. Gordon Wasson, Albert Hofmann, and Carl A. P. Ruck, *The Road to Eleusis: Unveiling the Secret of the Mysteries*, New York, Harcourt Brace Jovanovich, Inc., 1978; Ruck, "The Wild and the Cultivated: Wine in Euripides' *Bacchae*," *Journal of Ethnopharmacology*, 5 (1982), pp. 231-270; William Scott Shelley, *The Elixir: An Alchemical Study of the Ergot Mushrooms*, Notre Dame, Ind., Cross Cultural Publications, 1995; Shelley, *The Origins of the Europeans: Classical Observations in Culture and Personality*, San Francisco, International Scholars Publications, 1998, pp. 95–182.

[481] "The second edition of *The Sceptical Chymist* (1680) came out with an appendix (*The Producibleness of Chemical Principles*)...The Section of *The Producibleness* devoted to mercury testifies to his perplexities about its status. Boyle did not peremptorily deny that mercury was a simple and homogeneous substance. However, he did not decide whether mercury was produced or extracted. He seems to maintain that some mercury is contained in metals, though nobody had extracted it yet...Boyle's arguments for the existence of mercury in metals are particularly strong: the easy amalgamation of ordinary mercury with metals can be considered as evidence of a kind of 'cognation' between ordinary mercury and the one contained in metals. In addition, he suggests, mercury seems to be the principal cause of metals' gravity:

the ethereal Philosophers' Stone[482] but their expectation that it was attained through the transmutation of metals.[483] This was a common misjudgment, but in the case of Boyle and Newton it resulted in the Aristotelians retaining complete control of the University[484] and the subsequent decline of Hippocratic medicine and the Neoplatonic

---

The gravity of a metal cannot reasonably be supposed to proceed from the whole body of the metal, but only from some one ingredient heavier in specie than the rest, and than the metal itself. And this ingredient or principle can be no other than the most ponderous body, mercury... Boyle often repeated that mechanical philosophy was preferable to other theories of matter because it is based on principles that are simpler, more fundamental and more general than those of both Aristotelians and Paracelsians. Accordingly, it can explain a wider range of phenomena than the other two theories and does not have recourse to such entities as substantial forms, occult qualities, the universal spirit and the *Anima Mundi*." Clericuzio, *Elements, Principles and Corpuscles*, pp. 132, 135–136.

[482] "In advocating the value of chemical operations as a tool of natural philosophy, he [Boyle] divested chemical experiments of the Spagyristic interpretations. He believed that the troubles of chemistry coming from the Spagyrist doctrines could be avoided by sticking to mechanical principles. But the mechanical account of chemical processes that he attempted was not possible in his time; it had to wait for two and a half centuries." Yung Sik Kim, "Another Look at Robert Boyle's Acceptance of the Mechanical Philosophy: Its Limits and its Chemical and Social Contexts," *Ambix*, 38 (1991), p. 9.

[483] Dobbs reports that in his copy of the 1698 version of Lemery's *Chymistry*, Newton folded down the corner of a page in characteristic fashion to mark a point of special interest to him, directing attention to the following story:

> But the saddest consideration of all is, to see a great many of them [the alchemists], who have spent all the flower of their years, in this desparate concern, in which nevertheless they pertinaciously run on, and consume all they have, at last instead of recompense for their miserable fatigues, reduced to the lowest degree of poverty. *Penotus* will serve us for an instance of this nature, among thousands of others, he died a hundred years old wanting but two, in the Hospital of *Yverdon* in *Switzerland*, and he used to say before he died, having spent his whole life in vainly searching after the Philosopher's stone, *That if he had a mortal Enemy he did not dare to encounter openly, he would advise him above all things to give himself up to the study and practice of* Alchymy.

Nicholas Lemery, *A Course of Chymistry; containing An easie Method of Preparing those Chymical Medicines which are used in Physick...*, Third English Edition, J. Keill (trans.), London, W. Kettilby, 1698, pp. 62-63 [Trinity College, Cambridge, call no. NQ.8.118; Harrison, *Library of Newton*, p. 177]; in Dobbs, "Newton's Alchemy and Theory of Matter," p. 511.

[484] "As is well known he was forced by the criticism of Leibniz to acknowledge in the General Scholium to the *Principia mathematica* [1713, second edition] that he did not know the *cause* of gravity...Newton's speculations derived from four earlier traditions with which he was familiar. Drawing upon the Neoplatonic tradition of light metaphysics, he suggested that light might combine with matter to give it various active powers; the alchemical tradition linked ideas of light with ideas of an active spirit, present in all things, which again might be said to give rise to various unceasing activities of matter. This active spirit, in its turn, could be linked to more recent ideas, developed in the new mechanical philosophy, in which an all pervasive aether was used as a medium of transmitting impulse from one part of the universe to another. Newton's own aether speculations were by no means purely mechanistic, since his aether consisted of particles held apart from one another, and from particles of other matter, by repulsive forces operating between them, but they clearly owed something to the mechanical as well as the alchemical traditions. The fourth tradition was Christian theology: gravitation attraction being held to be brought about by God." John Henry, "'Pray Do Not Ascribe That Notion To Me': God and Newton's Gravity," *Books of Nature and Scripture*, p. 123.

philosophical tradition of Paracelsus, Jean Baptiste van Helmont, and George Starkey.[485]

The unwavering belief of Newton and Boyle in the existence of the numinous "immortal liquor"[486] and the aspiration to attain the *pharmakon tes athanasias*, the "drug of immortality" was a passion

[485] "From the time of his *Principia* until his death Newton was deeply troubled by the concept he had introduced: a universal gravity. He had been brought up in the 'received' philosophy, sometimes known as the 'mechanical philosophy,' centering around the ideas of Descartes, which held that all explanations in natural philosophy must be couched in terms of what Robert Boyle called 'those two grand and most catholick principles of bodies, matter and motion.' At first sight it would seem that to these Newton had added a third principle, force. Not only was this a departure from accepted norms, it also introduced a kind of force that was astonishing in its primary characteristics or qualities. For this force had to have the power of extending itself over many hundreds of millions of miles as a kind of grasping entity which could affect huge bodies. For instance, the gravitation force could extend far beyond the reaches of the solar system...to turn a comet around and cause it to return to the visible regions of the neighborhood of the sun. Newton again and again sought for some explanation of how universal gravity might act. That is, he attempted to reduce universal gravity to the action of something else, a shower of aether particles, electrical effluvia, variations in an all-pervading aether. All of these attempted 'explanations' or reductions of universal gravity to some accepted kind of mechanism failed—because none could fulfill two major requirements: that the resultant force vary inversely as the square of the distance and that this force act mutually on every pair of bodies so as to attempt to bring them together...The concluding General Scholium of the *Principia* was written essentially as a confession of failure, an expression of regret that Newton had not been able to find an explanation for gravity...It is enough (satis est), he concluded, that gravity and its laws lead us to retrodict and predict 'all the motions of the celestial bodies and of our seas' and much else. It is a paradoxical aspect of Newton's genius that even his failures should affect the future course of science. His public expression of his inability to find a 'cause' of gravity, and his boast that what he had achieved should be 'enough,' became eventually the declaration of the aim of natural science. It is today a commonplace that the goal of the sciences is not to find first causes and ultimate explanations, but rather—in the Newtonian manner—to establish a set of principles that enable us accurately and fully to predict and to retrodict the phenomena of the physical world, the results of experiment and critical observation...But we must not forget that, despite Newton's great interest in electricity and in an aethereal medium in relation to gravitational action, he was known to have said that universal gravity is caused directly by God." I. Bernard Cohen, "Newton's Third Law and Universal Gravitation," *Journal of the History of Ideas*, 48 (1987), pp. 587-588, 592-593 n. 49.

[486] "There is a widely held view that ether theories suffered a dramatic and sudden demise with the rise of the special theory of relativity, since a stationary ether permeating space seems incompatible with Einstein's two postulates...however, the view that ether theories did not survive 1905 is incorrect because there have been, and still are, many ether theories that, in principle, are perfectly compatible with special relativity and even general relativity. Moreover, quantum theory has led to new conceptions of ether, and not a few physicists have urged the *necessity* of some form of ether theory...Einstein, in an address delivered at the University of Leyden in 1920, stated:

> More careful reflection teaches us, however, that the special theory of relativity does not compel us to deny ether. We may assume the existence of an ether; only we must give up ascribing a definite state of motion to it, i.e. we must by abstraction take from it the last mechanical characteristic which Lorentz had still left in it...[There] is a weighty argument to be adduced in favour of the ether hypothesis. To deny ether is ultimately to assume that empty space has no physical qualities whatever. The fundamental facts of mechanics do not harmonize with this view...According to the general theory of relativity space without ether is unthinkable; for in such space there would not only be no propagation of light, but also no possibility of existence for standards of space and time (measuring-rods and -clocks), nor any space-time intervals in the physical sense.

held by many notable figures throughout history and an idea that reverberated through the Ages. In the nineteenth century, the English poet John Keats expressed this desire in the *Hyperion*: "And deify me, as if some blithe wine or bright elixir peerless I had drunk, and so become immortal."[487] These blissful words of Keats echo those in the *Pythian Odes* of the Greek poet Pindar in the fifth century B.C.: "they shall drip nectar and ambrosia on his lips and shall make him immortal."[488] These sentiments repeat those expressed in the *Rigveda* in the second millennium B.C.,[489] in which the Vedic priests proclaim after drinking the "Mead of the Gods,"[490] the ambrosial libation known as Haoma to the ancient Persian Magi: "We have drunk Soma and become immortal."[491]

Our second example, Dirac, submitted a short but famous letter to *Nature* in 1951, which included the following:

> Physical knowledge has advanced very much since 1905, notably by the arrival of quantum mechanics, and the situation has again changed. If one examines the question in the light of present-day knowledge, one finds that the aether is no longer ruled out by relativity, and good reasons can now be advanced for postulating an aether."

Cantor and Hodge, "Ether theories," pp. 53-54; Albert Einstein, "Ether and relativity," *Sidelights on Relativity*, G. B. Jeffrey and W. Perrett (trans.), London, W. Perrett, 1922; P. A. M. Dirac, "Is there an Æther?" *Nature*, 168 (1951), pp. 906-907.

[487] John Keats, *Hyperion* III.118-120; in H. W. Garrod (ed.), *Keats: Poetical Works*, Oxford, Oxford University Press, 1956, p. 242.

[488] Pindar, *Pythian Odes* IX.63; in William H. Race (trans.), *Pindar I*, Cambridge, Harvard University Press, 1997, p. 347.

[489] This is undoubtedly related to "holy Soma, the lord of the plants," the "wood-nymphs dispensing ambrosia," the "ambrosia dew," the "cups bristling with dew-besprent blades of barley," the "mingled scent of wine, ambrosia," the "elixir of life," and the "land of the philosopher's stone" mentioned in the *Harsacarita* (seventh century A.D.). Unfortunately, alchemy in India is beyond the scope of the present work. Bāna, *Harsacarita* 29, 82, 105, 107, 163, 223, 281; in E. B. Cowell and F. W. Thomas (trans.), *The Harsa-carita of Bāna*, Delhi, Motilal Banarsidass, 1961, pp. 20, 61, 80, 82, 130, 193, 251.

[490] The ingredients and method used by the Vedic priests to produce Soma was quite simple. Ergot infected cereal along with malt (which is easily produced by sprinkling the cereal with water and placing it in the shade) was boiled for a prolonged period and then allowed to ferment. In this process known as acid-catalyzed isomerization (cereals possess an acidic pH value) the poisonous alkaloids in ergot isomerize into their pharmacologically inactive isomers, and this also occurs in the amide derivatives of lysergic acid, the isomerization of lysergic acid hydroxyethylamide creating an hallucinogenic agent. These isomers are not water but alcohol soluble, and were extracted through the fermentation process. Soma was also consumed as a solid food, and subjecting ground ergot-infected cereal to prolonged boiling produced this form of Soma. The priesthood at Eleusis in Greece employed a similar method to produce the *Kykeon*, or "mixed drink." The ergot-infected barley was first boiled and then ground, and during the ceremony this was mixed into a fermented liquid composed of water and malted barley. William Scott Shelley, "Soma: Mead of the Gods," [unpublished]; Shelley, "Chemistry and the *Kykeon* of the Eleusinian Mysteries," [unpublished]; Shelley, "The 'Soma Vessels' of the Bactria-Margiana Archaeological Complex," [unpublished].

[491] *Rigveda* VIII.48.3; in Arthur Anthony Macdonell (trans.), *Hymns from the Rigveda*, London, Oxford University Press, 1922, p. 80.

Chapter 8. The Bubonic Plague Theory

Bubonic plague has become accepted as the principal cause of epidemic disease in Britain between 1348 and 1665.[492] It has also been generally accepted that the epidemics of the sixteenth and seventeenth centuries in England were the bubonic form of the disease.[493]

The bubonic plague is an infectious and fatal epidemic disease caused by the bacterium *Yersinia pestis*, discovered in the late-nineteenth century. There are two primary types of this disease, the better-known bubonic plague and pneumonic plague. The bubonic form is transmitted by the bite of fleas (*Xenopsylla cheopis*) from an infected host, including but not limited to the black rat (*Rattus rattus*), and is characterized by fever, chills, vomiting, diarrhea, and the formation of hard, red swellings called buboes that appear anywhere on the body. Bubonic plague cannot be transmitted from person to person.

---

[492] Susan Scott and Christopher J. Duncan, *Biology of Plagues: Evidence from Historical Populations*, Cambridge, Cambridge University Press, 2001; A. Lloyd Moote and Dorothy C. Moote, *The Great Plague: The Story of London's Most Deadly Year*, Baltimore, Johns Hopkins University Press, 2004. These works are among the latest papers that advocate bubonic plague as the pathogenic agent of the Great Plague in London. As with earlier works devoted to the bubonic plague hypothesis, and the multiple-disease hypothesis suggested by Shrewsbury, and those who identify "Fever" or "Ague" with malaria, all of these the authors either omit, or they are unaware of the medical literature pertaining to the Plague, and the diseases associated with the Plague that struck London in 1665. Other authors remain justifiably uncertain about the cause of the Plague. Stephen Porter, *Lord Have Mercy Upon Us: London's Plague Years*, Stroud, Eng., Tempus Publishing Ltd., 2005, pp. 9-21; John Theilmann and Frances Cate, "A Plague of Plagues: The Problem of Plague Diagnosis in Medieval England," *Journal of Interdisciplinary History*, 37 (2007), pp. 371-393.

[493] Leslie Bradley, "Some Medical Aspects of Plague," *The Plague Reconsidered: A new look at its origins and effects in 16th and 17th Century England*, Matlock, Eng., Local Population Studies, 1977, pp. 11–13. Bubonic plague can also progress into the extremely rare septicemic plague, which is a toxic poisoning of the blood stream.

Those who are afflicted with bubonic plague can also develop pneumonic plague. In the pneumonic form, the bacteria invade the lungs, and it is invariably accompanied by bloody sputum which is spread from person to person through infected droplets that are expelled from the lungs through talking, coughing, or sneezing.

However, there are fatal problems associated with the bubonic plague theory. The black rat, which has long been blamed for the epidemic, is a fairly gregarious animal, although it tends to live in the countryside. It is thought to be a less likely candidate than the brown rat (*Rattus norvegicus*), which did not arrive in Britain until the early eighteenth century.

Also, our knowledge of the modern *Yersinia pestis* does not square with the facts from the historical record. *Yersinia pestis* is not a human bacillus and is normally found in rats and other rodents, passing from animal to animal by their fleas. On occasion, the concentration of the bacillus will produce a massive mortality in the rodent population, termed an epizootic, and the fleas will jump to any warm body. Humans only become infected if the fleas are forced to abandon their rodent hosts. In other words, any outbreak of bubonic plague among humans is preceded by a massive mortality among the rodents, and no rodent mortality was recorded around the time of the Great Plague of London or noted by any of the earlier Western chroniclers of the disease. Because rat fleas do not stay on humans for very long, it is also difficult, if not impossible,[494] to explain how the disease moved from city to city if this was the culprit.[495]

Not only is there a rodent parameter to outbreaks of bubonic plague but also a climatic constraint and a time factor that determine the existence and survival of the disease. The eggs of fleas mature into adults in three weeks, and this will only occur in temperatures between 65°F (18°C) and 80°F (26°C) or 85°F (29°C), and the need for a great number of fleas is evident by the fact that in *Xenopsylla cheopis* only twelve per cent are "blocked," that is, capable of transmitting the bacillus. It is another hard fact that at no time between 1300 and 1850 were the average temperatures in July and August suitable for flea

[494] "[A]lthough in ideal conditions *X. cheopis* may live for a month without its host it is doubtful if at the end of that time it could transmit plague. As a rule, when not on a host, conditions are far from ideal and militate against the most effective transmission." Graham I. Twigg, *The Black Death: A Biological Reappraisal*, London, Batsford Academic and Educational, 1984, p. 85.

[495] William Naphy and Andrew Spicer, *The Black Death and the history of plagues 1345–1730*, Stroud, Eng., Tempus Publishing Ltd., 2000, pp. 54-55.

hatching. Moreover, the spread of bubonic plague from rat to rat is very slow; one epizootic in India travelled only three hundred feet in six weeks. In addition, in epidemics of bubonic plague, the time that it takes for the disease to kill the rat and then transfer to a human being and kill the individual is also very slow, between ten and fourteen days. This is all at odds with the accounts of the Black Death (a term that came into use after the seventeenth century) sweeping through Genoa in two days, which could not refer to bubonic plague, even in the pneumonic form.[496] The accounts of the Great Plague that struck London in 1665 also do not conform to the known parameters of the bubonic plague, even including the pneumonic form of the disease.

J. F. D. Shrewsbury attempted to explain the Plague epidemics in the British Isles from the fourteenth to the seventeenth centuries as a combination of bubonic plague and typhus, among other diseases.[497] To do this Shrewsbury postulated "a situation of plague entry and spread in London that ignored both the known logistics of plague diffusion and the data from the parish registers,"[498] as well as the medical literature from this period.[499] Typhus constitutes several forms of an infectious disease caused by microorganisms, and it will be demonstrated in later pages that the Plague of London in 1665 was not contagious.[500]

[496] "It has been suggested that the high medieval death rates were due to the fact that the plague bacillus at that time was a more virulent strain than the modern one...*Yersinia pestis* is a stable organism as shown by the fact that despite its modern day spread on all the continents...there is still only one serotype...Secondly, if the modern form is less virulent we would expect a low case mortality. Yet before antibiotics the case mortality of bubonic plague was between seventy and eighty per cent, pneumonic plague one hundred per cent and septicaemic probably the same." Graham I. Twigg, "The Black Death in England: An Epidemiological Dilemma," *Maladies Et Société (XIIe-XVIIIe siècles)*, N. Bulst and R. Delort (eds.), Paris, Éditions Du Centre National De La Recherche Scientifique, 1989, pp. 75, 77, 81, 84, 87, 93-94.

[497] J. F. D. Shrewsbury, *A History of the Bubonic Plague in the British Isles*, Cambridge, Cambridge University Press, 1970.

[498] Graham I. Twigg, "Plague in London: spatial and temporal aspects of mortality," *Epidemic Disease in London*, J. A. I. Champion (ed.), London, Centre for Metropolitan History, University of London, 1993, p. 1.

[499] "The sometimes acrid or cantankerous tone of voice he [Shrewsbury] uses, the amount of sheer assertion and reiteration of his argument, the over-emphasis of his points, the implication that he alone knows the aetiology of plague, the ill-disguised contempt for most earlier authorities are all suggestive of a certain obsessiveness. The obsession or mania might be called 'pesticidal'. All through his book he is out of minimize and if possible to scotch any alleged case of plague that he can and, whenever possible, to turn it into something else, preferably typhus. And this, as we have seen, will not always do...many of the views advanced with such confidence by Professor Shrewsbury are, to say the least, not proven..." Christopher Morris, "The Plague in Britain" [Book Review], *The Historical Journal*, 14 (1971), p. 215.

[500] Equally dubious is the haemorrhagic plague theory, which proposes on very little evidence that the Plagues of Europe throughout history were caused by a now extinct filovirus similar to the ebola pathogen. C. J. Duncan and S. Scott, "What caused the Black Death?" *Postgraduate Medical Journal*, 81 (2005), pp. 315-320.

The following thesis provides an alternative explanation to the highly problematic yet widely accepted bacterial theory of the Plague. Through the historical medical record and the accounts of those who were present in London during the epidemic in 1665, it will become apparent to the reader that the pathogen of the disease was fungal, and that the Plague was caused by poisonous cereal.[501]

[501] It is no coincidence that the British Isles began to see numerous epidemics of disease following the introduction of rye to the country by the Germanic Anglo-Saxon invaders in the fifth century. Wilfrid Bonser, "Epidemics During the Anglo-Saxon Period," *Journal of the British Archaeological Association*, Third Series, 9 (1944), pp. 48-71; Twigg, *Black Death: Biological Reappraisal*, p. 200.

CHAPTER 9. ERGOT AND *FUSARIUM* POISONING

Ergot poisoning occurs when a sufficient quantity of the ergot fungus (*Claviceps purpurea* (Fr.) Tul.) is consumed, producing the disease known as ergotism, which manifests in two forms: gangrenous and convulsive. Both types can occur together, with a wide range of symptoms.[502]

In France, where ergotism invariably occurred in the gangrenous form, the disease was called *Ignis Sacer*, or "Holy Fire," and *Ignis Plaga*, or "Fire Plague," and *Plaga Invisibilis*, or "Invisible Plague." The disease was also known as "St. Anthony's Fire," and *Ignis Sancti Martialis* after St. Martial.[503] Ergot poisoning was also called the "Peasants' Disease," as it was a malady that commonly afflicted the poor peasantry.[504]

Outbreaks of ergotism usually peak in late summer and fall, and they can be endemic or epidemic.[505] Ergot epidemics are common after the harvest, when the fungus is most toxic, although the alkaloids in ergot are quite stable and remain toxic for up to eighteen months.[506]

---

[502] Charles Creighton, *A History of Epidemics in Britain*, Volumes I-II, Cambridge, Cambridge University Press, 1891, vol. I, pp. 52-65; George Barger, *Ergot and Ergotism: A Monograph Based on the Dohme Lectures Delivered in Johns Hopkins University, Baltimore*, London, Gurney and Jackson, 1931, pp. 20-23, 80-81; Albert Hofmann, "Historical View on Ergot Alkaloids," *Pharmacology*, Supplement 1, 16 (1978), pp. 1–11; Klaus Lorenz, "Ergot in Cereal Grains," *CRC Critical Reviews in Food Science and Nutrition*, 11 (1979), pp. 311-354; Mervyn J. Eadie, "Convulsive ergotism: epidemics of the seratonin syndrome?" *The Lancet: Neurology*, 2 (2003), pp. 429-434.

[503] Frank James Bove, *The Story of Ergot: For Physicians, Pharmacists, Nurses, Biochemists, Biologists and Others Interested in the Life Sciences*, Basel, S. Karger, 1970, pp. 146–152.

[504] Ibid., pp. 153–154; Barger, *Ergot and Ergotism*, p. 25.

[505] Barger, *Ergot and Ergotism*, p. 52.

[506] Mary Kilbourne Matossian, *Poisons of the Past: Molds, Epidemics, and History*, New Haven, Yale University Press, 1989, p. 14.

In gangrenous ergot poisoning, the symptoms are generally mild until a critical level is reached and gangrene sets in. Individual susceptibility to gangrenous ergotism varies greatly, and rarely are entire families stricken by it.

Severe mental disturbances are limited to convulsive ergotism, a violent disease that commonly affects whole families who consume the same diet, and consequently it was considered to be infectious.[507]

Adolescents and infants who consume more food relative to bodyweight are especially prone to ergotism and other grain poisoning, and in some epidemics females are more susceptible than males, while elderly men can also be at higher risk.[508] This is also true of healthy adults who are generally more active and consume more food than their unhealthy counterparts.

Barger recounted an outbreak of convulsive ergotism in northern Bohemia in 1736 and 1737 that was carefully described by J. A. Scrinc:

> Out of 500 patients more than 100 died...About three-fifths of the patients were under 15 years of age. Two houses died out completely. The poor suffered from a very great and indescribable famine ; only one of the well-to-do was attacked, and he also had ergot among his corn. There was not a single case in the town of Niemes, where the bread was of good quality. The people believed the disease to be infectious, but Scrinc was convinced of the contrary and regarded the bad bread as the only cause. He considered that the toxicity of the latter was due to ergot...Henceforth convulsive ergotism became well known in Germany...[509]

Among the cereals, ergot of rye (*Secale cerealis*) produces the highest levels of alkaloids, primarily ergotamine and ergocristine with lower amounts of ergosine, ergocornine, ergokryptine, and ergonovine.[510] The alkaloid in ergot that produces hallucinations is the isomer of d-lysergic acid N-(1-hydroxyethyl) amide, the epimerization of the alkaloid occurring during normal food processing.[511]

In August 1951, an outbreak of ergot poisoning occurred in the French village of Pont St. Esprit,[512] which unlike other accounts of

[507] Barger, *Ergot and Ergotism*, pp. 21, 27, 37, 59.
[508] Ibid., pp. 39, 66, 71-72, 75, 78; Matossian, *Poisons of the Past*, p. 12.
[509] Barger, *Ergot and Ergotism*, pp. 71-72.
[510] P. M. Scott *et al.*, "Ergot Alkaloids in Grain Foods Sold in Canada," *Journal of AOAC International*, 75 (1992), pp. 773-779; Lorenz, "Ergot in Grains," p. 317.
[511] William Scott Shelley, "Chemistry and the *Kykeon* of the Eleusinian Mysteries," [unpublished].
[512] Bertram G. Katzung (ed.), *Basic & Clinical Pharmacology*, Seventh Edition, Stamford, Ct., Appleton & Lange, 1998, p. 278; Bove, *Story of Ergot*, p. 160.

ergotism that are generally repeated exposures to the toxin, this incident happened after a single poisoning and was fatal in some cases.[513] The first symptoms appeared 6 to 48 hours after consuming the toxic bread, and consisted of headache, nausea, abdominal pain, gastrointestinal disturbances, vomiting, diarrhea, a sensation of choking, pain at the nape of the neck, hot and cold fits, intense sweating with a fetid odor, excessive salivation, a weak pulse, muffled heart sounds, insomnia, depression, anguish, agitation, giddiness, fainting, trembling of the extremities, and vision problems; the more severe cases exhibited formication (numbness in the extremities), bloody urine, premature menstruation and hemorrhage, pains in the calves, an imperceptible pulse, fainting attacks, incoherent speech, tremor, twitching, convulsions, delirium, hallucinations, and gangrene.[514]

Ergot alkaloids remain relatively stable during normal food processing.[515] Rye was not only baked into bread but also made into soup, porridge, pancakes, dumplings, and other foods.[516] Toxic molds of the *Fusarium* genus also occur on rye, wheat, barley, oats, and other grains.[517] Belonging to the genus *Fusarium* are the toxic species of *Sporotrichiella*; the most prominent of these are *Fusarium sporotrichiodes* and *Fusarium poae*.[518] Species of *Fusarium* have been reported as contaminants of ergot and are known hyperparasites of *Claviceps* spp.[519] There can also be a synergistic interaction between mycotoxins, in which individual compounds become even more toxic.[520]

Of all the grains, rye is the most susceptible to ergot infection and provides the best substrate for the development of the *Fusarium* mold that produces lethal trichothecene toxins.[521] *Fusarium* toxins are not destroyed by normal cooking and remain stable at 125°C for a period

[513] This outbreak was caused by flour that was not government inspected and had been brought into the country illegally to avoid the grain tax. Roy E. Devitt *et al.*, "Ergot Poisoning," *Journal of the Irish Medical Association*, 63 (1970), p. 442.

[514] Drs. Gabbai, Lisbonne, and Pourquier, "Ergot Poisoning at Pont St. Esprit," *British Medical Journal*, 2 (1951), pp. 650-651.

[515] J. E. Fajardo *et al.*, "Retention of Ergot Alkaloids in Wheat During Processing," *Cereal Chemistry*, 72 (1995), pp. 291-298.

[516] Barger, *Ergot and Ergotism*, pp. 24, 78.

[517] Matossian, *Poisons of the Past*, p. 6.

[518] Abraham Z. Joffe, "The Genus Fusarium," *Mycotoxic Fungi, Mycotoxins, Mycotoxicoses: An Encyclopedic Handbook*, T. D. Wyllie and L. G. Morehouse (eds.), New York, Marcel Dekker, Inc., 1977, vol. I, pp. 59-82.

[519] R. L. Mower *et al.*, "Biological Control of Ergot by Fusarium," *Phytopatholgy*, 65 (1975), pp. 5–10.

[520] Abraham Z. Joffe, *Fusarium Species: Their Biology and Toxicology*, New York, John Wiley & Sons, Inc., 1986, p. 236; J. P. F. D'Mello *et al.*, "Fusarium Mycotoxins," *Handbook of Plant and Fungal Toxicants*, J. P. F. D'Mello (ed.), Boca Raton, Fl., CRC Press, 1997, pp. 287-301.

[521] Barger, *Ergot and Ergotism*, p. 1; Matossian, *Poisons of the Past*, p. 155.

of 30 minutes, and 110°C for 18 hours or longer.[522] The alkaloids in ergot are present in beer made of ergotized grain but not in distilled alcoholic beverages,[523] which is also the case with *Fusarium* toxins.[524]

Poisoning by *Fusarium* can be more dangerous than ergot poisoning, *Fusarium* poisoning was once designated as septic angina and called alimentary toxic aleukia (ATA), which is connected to the trichothecene compound named T-2 toxin.[525] *Fusarium* starts to form on the grain in fall, although it is in spring that it becomes toxic, and after the fungi disappear the toxin can remain poisonous for up to seven years.[526] *Fusarium* mold grows rapidly and also forms in the soil of the infected grain fields.[527] The levels of *Fusarium* toxins vary with environmental conditions, between regions, within a region, within the same field, and these levels are also affected by altitude, and are most toxic at lower elevations.[528] Moisture is necessary for the growth of *Fusarium*, whether rain, dew, honeydew, or the moisture of the host plant, and the fungus can produce toxins in temperatures up to 25°C.[529] Poisoning by *Fusarium* generally occurs in April, May, and mid-June, and it can take place during the fall harvest as well.[530] Entire families and villages can be affected by *Fusarium* poisoning, which can vary from region to region and between villages.[531]

The symptoms of *Fusarium* poisoning include a burning sensation in the mouth, tongue, palate, throat, esophagus and stomach, fever, headache, nausea, vomiting, diarrhea, excessive salivation, vertigo, fatigue, feeble pulse, red and purple spots on the skin, jaundice, bleeding from the nose, mouth, stomach, and intestines, necrotic ulcers in the mouth and throat, fetid breath, loss of voice, difficulty breathing, and swollen lymph nodes and cervical glands making it difficult to open the mouth, which can lead to strangulation.[532]

---

[522] Claude Moreau, *Moulds, Toxins and Food*, M. Moss (ed.), Chichester, Eng., John Wiley & Sons, 1979, p. 223.
[523] Matossian, *Poisons of the Past*, p. 14.
[524] Joffe, *Fusarium Species*, p. 11; Matossian, *Poisons of the Past*, p. 17.
[525] Joffe, *Fusarium Species*, p. 7.
[526] Ibid., p. 244; Moreau, *Moulds, Toxins and Food*, p. 221.
[527] Joffe, *Fusarium Species*, pp. 242, 246.
[528] Abraham Z. Joffe, "Environmental Conditions Conducive to *Fusarium* Toxin Formation Causing Serious Outbreaks in Animals and Man," *Veterinary Research Communications*, 7 (1983), pp. 187–193; Joffe, *Fusarium Species*, pp. 236, 239.
[529] Joffe, *Fusarium Species*, pp. 236-237, 247.
[530] Ibid., p. 230; Matossian, *Poisons of the Past*, p. 157.
[531] Joffe, *Fusarium Species*, pp. 228-229, 231.
[532] Moreau, *Moulds, Toxins and Food*, pp. 225-227.

Outbreaks of ATA occurred in the Soviet Union in the 1930s, and between 1942 and 1944 a widespread epidemic occurred in Orenburg and other districts near the Caspian Sea, causing an estimated 100,000 deaths between 1942 and 1948, the impact made worse by war and famine conditions that forced many to gather grain left in the fields throughout the winter. Because of the sudden outbreaks and high mortality rates associated with ATA, it was first thought by scientists to be an epidemic disease of infectious origin, and it was not until the 1960s that it was finally accepted that *Fusarium* poisoning was the cause of these devastating epidemics.[533]

[533] Abraham Z. Joffe, "Toxicity of Overwintered Cereals," *Plant and Soil*, 18 (1963), pp. 31-44; Abraham Z. Joffe, "Toxin Production by Cereal Fungi Causing Alimentary Toxic Aleukia in Man," *Mycotoxins in Foodstuffs*, G. N. Wogan (ed.) Cambridge, MIT Press, 1965, pp. 77-85; Joffe, *Fusarium Species*, pp. 225-229.

Chapter 10. The English Diet in the Seventeenth Century

Adam Lonicer was the first author to unambiguously mention ergot in his book published in Germany in 1582.[534] Following Lonicer, nearly all botanists up through the eighteenth century believed that ergot was a rye kernel that transmuted in place of a seed.[535] Ergot was finally proved to be a separate plant by Tulasne in 1853.[536] In France, Denis Dodart recognized ergot as the cause of gangrenous ergotism in 1676, and in Germany in 1695 Brunner published the first work that identified ergot as the cause of the convulsive disorder.[537] Yet because of the remarkable inability of physicians to recognize ergot as the cause of the disease, the controversy over this identification continued in Germany through the eighteenth century.[538] The confusion was furthered in a dissertation by Rothman (1763), a pupil of the Swedish botanist Linnaeus, who attributed ergotism to charlock (*Raphanus raphanistrum*), and on the authority of Linnaeus the name raphania was at one time used for convulsive ergot poisoning.[539]

English physicians and botanists in 1665 appear to have been completely unaware of ergot, for which there is no Anglo-Saxon name.[540] The first mention of ergot in English is from the 1677

[534] Barger, *Ergot and Ergotism*, pp. 7-9.
[535] Ibid., p. 85; Bove, *Story of Ergot*, p. 31.
[536] The belief that ergot was malformed rye survives in the name *Secale cornutum*. Barger, *Ergot and Ergotism*, p. 85; Bove, *Story of Ergot*, pp. 28, 36.
[537] Barger, *Ergot and Ergotism*, pp. 11, 59-60.
[538] Ibid., pp. 69-77.
[539] Ibid., pp. 29, 79.
[540] Ibid., pp. 4-5, 7, 40.

publication of John Ray's *Catalogus Plantarum Angliæ*.[541] In 1765, the Royal Society published the first paper in England that identified ergot as the cause of ergotism.[542] Sporadic outbreaks of ergot poisoning have continued in England into the twentieth century, and according to Barger, this shows "how little is known about the subject in this country. English analysts do not seem to be aware of the many papers published in Germany, Austria and Russia on the detection and estimation of ergot in flour."[543]

By the 1670s Londoners were consuming an estimated 1.3 million quarters of grain annually, much of it arriving in the city by wagons, river barges, and coastwise shipments from the eastern counties and southeastern ports of England.[544] In 1650, the English imports of Baltic rye fell dramatically, and for the next century the English relied entirely upon their own cereal production.[545] Between 1650 and 1682, wheat prices were high, causing much of the population, particularly those of limited means, to depend upon the cheaper rye as their cereal staple. Improvements in agricultural techniques after 1682 permitted an increase in the production and decrease in the price of wheat, and throughout the following century grains less susceptible to ergot infection—wheat, barley, and oats, replaced rye production.[546] Consequently, the consumption of rye bread in England fell throughout the seventeenth and eighteenth centuries.[547]

Large acreages of rye were cultivated in the northwestern county of Cheshire,[548] the southeastern counties of Norfolk, Suffolk, and

[541] Ibid., p. 10; Bove, *Story of Ergot*, p. 162.

[542] S. A. D. Tissot, "An Account of the Disease, called *Ergot* in *French*, from its supposed Cause, *viz.* vitiated Rye," *Philosophical Transactions*, 55 (1765), pp. 106–126.

[543] Barger, *Ergot and Ergotism*, pp. 63-65.

[544] Joan Thirsk (ed.), *The Agrarian History of England and Wales 1640–1750*, Cambridge, Cambridge University Press, 1985, vol. V.II, p. 22.

[545] Norman Scott Brien Gras, *The Evolution of the English Corn Market: From the Twelfth to the Eighteenth Century*, Cambridge, Harvard University Press, 1915, pp. 102–109; J. A. Faber, "The Decline of the Baltic Grain-Trade in the Second Half of the 17th Century," *Acta Historiae Neerlandica*, 1 (1966), p. 128.

[546] William Ashley, "The Place of Rye in the History of English Food," *Economic Journal*, 31 (1921), pp. 285-308; J. A. Yelling, "Changes in Crop Production in East Worcestershire 1540-1867," *Agricultural History Review*, 21 (1973), pp. 18-34; D. C. Coleman, *The Economy of England, 1450–1750*, London, Oxford University Press, 1977, pp. 113–125.

[547] After 1770 potatoes became more important in the diet of the English poor. Redcliffe N. Salaman, *The History and Social Influence of the Potato*, Cambridge, Cambridge University Press, 1949, p. 537; Capel Lofft, "Food of the Poor of Ingleton," *Annals of Agriculture*, 26 (1796), p. 226; Eric Kerridge, *The Farmers of Old England*, Totowa, N.J., Rowman and Littlefield, 1973, p. 163; E. J. T. Collins, "Dietary Change and Cereal Consumption in Britain in the Nineteenth Century," *Agricultural History Review*, 23 (1975), pp. 97–115.

[548] The Plague occurred in Cheshire and struck the poor farming and mining community around Eyam in the neighboring county of Derbyshire, situated in a rye-growing region. In a fourteen-

Essex in East Anglia, as well as Sussex, and Dartmoor in Devon in the West Country.[549] By 1665 the marshes of the Vale of London, western Essex, and the Essex coast had been drained for agricultural use,[550] and many of the heaths of the Vale of London were situated on poor, sandy soils.[551] Rye is grown in sour soils where other cereals are unable to thrive; it was consequently sown on recently drained marshland and sandy soils, and was a predominant crop in the region of London.[552] Rye grown on low-lying wetland is most susceptible to ergot infection, and these regions are most likely to produce lethal levels of ergot alkaloids.[553]

Rye bread and bread of rye sown or mixed with wheat, called maslin, was a food eaten by the English poor.[554] Although it was widely cultivated in England, the gentry considered rye to be inferior to wheat and more difficult to digest, and believed maslin healthier than rye.[555] In the early seventeenth century Tobias Venner wrote:

> Bread made of Rie, is in wholsom[e]nesse much inferiour to that which is made of Wheat : it is cold, heavie, and hard to digest, and by reason of the massivinesse thereof, very burdensome to the stomack. It breedeth a clammie, tough, and melancholick juyce ; it is most meet for rustick laborers : for such, by reason of their great travaile, have commonly very strong stomackes. Rie in divers places is mixed with Wheat, & a kind of bread made of them, called *Messeling-bread*, which is whol[e]somer than that which is made of Rie...[556]

During the same period Thomas Cogan writes:

month period there were 266 Plague deaths in Eyam with an estimated population of 350, that is, seventy-six percent of the inhabitants perished. Leslie Bradley, "The Most Famous of All English Plagues: A Detailed Analysis of the Plague at Eyam, 1665-6," *Plague Reconsidered*, pp. 63-94; Twigg, "Black Death in England," p. 94. On the Plague in the county of Devon, see Paul Slack, *The Impact of Plague in Tudor and Stuart England*, London, Routledge & Kegan Paul, 1985, pp. 80-99, 107–110. Rye was also grown in the region of the Tyne River, where the Plague also occurred. William Ashley, *The Bread of our Forefathers: An Enquiry in Economic History*, Oxford, Clarendon Press, 1928, pp. 7, 13.

[549] Joan Thirsk (ed.), *The Agrarian History of England and Wales 1500–1640*, Cambridge, Cambridge University Press, 1967, vol. IV, pp. 33, 104, 132, 169; Eric Kerridge, *The Agricultural Revolution*, New York, Augustus M. Kelley, 1967, pp. 59, 69, 71-73, 77, 79, 83, 85.

[550] Kerridge, *Agricultural Revolution*, pp. 223-227.

[551] Ibid., p. 173.

[552] Ibid., p. 176.

[553] Oliver Prescott, *A Disseration on the Natural History and Medicinal Effects of the Secale Cornutum, or Ergot*, Boston, Cummings & Hilliard, 1813, p. 5; Barger, *Ergot and Ergotism*, p. 14; Matossian, *Poisons of the Past*, p. 120.

[554] William Harrison, *The Description of England*, Ithaca, Cornell University Press, 1968, p. 135; Barger, *Ergot and Ergotism*, p. 2.

[555] Tobias Venner, *Via Recta ad Vitam Longam...*, London, Printed by *R. Bishop...*, 1638, pp. 19-20.

[556] Ibid., p. 21.

> *Secale*, commonly called Rye, a graine much used in bread, almost thorowout this Realm, though more plentifull in some places than in other : yet the bread that is made thereof is not so wholesome as wheate-bread, for it is heavy and hard to digest, and therefore most meet for labourers, and such as worke or travaile much, and for such as have good stomacks. There is made also of Rie mixed with Wheate, a kind of bread named misseling or masseling bread, much used in divers Shires, especially among the family.[557]

In 1587, William Harrison wrote: "[T]he gentility commonly provide themselves sufficiently of wheat for their own tables, whilst their household and poor neighbors in some shires are enforced to content themselves with rye or barley."[558] Fynes Moryson wrote in the early-seventeenth century: "The English Husbandmen eate Barley and Rye browne bread, and preferre it to white bread as abiding longer in the stomack, and not so soone digested with their labour, but Citizens and Gentlemen eate most pure white bread, England yeelding (as I have said) all kinds of Corne in plenty."[559]

Maslin was believed to keep better than wheat, and even the well off would mix rye with wheat for this purpose.[560] While traveling in Leicestershire, the clergyman Ralph Josselin was repelled by the thought of eating black bread and pocketed two white loaves "against the worst," yet he continued to produce rye and maslin at his home in Essex, and up until 1674 he and his family suffered from "Ague" every fall harvest.[561] By the 1620s, those farmers in southeastern England who could afford to were subsisting on wheat or maslin while producing rye for the London market.[562] The poverty caused by the flight of the wealthy from London certainly would have forced the abandoned population into greater rye consumption, considered an inferior bread cereal and a cheaper source of food.[563]

[557] Thomas Cogan, *The Haven Of Health*, London, Printed by Anne Griffin..., 1636, pp. 28-29.
[558] Harrison, *Description of England*, p. 133.
[559] Fynes Moryson, *An Itinerary*, Glasgow, James MacLehose and Sons, 1908, vol. IV, p. 171.
[560] Thirsk, *Agrarian History of England*, vol. IV, p. 169.
[561] Ralph Josselin, *The Diary of Ralph Josselin 1616–1683*, A. MacFarlane (ed.), London, Oxford University Press, 1976, p. 46; Matossian, *Poisons of the Past*, p. 63.
[562] Kerridge, *Farmers of Old England*, p. 163; Kerridge, *Agricultural Revolution*, p. 345.
[563] Thirsk, *Agrarian History of England*, vol. V.II, pp. 41-44.

Chapter 11. Environmental Conditions and the Plague

The growth of ergot is divided into three stages. The first is the sphacelia, or conidia, or honeydew stage, in which the fungus forms on the host plant. The growth of ergot on rye and other grains is favored by extremes in weather conditions that traumatize the host plant. The second stage is the sclerotial or resting stage when the fungus is on the ground, and the third is the germinating stage, in which the sclerotium produce small, mushroom-like projections called stromata, which release acospores, each sclerotium producing up to one million acospores capable of germination. The stromata are indistinguishable from other species of mushrooms that grow in the fields.[564] The host plant is inoculated by the acospores that are distributed by the wind, and it is further transmitted by rain splashing onto the honeydew that also drips into the florets below and onto flies and other insects that are attracted by the honeydew, and by contact between stalks.[565]

According to William Boghurst, all wild and domestic animals were unaffected by the Plague in 1665.[566] However, Edward Baynard, a member of the College of Physicians who fled London, observed that the birds seemed to pant for breath and flew heavier than usual,[567]

[564] Bove, *Story of Ergot*, pp. 16-24, 36.

[565] Ibid., pp. 17, 30; Barger, *Ergot and Ergotism*, p. 97; Lorenz, "Ergot in Grains," p. 314; Klaus B. Tenberge, "Biology and Life Strategy of the Ergot Fungi," *Ergot: The Genus Claviceps*, V. Křen and L. Cvak (eds.), Amsterdam, Harwood Academic Publishers, 1999, p. 48.

[566] William Boghurst, *Loimographia, An Account of the Great Plague in London in the Year 1665*, J. F. Payne (ed.), London, Shaw and Sons, 1894, p. 96.

[567] "According to the rat-flea theory, in order for fleas to transfer from rat to human first the rat population must suffer an epizootic, there are few (if any) references to an increased number

and there was an abundance of mildews that summer.[568] According to William Kemp, the signs that the Plague was approaching included a "more than ordinary encrease of Mushro[o]mes, if there hath been a Murrain among Sheep or Cattel ; for though the same *Plague* that destroys Man, doth not hurt Sheep, neither doth the same Disease that kills Sheep, presently assault Men."[569]

An outbreak of ergotism called "lethargy" in fourteenth-century England was associated with murrain of cattle,[570] and Barger believed that the epidemic of convulsive ergotism in 1749 that occurred near Lille, France, was likely connected with an epizootic of cattle that preceded the outbreak.[571] Nathaniel Hodges writes:

> [I]t is not at all foreign to our Purpose here to take Notice, that on the Year before the last pestilential Sickness, there was a great Mortality amongst the Cattel, from a very wet Autumn...And the Conjecture that a Sickness amongst Cattle is transferrable to the humane Species, hath not yet appeared on any good Foundation...no one doubts but that a Plague amongst Cattle, from some common Cause, as a Corruption of the aerial Nitre, and which differs from the Plague amongst Men but in Degree...[572]

Hodges writes:

> Furthermore, this nitrous Principle may be sometimes changed... there may arise several Sorts of Distemperature ; as Blasts upon Trees [i.e., *Fusarium* spp.], and Diseases amongst Cattle ; and at last end in Pestilence amongst Mankind."[573] Hodges also writes: "in this Regard we may consider the frequent Mortalities amongst Cattle, which foregoe an Infection amongst Mankind ; for these Creatures living...in

of rat carcasses either in printed works or in private diaries." J. A. I. Champion, *London's Dreaded Visitation: The Social Geography of the Great Plague in 1665*, London, Centre for Metropolitan History, University of London, 1995, pp. 7-8. In the summer of 1664 there was said to be many flies, ants, and frogs, and the year following the Plague there were very few flies and frogs. Boghurst, *Loimographia*, pp. 26, 98; Barger, *Ergot and Ergotism*, p. 71.

[568] Laurence Echard, *The History of England. From the First Entrance of Julius Cæsar and the Romans, to the Conclusion of the Reign of King James the Second, and the Establishment of King William and Queen Mary upon the Throne, in the Year 1688*, Third Edition, London, Printed for Jacob Tonson..., 1720, p. 823.

[569] "But as the same Plague and Murrain that kills Sheep and Beasts will not hurt men, so will not the Plague that kills men, hurt Sheep or Cattel." William Kemp, *A Brief Treatise Of the Nature, Causes, Signes, Preservation From, and Cure Of The Pestilence*, London, Printed for...*D. Kemp*..., 1665, pp. 28, 44.

[570] Creighton, *Epidemics in Britain*, vol. I, p. 62.

[571] Barger, *Ergot and Ergotism*, p. 26.

[572] Nathaniel Hodges, *Loimologia: Or, An Historical Account of The Plague in London in 1665*..., J. Quincy (trans.), Third Edition, London, Printed for *E. Bell*..., 1721, pp. 58-60.

[573] Ibid., p. 41.

> the open air, not only are more influenced by it when tainted, but are also hurt by the infectious Venom which gathers upon the Herbage."[574]

Like the Plague,[575] the outbreak of convulsive ergotism that occurred in Holstein in 1716 and 1717 was attributed to a peculiar constituent in the air.[576] Hodges writes of the Plague: "It is said to be poisonous also, from its Similitude to the Nature of a Poison, both being equally destructive to Life, and killing Persons much after the same Manner, so that they seem to differ in Degree only ; for the deadly Quality of a Pestilence vastly exceeds either the arsenical Minerals, the most poisonous Animals or Insects, or the killing Vegetables ; nay, the Pestilence seems to be a Composition of all the other Poisons together.[577]

The winter of 1664 was very severe, with a hard frost covering the ground from December almost until the middle of April.[578] According to the *Diary* of Ralph Josselin, the grain harvest in 1664 was very wet, the fall afterwards rainy, and in February and March there was alternate thawing and freezing.[579] These are conditions that are highly favorable to the formation of *Fusarium* T-2 toxin on rye that is stored or overwintered in the fields.[580] Cold winters also favor the formation

[574] Ibid., p. 139.

[575] "The notion that pestilence was a 'vapore velonoso', which emanated from the air was widespread in the fourteenth and fifteenth centuries and can be found in all the surviving plague Consigli from...the Neoplatonic doctor Marsilio Ficino writing in response to the epidemics in the last decades of the fifteenth century...For Ficino...Hippocrates and even more Galen figure prominently, particularly in his descriptions of poisonous vapours...Ficino also cites...the somewhat recondite medical works of Raymond Lull. Most historians who have written about these Consigli also suggest that one of the main reasons for their similarity was that they dealt with the same disease, namely bubonic plague. But whether we can seriously associate this illness with the one we know today from the third pandemic of the 1920s must remain open to debate. However, here we enter slightly muddied waters because the terms employed by most treatise-writers were generic : 'pestilenza', 'morbo', or 'peste'. Some tracts also describe symptoms. Ficino, for example, lists breathlessness, slow pulse, general heaviness, deliriousness, and blood-shot eyes. But these are fairly general symptoms of any acute disease and do not help us to distinguish this 'pest' from any other. More specific is the appearance of the 'segno' or carbuncle which, according to him and many other writers, appears in specific locations : behind the ear, in the groin, and under the arm...even if as well-trained an observer as Ficino was aware that the expectoration of blood could indicate the presence of another type of 'pest', the layman might easily have continued to look for the tell-tale 'segno'." John Henderson, "Epidemics in Renaissance Florence: Medical Theory and Government Response," *Maladies Et Société*, pp. 168–169.

[576] Barger, *Ergot and Ergotism*, p. 70.

[577] Hodges, *Loimologia*, p. 33.

[578] Ibid., p. 5; Symon Patrick, *The Auto-Biography of Symon Patrick, Bishop of Ely*, Oxford, J. H. Parker, 1839, p. 51.

[579] *Diary of Josselin*, pp. 510-531.

[580] Joffe, *Fusarium Species*, p. 231; Moreau, *Moulds, Toxins and Food*, p. 221; Matossian, *Poisons of the Past*, p. 51.

of ergot on rye, and the winter rye that is planted in fall is more likely to be infected with ergot after a severe winter.[581]

Theophilus Garencieres writes: "by the long continuation of the late Winter, we had little or no Spring, but the Winter ended almost with the beginning of Summer."[582] A drought occurred throughout much of the year 1665, and southeastern England suffered an extended drought from November of that year until September 1666.[583] The formation of ergot is also associated with drought which traumatizes the grain.[584] Boghurst noted that the spring of 1665 was dry for six or seven months, with a few showers at the end of April.[585] According to Richard Baxter, the winter, spring, and summer were the driest in memory.[586] Samuel Pepys recorded rain on 22 and 23 April, around the onset of the epidemic, as well as other rainfall before the end of June.[587] According to Pepys, the middle of May was hot, and in early June he wrote of "the mighty heat of the weather" and called the seventh of that month "The hottest day that ever I felt in my life," which was followed by rain that night.[588] Spring rains followed by a warm and dry summer create optimal conditions for the formation of ergot, the growth of which is also affected by rain during this period.[589] The growth of ergot is also favored by heat, which is thought to increase the alkaloid levels and toxicity of the fungus.[590]

The west winds blew strong from Christmas until July in 1665,[591] and moderate breezes extended through the summer.[592] Temperatures in the months of July and August were average.[593] Pepys noted in his *Diary* on 8 July, "the mightiest storm came of wind and rain that almost

[581] Prescott, *Medicinal Effects of Ergot*, p. 5; Matossian, *Poisons of the Past*, pp. 13, 67, 80.

[582] Theophilus Garencieres, *A Mite Cast into the Treasury of the Famous City of London...*, London, Printed by *Tho. Ratcliffe...*, 1666, p. 8.

[583] Bell, *Plague in London*, pp. 251-252; D. J. Schove, "Fire and Drought, 1600–1700," *Weather*, 21 (1966), p. 313.

[584] Barger, *Ergot and Ergotism*, p. 73.

[585] Boghurst, *Loimographia*, p. 29.

[586] Richard Baxter, *Reliquiæ Baxterianæ...*, M. Sylvester (ed.), London, Printed for *T. Parkhurst...*, 1696, p. 448.

[587] Samuel Pepys, *The Diary of Samuel Pepys*, R. Latham and W. Matthews (eds.), London, G. Bell and Sons Ltd., 1972, vol. VI, pp. 86-87.

[588] Ibid., vol. VI, pp. 106, 120.

[589] Prescott, *Medicinal Effects of Ergot*, p. 5; Barger, *Ergot and Ergotism*, pp. 10, 98; Bove, *Story of Ergot*, p. 27.

[590] Barger, *Ergot and Ergotism*, p. 98.

[591] Boghurst, *Loimographia*, p. 28.

[592] Hodges, *Loimologia*, p. 13.

[593] H. H. Lamb, *Climate: Present, Past and Future*, London, Methuen & Co. Ltd., 1977, vol. II, pp. 571-572; Gordon Manley, "Central England Temperatures: Monthly Means, 1659 to 1973," *Quarterly Journal of the Royal Meteorological Society*, 100 (1974), p. 393.

could be for a quarter of an hour, and so left."[594] On 16 July Pepys wrote: "it was most extraordinary hot that ever I knew."[595] Josselin noted the beginning of the grain harvest in late July, and heavy rain in early August.[596] In mid-August a day or two of heavy rain fell,[597] and the temperatures fell at the end of the month; Pepys mentioned the "coolness of the last 7 or 8 days" to Lady Carteret on 4 September.[598] Heavy rain again fell on 9 September, the first since mid-August.[599]

The mortality rates of the Plague were closely tied to the rainfall. The weekly death rates continued to climb throughout the summer until the week ending 5 September, and declined the week ending 12 September, then peaked the week ending 19 September following the heavy rain.[600] Boghurst observed: "If very hott weather followed a shower of raine, the disease encreased much."[601] According to Jean-Noël Biraben, "We have found innumerable examples where rain has provoked the recrudescence of the plague, beginning with the year 1348...We have never found a contrary example, where the plague ceased following rains."[602]

Symon Patrick wrote to Elizabeth Gauden on 8 September 1665:[603] "the Wind is changed againe, & brings Abundance of Raine with it : & indeed we have no settled weather since I saw you, which hath made the Sicknesse, I believe, rage more : for *South Winds* are alwayes observed to be *bad* in such *Times* : & the Wind stays not long out of that Quarter." Patrick then repeated a suspicion shared by others : "And how should it be more dangerous then to receive Bear & Wine, the Vessells being capable of Infection? but especially Bread, they say, is the most attractive of it, which I am forced to buy : for I know not

[594] *Diary of Pepys*, vol. VI, p. 152.
[595] Ibid., vol. VI, p. 160.
[596] *Diary of Josselin*, p. 519.
[597] Ibid., p. 520; *Diary of Pepys*, vol. VI, p. 189.
[598] Pepys to Lady Carteret, 4 September 1665; in Samuel Pepys, *Memoirs of Samuel Pepys, Esq. F. R. S.*, Second Edition, R. L. Braybrooke (ed.), London, Frederick Warne and Co., 1870, p. 598.
[599] *Diary of Pepys*, vol. VI, p. 218; *Diary of Josselin*, p. 520.
[600] John Graunt, *London's Dreadful Visitation: Or A Collection of All the Bills of Mortality for this Present Year...*, London, Printed and are to be sold by *E. Cotes...*, 1665, pp. 37-39; Watson Nicholson, *The Historical Sources of Defoe's Journal of the Plague Year*, Boston, Stratford Co., 1919, pp. 170–171.
[601] Boghurst, *Loimographia*, p. 24.
[602] Jean-Noël Biraben, *Les hommes et la peste en France et dans les pays européens et méditerranéens, Tome I*, Paris, Mouton, 1975; in Matossian, *Poisons of the Past*, p. 50.
[603] "On the 3rd my brother was taken very ill, and vomited forty or fifty times, and my servant also had a swelled face, and I myself on the 7th had a sore pain in my leg, which broke my sleep, and made me suspect some touch of the plague, which was now come to its height, there dying ten thousand in one week. But, blessed be God, all these maladies went well over without danger." *Autobiography of Patrick*, pp. 53-54.

otherwayes to have it."[604] A few weeks later he informed Gauden: "I have quite changed my Diet. I eat boiled Meats & Broth more then I used : something at Supper also."[605]

[604] Patrick to Gauden, 8 September 1665; in Nicholson, *Sources of Defoe's Journal*, pp. 155, 157.
[605] Patrick to Gauden, 30 September 1665; in Ibid., p. 160.

Chapter 12. The Great Plague: An Overview of the Epidemic

The Plague appeared in Holland in 1663 and 1664, which affected the towns and small villages in the countryside, and killed 35,000 in Amsterdam.[606] Nathaniel Hodges, a member of the College of Physicians who remained in London during the Plague, believed that the disease came to England from Holland: "After a most strict and serious Inquiry, by undoubted Testimonies, I find that this *Pest* was communicated to us from the *Netherlands* by way of Contagion ; and if most probable Relations deceive me not, it came from *Smyrna* to *Holland* in a parcel of infected Goods."[607] Hodges writes:

> And for what concerns that Pestilence now under Enquiry, this we have as to its Origin, from the most irrefrigable Authority, that it first came into this Island by Contagion, and was imported to us from *Holland*, in Packs of Merchandize ; and if any one pleases to trace it further, he may be satisfied by common Fame, it came thither from *Turkey* in Bails of Cotton, which is a strange Preserver of the Pestilential Steams. For that Part of the World is seldom free from such Infections, altho' it is sometimes more severe than others, according to the Disposition of Seasons and Temperature of Air in those Regions...[608]

William Boghurst, an apothecary who also remained in London throughout the Plague, reported that the disease was present in the

[606] Jonathan Israel, *The Dutch Republic: Its Rise, Greatness, and Fall, 1477–1806*, Oxford, Clarendon Press, 1995, p. 625.
[607] Nathaniel Hodges, [8 May 1666], "Letter from Dr. Hodges to a Person of Quality," *A Collection of Very Valuable and Scarce Pieces Relating to the Last Plague in the Year 1665*, Second Edition, London, Printed for *J. Roberts*..., 1721, pp. 14–15.
[608] Hodges, *Loimologia*, p. 30.

city in the years preceding the epidemic: "The Plague hath put itselfe forth in St. Giles's, St. Clement's, St. Paul's, Covent Garden, and St. Martin's this 3 or 4 yeares, as I have beene certainly informed by the people themselves that had it in their houses in those Parishes."[609]

London was no stranger to the Plague, and with knowledge the preceding year of the Plague in Holland, measures were taken to prevent the spread of the disease to Britain. Between 1603 and 1665 there were only four separate years in which no Plague deaths were recorded in London, and in 1664 there were five deaths recorded as Plague, and nine in 1663.[610] In 1603, the year Queen Elizabeth died, the *Bills of Mortality* recorded a total of 33,347 deaths by Plague, and in the ill-omened year of 1625 when Charles I ascended to the throne, there were 41,313 who perished by Plague. There were many Londoners in 1665 that could still remember the Plague that raged between 1640 and 1647, when 14,420 deaths in London were attributed to the disease.

The population of London in 1665 has been roughly estimated at between 400,000 and 500,000, and according to the *Bills of Mortality* for 1665 London officially lost 97,366 lives; of these, 68,956 were recorded as death by Plague.[611] Because the Jews, Quakers, Anabaptists, and other sectaries did not allow their deaths to be reported to a church and had separate burial grounds, it is thought that the actual number of Plague deaths far exceeds 100,000.[612] It is believed that London lost at least one-quarter and possibly more than one-third of its permanent resident population in 1665.[613]

The Plague lingered in London through 1666, causing an additional 1,998 deaths.[614] Although some have suggested that the Great Fire in London in 1666 destroyed the agent of the epidemic, the fire was confined to the old walled city and would have had no effect on the disease, regardless of the pathogen.[615] The Plague returned intermittently in London until 1679, and this was the last year the disease was listed as a cause of death in the *Bills of Mortality*.[616]

The population in the poverty-stricken suburbs south of the Thames, including Southwark, suffered excessive mortality rates

[609] Boghurst, *Loimographia*, p. 26.
[610] Creighton, *Epidemics in Britain*, vol. I, p. 533.
[611] Graunt, *London's Dreadful Visitation*.
[612] Bell, *Plague in London*, pp. 3-7, 12.
[613] Shrewsbury, *History of Plague*, p. 454.
[614] Creighton, *Epidemics in Britain*, vol. I, p. 533.
[615] Shrewsbury, *History of Plague*, p. 485.
[616] Ibid., p. 486.

during the Plague that were comparable to the disproportionally high death rates in the impoverished suburbs surrounding the walled city.[617] Farther from London the Plague was also found throughout southeastern England, including the coastal townships of Dover, Ipswich, Yarmouth, Poole, Colchester, Southampton, and Newport on the Isle of Wight, and it also struck the northwestern county of Cheshire and the northeastern region of the Tyne River.[618] Dover lost between one-quarter and one-third of its inhabitants to Plague, and in Southampton nearly one-half of the population perished between 1665 and 1666.[619] In Colchester the Plague continued its ravages through 1666, and when it had passed nearly one-half of the population was deceased.[620] Graham Twigg writes: "The overwhelming part that plague has been believed to play in London epidemics is curious, for at no time, even during the plague of Canton, has it produced such high death rates. For example, in India plague killed only 0.135 per cent of the population *per annum* between 1896 and 1917 and death rates in other parts of the world were comparable."[621]

In the small and poor East Anglian town of Braintree in Essex, the Plague claimed nearly one-half of the population in 1665 and 1666, and the fatality rate among the afflicted reached 97 per cent between Septembers of those years.[622] The marshland parishes of southeastern England suffered greatly during the Plague,[623] although within the most heavily struck regions there were many communities that were untouched by the epidemic. Seven miles from Braintree stood Witham, a town that was also located on the main road between Chelmsford and Colchester, which escaped the mass death suffered by neighboring townships.[624] The Plague also struck Ipswich and Needham Market eight miles away, yet four miles further inland Stowmarket was unaffected, as was the port town of Lowestoft.[625]

[617] Champion, *London's Dreaded Visitation*, pp. 43, 104–106.

[618] Bell, *Plague in London*, pp. 140–143.

[619] A. Temple Patterson, *A History of Southampton 1700–1914*, Southampton, Southampton University Press, 1966, vol. I, pp. 2, 6; Shrewsbury, *History of Plague*, pp. 491-493; Mary J. Dobson, *Contours of Death and Disease in Early Modern England*, Cambridge, Cambridge University Press, 1997, p. 409.

[620] Creighton, *Epidemics in Britain*, vol. I, pp. 688-689; I. G. Doolittle, "The Plague in Colchester—1579–1666," *Transactions of the Essex Archeological Society*, Third Series, 4 (1972), pp. 137–138; Shrewsbury, *History of Plague*, pp. 499-502; Dobson, *Contours of Death*, p. 408.

[621] Twigg, "Plague in London," p. 15.

[622] Slack, *Impact of Plague*, pp. 106–107, 177; Dobson, *Contours of Death*, pp. 194, 282, 411.

[623] Dobson, *Contours of Death*, p. 194.

[624] Janet Gyford, *Witham 1500–1700: Making a Living*, Witham, Eng., Privately Printed, 1996, pp. 7-8.

[625] A. G. E. Jones, "Plagues in Suffolk in the Seventeenth Century," *Notes and Queries*, 198 (1953), pp. 384-386.

These patterns are not typical of an infectious disease. Oxford also escaped the Plague, and in the communities between London and Oxford the disease was not widespread, with confined outbreaks occurring at Uxbridge and High Wycombe.[626] Northeast of Oxford most Midland towns also escaped the epidemic, as did most of the communities in the West Country, Wales, and northern England.[627]

[626] Shrewsbury, *History of Plague*, pp. 481, 517.

[627] William Durrant Cooper, "Notices of the Last Great Plague, 1665-6; from the Letters of John Allin to Philip Fryth and Samuel Jeake," *Archaeologia*, 37 (1857), pp. 19-20; Bell, *Plague in London*, pp. 140–143; A. W. Langford, "The Plague in Herefordshire," *Transactions of the Woolhope Naturalists' Field Club*, 25 (1955–1957), pp. 152–153.

Chapter 13. The Great Plague in London

The Reverend Symon Patrick observed that at the end of 1664 there "was a very hard frost, which lasted from Christmas till near the middle of April in the year 1665, when the plague began to break out, a little after the breaking of the frost."[628] Spring is extremely early for the onset of bubonic plague, a tropical disease that requires temperatures of at least 65°F (18°C) for a period of several weeks. Boghurst said that the Plague started on the highest ground, and first struck the parish of St. Giles, located on the highest ground in London and some distance from the river Thames.[629] St. Giles-in-the-Fields was a larger out-parish of 1,500 households located west of the city of London, essentially a town area that was inhabited by the poor and destitute.[630] Boghurst noted that Southwark across the Thames was infected almost as soon as the West end.[631] The earliest reports of Plague were "only in the outskirts of the town, and in the most obscure alleys, amongst the poorest people," according to the Earl of Clarendon, the Lord Chancellor.[632] This is an unlikely place to find the onset of an infectious disease allegedly brought into the city in merchandise imported from Holland. According to Boghurst, the Plague did not start in one area and then spread throughout the

[628] *Autobiography of Patrick*, p. 51.
[629] Boghurst, *Loimographia*, p. 26.
[630] Bell, *Plague in London*, p. 28.
[631] Boghurst, *Loimographia*, p. 28.
[632] Edward Hyde, Earl of Clarendon, *Selections from Clarendon*, G. Huehns (ed.), London, Oxford University Press, 1955, p. 410.

region, but "fell upon severall places of the City and Suburbs like raine, even at the first at St. Giles's, St. Martin's, Chancery Lane, Southwark, Houndsditch, and some places within the City, as at Proctor's House."[633]

The *Bills of Mortality* recorded two deaths by Plague in late April and another forty-four Plague deaths in May.[634] In May and early June Plague deaths were recorded east of the city in the out-parishes of St. Botolph Bishopgate and St. Mary Whitechapel.[635] On 10 June, Pepys recorded the death of his physician's servant within the walls of London: "to my great trouble hear that the plague is come into the City (though it hath these three or four weeks since its beginning been wholly out of the City); but where should it begin but in my good friend and neighbour's, Dr. Burnett in Fanchurch-street—which in both points troubles me mightily."[636] On 15 June Pepys wrote: "The town grows very sickly, and people do be afeared of it—there dying this last week of the plague 112, from 43 the week before—whereof, one in Fanchurch-street and one in Broadstreete by the Treasurer's office."[637] On 20 June Pepys optimistically wrote: "people do think that the number will be fewer in the town then it was the last week."[638]

However this was not the case, and on the week ending 27 June there were 684 burials with 267 deaths recorded as Plague.[639] On 29 June Pepys wrote: "This end of the town every day grows very bad of the plague. The Mortality bill is come to 267—which is about 90 more then the last; and of these, but 4 in the City—which is a great blessing to us."[640] On 19 July Pepys wrote: "I hear the sickness, is, and endeed is scattered almost everywhere—there dying 1089 of the plague this week."[641] On 22 July Pepys mentioned the Plague Pits, the mass graves in which the dead were buried after the cemeteries in the churchyards

---

633 Boghurst, *Loimographia*, p. 29. "According to the rat-flea vector theory, the disease is dependent for its movement upon the existence of infected rat populations, since technically an epizootic must precede an epidemic. The disease, therefore, should move slowly and regularly from identifiable foci. However, the incidence of plague deaths in the parishes week by week throughout the epidemic, as indicated by the *Bills of Mortality* does not conform to this epidemiological model." Champion, *London's Dreaded Visitation*, pp. 81-82.

634 "Bills of Mortality"; in John Bell, *Londons Remembrancer...*, London, Printed and are to be sold by *E. Cotes...*, 1665.

635 Graunt, *London's Dreadful Visitation*, pp. 23-24.

636 *Diary of Pepys*, vol. VI, p. 124.

637 Ibid., vol. VI, p. 128.

638 Ibid., vol. VI, p. 133.

639 Creighton, *Epidemics in Britain*, vol. I, p. 662.

640 *Diary of Pepys*, vol. VI, p. 142.

641 Ibid., vol. VI, p. 163.

were unable to hold more victims.[642] John Allin gauged the extent of the catastrophe by the "dolefull and almost universall and continuall ringing and tolling of bells" by the parishes.[643] Hodges writes: "the Bells seemed hoarse with continual tolling, until at last they quite ceased ; the burying Places would not hold the Dead, but they were thrown into large Pits dug in waste Grounds, in Heaps, thirty or forty together."[644]

Referring to the week ending 25 July, Pepys wrote on 31 July that the disease "grows mightily upon us, the last week being about 1700 or 1800 of the plague."[645] On 3 August Pepys wrote: "the Bill...I had heard was 2010 of the plague, and 3000 and odd of all diseases."[646] On 10 August he wrote: "we sat all the morning, in great trouble to see the Bill this week rise so high, to above 4000 in all, and of them, about 3000 of the plague."[647] According to Pepys' *Diary* entry of 12 August, the deaths by this time had become so numerous that the authorities were forced to bury the dead during daylight because the night was not long enough.[648] On 31 August Pepys wrote: "the plague having a great encrease this week beyond all expectation, of almost 2000...In the City died this week 7496; and of them 6102 of the plague. But it is feared that the true number of the dead this week is near 10000—partly from the poor that cannot be taken notice of through the greatness of the number, and partly from the Quakers and others that will not have any bell ring for them."[649] Thomas Vincent writes:

> In *August*...the people fall as thick as the leaves from the Trees in *Autumn*, when they are shaken by a mighty wind...and there is a dismal solitude in *London-streets*...Now shops are shut in, people rare and very few that walk about, in so much that the grass begins to spring up in some places, and a deep silence in every place, especially within the Walls ; no rat[t]ling Coaches, no prancing Horses, no calling in Customers, nor offering Wares ; no *London-cries* sounding in the ears ; if any voice be heard, it is the groans of dying persons, breathing forth their last ; and the Funeral-knells of them that are ready to be carried to their Graves. Now shutting up of Visited-Houses (there being so many) is at an end, and most of the well are mingled among the sick,

[642] Ibid., vol. VI, p. 165.
[643] Allin to Fryth, 2 September 1665; in Cooper, "Letters of Allin," p. 9.
[644] Hodges, *Loimologia*, p. 18.
[645] *Diary of Pepys*, vol. VI, pp. 173, 178.
[646] Ibid., vol. VI, p. 180.
[647] Ibid., vol. VI, p. 187.
[648] Ibid., vol. VI, p. 189.
[649] Ibid., vol. VI, pp. 207-208.

> which otherwise would have got no help. Now in some places where the people did generally stay, not one house in an hundred but is infected ; and in many houses half the Family is swept away ; in some the whole...Now the nights are too short to bury the dead, the whole day (though at so great a length) is hardly sufficient to light the dead that fall therein, into their beds. Now we could hardly go forth, but we should meet many Coffins, and see many with sores, and limping in the streets...[650]

On 7 September Pepys wrote: "sent for the Weekly Bill and find 8252 dead in all, and of them 6978 of the plague—which is a most dreadfull Number."[651] John Evelyn noted in his *Diary* on 7 September that there were "perishing now neere ten-thousand poore Creatures weekely : however I went all along the Citty & suburbs from *Kent streete* to *St. James's*, a dismal passage & dangerous, to see so many Cof[f]ines expos[e]d in the streetes & the streete thin of people, the shops shut up, & all in mournefull silence, as not knowing whose turne might be next."[652] Vincent writes: "in *September*...the Grave doth open it's mouth without measure...the Church-yards now are stuft so full of dead Corpses, that they are in many places swell'd two or three feet higher then they were before ; and new ground is broken up to bury the dead."[653] John Tillison wrote to Dean William Sancroft on 14 September:

> Death stares us continually in the face in every infected person that passeth by us ; in every coffin which is daily and hourly carried along the streets. The bells never cease to put us in mind of our mortality. The custom was, in the beginning, to bury the dead in the night only ; now, both night and day will hardly be time enough to do it. For the last week, mortality did too apparently evidence that, that the dead was piled in heaps above ground for some hours together, before either time could be gained or place to bury them in.[654]

On the week ending 12 September the number of Plague deaths fell to 6,544 with 7,690 burials, according to Pepys, a "decrease of 500 and more, which is the first decrease we have yet had in the

[650] Thomas Vincent, *Gods Terrible Voice in the City of London...*, Cambridge, Printed by *Samuel Green*, 1667, pp. 8-9.

[651] *Diary of Pepys*, vol. VI, p. 214.

[652] John Evelyn, *The Diary of John Evelyn*, E. S. de Beer (ed.), London, Oxford University Press, 1959, pp. 479-480.

[653] Vincent, *Gods Terrible Voice*, p. 10.

[654] Tillison to Sancroft, 14 September 1665, Harleian MSS. 3785, fol. 50; in Henry Ellis (ed.), *Original Letters, Illustrative of English History...*, Fourth Series, London, Harding and Lepard, 1827, vol. IV, p. 36.

sickness since it begun—and great hopes that the next week it will be greater."[655] However, during the week ending 19 September the Plague rebounded, and the weekly death toll climbed to its infamous peak. Hodges estimated the deaths that week at 12,000,[656] and the Duke of Albemarle told the French ambassador the figure was 14,000, not counting the Quakers.[657] On 20 September Pepys wrote:

> But Lord, what a sad time it is, to see no boats upon the River—and grass grow up all up and down Whitehall-court—and nobody but poor wretches on the streets. And which is worst of all, the Duke showed us the number of the plague this week, brought in the last night from the Lord Mayor—that it is encreased about 600 more then the last, which is quite contrary to all our hopes and expectations from the coldness of the late season: for the whole general number is 8297; and of them, the plague 7165—which is more in the whole, by above 50, then the biggest Bill yet—which is very grievous to us all.[658]

Hodges writes:

> In the Months of *August* and *September*, the Contagion chang'd its former slow and languid Pace, and having as it were got Master of all, made a most terrible Slaughter, so that three, four, or five Thousand died in a Week, and once eight Thousand ; who can express the Calamities of such Times? The whole *British* Nation wept for the Miseries of her Metropolis. In some Houses Carcases lay waiting for Burial, and in others, Persons in their last Agonies ; in one Room might be heard dying Groans, in another the Raving of a Delirium, and not far off Relations and Friends bewailing both their Loss, and the dismal Prospect of their own sudden Departure ; Death was the sure Midwife to all Children, and Infants passed immediately from the Womb to the Grave ; who would not melt with Grief, to see the Stock for a future Generation hang upon the Breasts of a dead Mother? Or the Marriage-Bed changed the first Night into a Sepulchre, and the unhappy Pair meet with Death in their first Embraces? Some of the infected run about staggering like drunken Men, and fall and expire in the Streets ; while others lie half-dead and comatous, but never to be waked but by the last Trumpet ; some lie vomiting as if they had drank Poison ; and others fall dead in the Market, while they are buying Necessaries for the Support of Life.[659]

[655] *Diary of Pepys*, vol. VI, pp. 224-225.
[656] Hodges, *Loimologia*, p. 19.
[657] Bibliothèque Nationale MSS. Fr. 15889; in Bell, *Plague in London*, pp. 243-244.
[658] *Diary of Pepys*, vol. VI, pp. 233-234.
[659] Hodges, *Loimologia*, pp. 16–17.

On 27 September Pepys wrote: "I saw this week's Bill of Mortality, wherein, blessed be God, there is above 1800 decrease, being the first considerable decrease we have had."[660] Walking past the Tower on 16 October, Pepys encountered "so many poor people sick people in the streets, full of sores,"[661] and riding through Kent-street in Southwark on 27 October, Pepys described the neighborhood as "a miserable, wretched, poor place, people sitting sick and muffled up with plasters at every four or five door."[662] The Plague continued a steady decline through autumn, and on 31 December Pepys noted, "the plague is abated almost to nothing."[663]

Entire families were destroyed by the epidemic, and Allin observed that "many whole familyes of 7, 8, 9, 10, 18 in a family" perished together.[664] When the plague was determined to exist within a home, the entire household was confined to the premises for forty days, and in many cases the entire family perished.[665] In London's alleys where crowded tenements often lived 4 to 6 families, all were shut in together, often a death sentence for all.[666] This policy received the approval of the College of Physicians, operating under the theory that the Plague was contagious, and spread through contact.[667] Hodges writes:

> In Order whereunto, it is to be observ'd, that a Law was made for marking the Houses of infected Persons with a Red Cross, having with it this Subscription, LORD HAVE MERCY UPON US: And that a Guard should attend there continually, both to hand to the Sick the Necessaries of Food and Medicine, and to restrain them from coming Abroad until Forty Days after their Recovery. But although the *Lord Mayor* and all inferior Officers readily and effectually put these Orders in Execution, yet it was to no Purpose, for the Plague more and more increased ; and the Consternation of those who were separated from all Society, unless with the infected, was inexpressible...[668]

[660] *Diary of Pepys*, vol. VI, p. 243.

[661] Ibid., vol. VI, p. 268.

[662] Ibid., vol. VI, p. 279.

[663] Ibid., vol. VI, p. 341.

[664] Allin to Fryth, 26 July 1665; in Cooper, "Letters of Allin," p. 7.

[665] Walter George Bell, *The Great Plague: Grim Tragedy Told in Church's Records of 1665*, London, Published by the Guild of St. Bride, 1958, pp. 2-8.

[666] Shrewsbury, *History of Plague*, p. 448. Statistics from a sample of London parishes show that the majority of infected homes experienced one or two deaths, and that the households that suffered multiple casualties were those of the poor. "It is clear from the parish studies that the pattern of infection within the household does not match the predictions of the rat/flea theory." Champion, *London's Dreaded Visitation*, pp. 82-87.

[667] Bell, *Plague in London*, pp. 114–116; Shrewsbury, *History of Plague*, p. 449.

[668] Hodges, *Loimologia*, p. 7.

The majority of London's elite College of Physicians escaped to the country early in the epidemic, including the College president Sir Edward Alston; and the great Thomas Sydenham also fled the city.[669] Following a special session of the College on 26 June attended by only twenty members,[670] their building at Amen Corner was completely deserted by the faculty, and was robbed by thieves in late June.[671] In early July, King Charles II, his mother, and the Royal court also fled London.[672] Referring to those who abandoned the city, Hodges writes: "in *December* they crowded back as thick as they fled."[673] Six members of the College of Physicians met to elect a president on 17 March 1665/66, and the College held its first meeting in late June, attended by twenty-seven members.[674]

Grain and other foods were plentiful during the Plague.[675] Hodges writes: "yet although the more opulent had left the Town, and it was almost left uninhabited, the Commonalty that were left felt little of Want ; for their Necessities were relieved with a Profusion of good Things from the Wealthy, and their Poverty was supported with Plenty."[676] Hodges added, "Nor ought we here to pass by the beneficent Assistances of the Rich, and the Care of the Magistrates ; for the Markets being open as usual, and a great Plenty of all Provisions, was a great Help to support the Sick ; so that there was the Reverse of a Famine, which hath been observed to be so fatal to pestilential Contagions."[677] However, the parishes were soon unable to shoulder the financial burden of the sick,[678] and with the flight of the gentry the economy of London suffered, and the working poor were left without a means of support. Without employment and the inability of the parishes to provide assistance, hunger set it, and desperation forced many into begging.[679] In his *Diary* entry of 6 July, Pepys bemoaned the

[669] Sir George Clark, *A History of the Royal College of Physicians of London*, Oxford, Clarendon Press, 1964, vol. I, p. 319.

[670] Harold J. Cook, *The Decline of the Old Medical Regime in Stuart London*, Ithaca, Cornell University Press, 1986, p. 156.

[671] Bell, *Plague in London*, pp. 62-63.

[672] Shrewsbury, *History of Plague*, p. 448.

[673] Hodges, *Loimologia*, p. 27.

[674] Clark, *College of Physicians*, vol. I, p. 326; Cook, *Old Medical Regime*, p. 160.

[675] Boghurst, *Loimographia*, p. 30.

[676] Hodges, *Loimologia*, p. 15.

[677] Ibid., p. 21.

[678] Tillison to Sancroft, 15 August 1665, Harleian MSS. 3785, fol. 48; in Ellis, *Original Letters*, p. 31; Nicholson, *Sources of Defoe's Journal*, p. 85.

[679] *Diary of Pepys*, vol. VI, pp. 255, 268, 279, 288, 297; *Diary of Evelyn*, p. 481; William Taswell, "Autobiography and Anecdotes," *Camden Miscellany*, G. P. Elliott (ed.), London, Printed for

lack of finances in the Kingdom during this difficult period: "we have not enough to stop the mouths of poor people and their hands from falling about our eares here, almost in the office."[680]

The Plague was erratically distributed throughout the city, and the deaths by the disease were a reflection of the distribution of wealth. The parishes that were spared high death rates were generally clustered around Cheapside, the wealthiest part of the city.[681] In the impoverished city of Norwich in East Anglia, the second most populous city in the Kingdom, the distribution of the disease also reflected the distribution of wealth, and the mortality rates were three times higher in the poor parishes than those in the wealthy parishes.[682] On 9 July Pepys wrote in his *Diary*:

> The most observable thing I found there, to my content, was to hear him and his clerk tell me that in this parish of Michells Cornhill, one of the middlemost parishes and a great one of the town, there hath, notwithstanding this sickliness, been buried of any disease, man, woman, or child, not one for thirteen months past; which very strange. And the like in a good degree in most other parishes, I hear, saving only of the plague in them. But in this, neither the plague nor any other disease.[683]

Within the walls of London the parishes of St. John Evangelist recorded no deaths by Plague in 1665, St. Bennett Sherehog only one death, St. Alhallows Honey-lane five deaths, and both St. Mary Colechurch and St. Matthew Friday-Street six deaths by Plague. Within the city walls the most severely affected parishes were St. Anne Black-Friers and Christs Church, both with 467 deaths by Plague. In the poor suburbs located outside of the walled city, the parishes of St. Botolph Aldgate recorded 4,051 Plague deaths, St. Giles Cripplegate 4,838 Plague deaths and 8,069 burials, and Stepney 6,583 Plague deaths and 8,598 burials.[684] These are clearly not mortality distribution patterns that occur with an infectious disease. As Twigg has pointed out, the intensity of the disease in the poor out-parishes of London is at odds with what is known of rat biology, which in modern times has selected the wealthier parts of the city where

the Camden Society, 1853, vol. II, p. 9; Creighton, *Epidemics in Britain*, vol. I, p. 663; Champion, *London's Dreaded Visitation*, p. 91.

680 *Diary of Pepys*, vol. VI, p. 149.

681 Champion, *London's Dreaded Visitation*, pp. 104–106; Slack, *Impact of Plague*, pp. 156–158.

682 Slack, *Impact of Plague*, pp. 137–139.

683 *Diary of Pepys*, vol. VI, pp. 152–153.

684 Graunt, *London's Dreadful Visitation*.

warmth is greater and food more plentiful. There is no reason to believe that these same features did not control the distribution of the animal in the seventeenth century.[685]

Furthermore, there were no fatalities by Plague among the nobility, statesmen, the House of Commons, the House of Lords, judges, the higher ranks of the clergy, or financiers, most of whom fled London. Among those who stayed behind were the City Aldermen, and there were no deaths reported among the Court of the Aldermen. There were also no casualties among the nine magistrates in the out-parishes who were directed by King Charles to remain in London during the epidemic.[686] Among those who remained behind were the Duke of Albemarle, the sole representative of the Government; his most active lieutenant, the Earl of Craven, who opened his home on Drury Lane to the afflicted; and the Lord Mayor, Sir John Lawrence, all of whom survived the epidemic.[687]

Boghurst said that the strong and young died sooner in the Plague than the weak and old,[688] and writes: "Strength of constitution of body was noe protection against the disease nor death, for it made the hottest assault uppon strong bodyes and determined soonest, for they dyed sooner than people of weake constitution, and men dyed sooner than women."[689] According to Boghurst, the Plague took few prostitutes and drunkards, and the disease "left the rotten bodyes, and took the sound."[690] Children were especially susceptible to the disease.[691] Hodges writes: "of the Female Sex most died ; and hardly any Children escaped."[692] Kemp writes: "Children are most in danger, Women with Child next, and young Maids that are marriageable, more than Elder or aged persons."[693] Pepys recorded an incident that later came to symbolize the epidemic:

---

[685] Twigg, "Plague in London," p. 15. "The crucial determinants in the growth of a flea population are temperature and humidity: outside of certain very circumscribed parameters the flea population cannot breed rapidly enough to sustain the epidemic patterns of infection. From most of the surviving accounts the climate in England between 1664 and 1665 does not seem suitable to have provided the environmental infrastructure of an infected flea population... and so in terms of the rat-flea theory could not have determined the nature of the epidemic." Champion, *London's Dreaded Visitation*, pp. 8-9.

[686] Bell, *Plague in London*, pp. 312-313; Shrewsbury, *History of Plague*, p. 448.

[687] Bell, *Plague in London*, pp. 69-70.

[688] Boghurst, *Loimographia*, p. 99.

[689] Ibid., p. 25.

[690] Ibid., p. 96.

[691] Ibid., p. 97.

[692] Hodges, *Loimologia*, p. 18.

[693] Kemp, *Treatise Of The Pestilence*, p. 33.

> Among other stories, one was very passionate methought—of a complaint brought against a man in the town for taking a child from London from an infected house. Alderman Hooker told us it was the child of a very able citizen in Gracious-street, a saddler, who had buried all the rest of his children of the plague; and himself and wife now being shut up, and in despair of escaping, did desire only to save the life of this little child; and so prevailed to have it received stark-naked into the arms of a friend, who brought it (having put it into new fresh clothes) to Grenwich; where, upon hearing the story, we did agree it should be [permitted to be] received and kept in the town.[694]

Henry Oldenburg remained at his home in Pall Mall during the Plague, and with others observed that the disease affected very few who lived "orderly and comfortably," while those "wanting necessaries and comfortable relief" suffered enormously.[695] The Earl of Clarendon writes: "The greatest number of those who died consisted of women and children, and the lowest and poorest sort of the people...not many of wealth or quality or of much conversation."[696] John Gadbury concluded that the "dregs" were destroyed by Plague while the noble survived.[697] Hodges writes: "But it is incredible to think how the Plague raged amongst the common People, insomuch that it came by some to be called the *Poors Plague*."[698]

[694] *Diary of Pepys*, vol. VI, pp. 211-212.

[695] Oldenburg to Hooke, 23 August 1665; in A. Rupert Hall and Marie Boas Hall (eds.), *The Correspondence of Henry Oldenburg*, Madison, University of Wisconsin Press, 1965, vol. II, pp. 449, 479.

[696] Hyde, *Selections of Clarendon*, p. 413.

[697] Bernard Capp, *Astrology and the Popular Press: English Almanacs 1500–1800*, London, Faber & Faber, 1979, p. 135.

[698] Hodges, *Loimologia*, p. 15. "One common assumption is that the flea only attacked the poor. Interestingly most of the very few references to fleas survive in the literary accounts of the upper orders." "One of the central problems with the dominance of the rat flea theory is that it has meant that historians have accepted a very static medical model of the interaction between disease and social structure. Part of the difficulty lies in the implicit connection between a definition of both disease and poverty that relies upon ideas of cleanliness and dirt. For example, William Bell illustrates the worst of this presumption...He insisted that Stuart England was as filthy as 'Old Cairo,' and described the 'squalid and overcrowded quarters where the poor herded' that attracted fleas and 'became a focus for the accumulation and dissemination of poison.' In this manner the understanding of the relationship between poverty and disease is static: poor households are dirty and therefore they attract the plague...undoubtedly men were as dirty as women and rich as poor, but each of these groups experienced the epidemic in different manners." Champion, *London's Dreaded Visitation*, pp. 9 n. 20, 87; Bell, *Plague in London*, pp. 250-252.

CHAPTER 14. THE MEDICAL EVIDENCE: DISEASES PRECEDING THE PLAGUE

Ergot infection continues to be prevalent among the cereals and wild grasses in England today, and it occurs on a regular basis.[699] Following a wet summer in 1927, a mild but extensive epidemic occurred among Jewish immigrants in Manchester after consuming ergot-infected rye.[700] Epidemics of either convulsive or gangrenous ergotism usually recurred within a specific geographical region.[701] There has only been one recorded occurrence of gangrenous ergot poisoning in England, and it attacked the family of a poor agricultural laborer in Wattisham in 1762, apparently caused by ergot-infected wheat.[702] Creighton has identified two eighteenth century outbreaks

[699] G. Wood and J. R. Coley-Smith, "Observations on the Prevalence and Incidence of Ergot Disease in Great Britain with Special Reference to Open-flowering and Male-sterile Cereals," *Annals of Applied Biology*, 95 (1980), pp. 41-46.

[700] Barger, *Ergot and Ergotism*, pp. 64-65.

[701] "Most mixed epidemics appear to have occurred in Russia ; generally however nervous ergotism predominated in the north, while the gangrenous type has been recorded from the south of that country. In France gangrenous ergotism was the rule and of 41 references to ergotism, which I have collected from French chronicles, only two mention convulsions. Both references refer to the same epidemic (or to two epidemics in rapid succession) in Lorraine, bordering on Germany, where convulsive ergotism was the universal type. A small Swiss epidemic of 1709, accurately described by LANG, is often quoted as an example of the mixed type, but was essentially gangrenous...In Germany gangrene was extremely rare. The chronicle of Meissen in Saxony for 1486 mentions it. Apart from a few sporadic cases, such as that of BRUNNER, there was no ergot gangrene in Germany (TAUBE is very definite on this point) ; none is recorded for Bohemia, Hungary, Sweden or Finland. East of the Rhine gangrene was only met with in a few of the many Russian epidemics, in which convulsive symptoms greatly predominated..." George Barger, "The Alkaloids of Ergot," *Handbuch der Experimentellen Pharmakologie*, W. Heubner and J. Schüller (eds.), Berlin, Verlag von Julius Springer, 1938, vol. VI, pp. 205-206.

[702] Barger, *Ergot and Ergotism*, pp. 27-28, 63-64.

in England as ergot poisoning.[703] The first occurred in the village of Blackthorn in Oxfordshire in 1700, investigated by Dr. Friend, who found the victims making barking sounds: "He found that this *pestis* or plague had invaded two families in the village, on terms of close intimacy with each other. Two or three girls in each family are specially referred to : They were seized at intervals of a few hours with spasms of the neck and mouth, attended by vociferous cries ; the spasmodic movements increased to a climax, when the victims sank exhausted."[704] The second outbreak occurred in Lancashire and Cheshire in 1702, reported to the Royal Society by Dr. Charles Leigh:

> We have this year [1702] had an epidemical fever, attended with very surprising symptoms. In the beginning, the patient was frequently attacked with the colica ventriculi ; convulsions in various parts, sometimes violent vomitings, and a dysentery ; the jaundice, and in many of them, a suppression of urine ; and what urine was made was highly saturated with choler. About the state of the distemper, large purple spots appeared, and on each side of'em two large blisters... But the most remarkable instance I saw in the fever was in a poor boy... Upon the crisis or turn of the fever, he was seized with an aphonia, and was speechless six weeks [? days], with the following convulsions : the distemper infested the nerves of both arms and legs which produced the Chorea Sancti Viti, or St Vitus's dance...he barked in all the usual notes of a dog, sometimes snarling, barking, and at the last howling like an hound...These symptoms were so amazing that several persons about him believed he was possessed.[705]

Grain poisoning is thought by Mary Matossian to be involved in the epidemics that occurred in East Anglia between the years 1430 and 1480.[706] During this period the mortality rates peaked most frequently

---

[703] Creighton has also identified a number of other epidemics in England that he believed were caused by ergot poisoning. In 1340, a lethal epidemic appeared throughout England and especially in Leicestershire, causing convulsions, pains, and noises similar to the bark of a dog. In 1355, an outbreak of "madness" occurred throughout Worcester that caused many people to believe they were seeing "demons." A lethal epidemic of "lethargy" occurred around 1362 causing many women to die by the "flux," and in 1389 a lethal epidemic of "phrensy of the mind" struck Cambridge, characterized by sudden and violent disturbances of the central nervous system. Creighton, *Epidemics in Britain*, vol. I, pp. 59-63. Cf. George Lyman Kittredge, *Witchcraft in Old and New England*, Cambridge, Harvard University Press, 1929, pp. 125–126.

[704] D. Johannis Friend, "Epistola D. Johannis Friend ad Editorem missa, de Spasmi Rarioris Historia," *Philosophical Transactions*, 22 (1701), p. 799; in Creighton, *Epidemics in Britain*, vol. I, p. 61.

[705] Charles Leigh, "Part of a Letter from *Dr. Charles Leigh* of *Lancashire* to the Publisher, giving an account of strange Epileptic Fits," *Philosophical Transactions*, 23 (1702), pp. 1174–1175; I. F. C. Hecker, *The Epidemics of the Middle Ages: No. II. The Dancing Mania*, B. G. Babington (trans.), Philadelphia, Haswell, Barrington, and Haswell, 1837; Creighton, *Epidemics in Britain*, vol. I, p. 60.

[706] "In 1348, following a century of wet and cold weather, and fifty years of famine and subsistence crisis, England was struck by the Black Death...The Black Death reached London late in September 1348...With 50,000 or more inhabitants, London was England's largest town,

at harvest time in spring and autumn, varied greatly between regions and communities, and these rates were much higher among the working poor than those of the upper class.[707] England suffered a lethal epidemic in 1513 that also appears to have been caused by poisonous grain:

> This Year a Great Mortality prevailed in *England*, say our Historians ; they call it (as indeed all Diseases) the Plague ; but to know what it was, we must consult Foreigners. Says *Cole*, when Dearth, Scarcity of Corn, Famine, rainy Seasons, and severe cold ones had afflicted *Italy* for two Years, and People were forced to eat uncommon and unwholsome Food, arose an epidemic contagious Fever, with a Dysentery, and black Spots over the whole Body. And from this want of Food, great Weakness and unhealthy Juices, they had a pale cacochimic and depraved Countenance, a Swelling of their Feet, and Difficulty breathing. Their Excrements were black, and corroded their Bowels ; their Urine black with a Strangury ; their Breath, Urine, and Spittle, stunk intolerably : All forsook the Sick, and fled.[708]

Following a hot and dry spring and summer in 1657, a mild "Epidemick Fever" affecting entire families attacked the farmers and rural inhabitants of England, the outbreak conforming to the epidemiological model of ergot poisoning. At the end of July the disease began to appear sporadically, characterized by fever, sweating, chills, shivering, thirst, vomiting, cramps, loss of appetite, discolored urine, weak pulse, lethargy, fainting, swooning, watching, restlessness, delirium, and convulsions. The disease continued into August, and

---

the only one comparable to the great cities of the continent...In all, the Black Death lingered until late spring 1350, and killed at least a third of London's population...Estimates of mortality in England range between fifteen and fifty percent; a third to forty percent seems likely...But in the spring of 1361 England was struck by the *pestis secunda*. This second epidemic began a cycle in which plague would recur every few years until 1480, and continue the depopulation that, in some places, lasted until the sixteenth century...Plague returned in 1369. Thereafter the most important feature of plague was not the death rate in any given epidemic but the frequency with which epidemics occurred. From 1369 to 1479 no epidemic killed more than ten to fifteen percent, save perhaps in a few locales. Some epidemics probably killed five percent. But plague entered a cycle in which it recurred from five or six, to ten or twelve years...Parts of England experienced plague epidemics in eleven of the years between 1442 and 1459. London was particular hard hit, suffering on six occasions. From 1463 to 1465 another major epidemic struck the kingdom, followed by one more in 1467. This set the stage for the terrible 1470s." Robert S. Gottfried, "Plague, Public Health and Medicine in Late Medieval England," *Maladies Et Société*, pp. 335, 343-346.

[707] Robert S. Gottfried, *Epidemic Disease in Fifteenth Century England: The Medical Response and the Demographic Consequences*, New Brunswick, Rutgers University Press, 1978, pp. 99–117, 127–138; Matossian, *Poisons of the Past*, pp. 55-57.

[708] Thomas Short, *A General Chronological History of the Air, Weather, Seasons, Meteors, &c. in Sundry Places and Different Times...*, London, Printed for T. Longman..., 1749, vol. I, pp. 204-205.

the mortality rates were estimated at one in a thousand.[709] According to John Graunt, the following years of 1658 and 1661 were both "sickly Years" in London: "We mean by a sickly Year, such wherein the Burials exceed those both of the precedent and subsequent Years, and not above two hundred dying of the *Plague* : for Years exceeding that number of the *Plague*, we call *Plague* Years."[710]

In 1658, following a severely cold winter, England was struck by *Fusarium* poisoning in spring and ergot poisoning in late summer. Snow covered the ground from early December until the start of spring, with only an occasional day or two of hot weather until the first of June. In late April a lethal "Epidemick Fever" suddenly appeared, striking both the rural and urban communities. The victims displayed a variety of symptoms: fever, cough, headache, excessive spitting and discharge from the nose and mouth, thirst, lassitude, burning of the precordial region, pain in the back, joints, loins, and limbs, loss of appetite, hoarseness, difficulty breathing, headache, watching, spitting blood, nose bleed, and bloody stool. The disease disappeared six weeks later, and during this time some villages had one thousand people affected in a week.[711]

That year a continual north wind produced an unusually cool summer that traumatized the grain, causing farmers to fear there would be very little of their crops that would reach maturity. In early August a heat wave struck, and in late August a lethal "Epidemick Fever" appeared that peaked in September and particularly affected the rural communities. Primarily attacking the central nervous system in a manner consistent with ergot poisoning, the disease produced the following symptoms: fever, violent headache, shivering, sweating, hot and cold fits, thirst, roughness of the tongue, nausea, loss of appetite, vomiting, diarrhea, bloody urine, difficulty breathing, skin eruptions, deafness, ringing in the ears, speechlessness, loss of the senses, pain in the loins, trembling, twitching, spasms, a weak, uneven and intermittent pulse, weakness, dizziness, vertigo, fainting, drowsiness, stupor, coma, catalepsy, stupidity, distraction, idle talk, watching, restlessness, insomnia, delirium, convulsions, and frenzy. These symptoms were intermittent for some and continual for others,

---

[709] Thomas Willis, *The London Practice Of Physick...*, London, Printed for *Thomas Basset...*, 1685, pp. 648-655.

[710] John Graunt, "Reflections on the Weekly Bills of Mortality...," *Collection of Very Valuable and Scarce Pieces*, p. 81.

[711] Willis, *Practice Of Physick*, pp. 657-662.

and those who recovered did so only after a long convalescence. Thomas Willis attributed the disease to the "ill and course" diet of the farmers and inhabitants of the rural countryside.[712] Oliver Cromwell died on 3 September of that year after suffering from insomnia, severe pains in the bowels and back, and convulsions.[713]

In the spring of 1661 London suffered a food shortage, and a lethal "Epidemick Fever" appeared that struck entire families and primarily affected children and young people, displaying symptoms consistent with *Fusarium* and ergot poisoning, causing fever, nausea, vomiting, diarrhea, loss of appetite, bloody urine, difficulty breathing, coughing and lung problems, spots and ulcers on the skin, deafness, ringing in the ears, speechlessness, loss of the senses, thirst, roughness of the tongue, weakness, giddiness, vertigo, fainting, stupor, catalepsy, insomnia, twitching, delirium, frenzy, convulsions, and swelling of the lymphatic glands of the neck, which broke and oozed a "thin and stinking Ichor." The disease often turned into "Consumption," and recovery from the affliction took months.[714] Thomas Sydenham writes:

> The autumnal intermittent fevers which had reigned for several years backwards, appeared with new force in the year 1661, especially a bad kind of tertian, about the beginning of July, which continually increased, so as to prove extremely violent in August, seizing almost whole families in many places with great devastation...A few quartans accompanied these tertians...(for they seized upon none that were unaffected by them before) and were followed by a continued fever... How long this continued fever had prevailed, I cannot say...This, however, I know, that there was only one species of continued fevers in the year 1665...The above-mentioned tertian fever, which spread very wide in 1661...contracted itself in the succeeding year ; for, in the following autumns, quartans prevailed over the other epidemics...at the beginning of May, the small pox appeared a little, but disappeared again upon the coming in of the autumnal epidemics ; viz. the continued fever and quartans, which then reigned. In this order did the epidemic diseases appear and succeed each other...[715]

Sydenham adds:

> The continued fever and intermittents...of the years 1661, 1662, 1663, and 1664...likewise of the small pox of that constitution...in those

[712] Ibid., pp. 662-672; Creighton, *Epidemics in Britain*, vol. I, pp. 572-574; Matossian, *Poisons of the Past*, pp. 65-68.

[713] Antonia Fraser, *Cromwell, The Lord Protector*, New York, Alfred A. Knopf, 1973, p. 670.

[714] Willis, *Practice Of Physick*, pp. 273-278.

[715] Thomas Sydenham, *The Works of Thomas Sydenham, M.D. on Acute and Chronic Diseases; with their Histories and Modes of Cure*, B. Rush (ed.), Philadelphia, Published by B. & T. Kite..., 1815, pp. 50-52.

> years they prevailed much in the beginning of May, but went off, upon the coming of the autumnal epidemics, namely, the continued and intermittent fevers. The tops of the eruptions had small pits for the most part, about the size of the head of a small pin, and in the distinct kind the eighth day was attended with most danger...a delirium, great restlessness, pain and sickness, a frequency of making urine in small quantities succeeded, and the patient died in a few hours very unexpectedly...the fever prevailed...and continued to the spring of the year 1667...the fevers which prevail for a year or two after a severe plague, are generally pestilential ; and though some have not the genuine signs of the plague, yet they are much of the same nature...[716]

In the 1650s and 1660s the summer temperatures in England were favorable to the development of ergot,[717] and there is evidence for other epidemics of ergot poisoning occurring in some of the rye-growing regions during the 1650s.[718] The "Epidemick Fevers" of 1657, 1658, 1661, 1662, 1663, and 1664 were antecedent epidemics that culminated with the Great Plague in 1665.

[716] Ibid., pp. 90-93.

[717] The optimum temperature for the growth of ergot is 17.4°–18.9°C (63.3°– 66°F). B. Košir, P. Smole and Z. Povšič, ["Factors affecting the yield and quantity of sclerotia from *Claviceps purpurea*"], (in Slovenian, with English abstract), *Farmacevtski Vestnik*, 32 (1981), pp. 21-25; Zdeněk Řeháček and Přemysl Sajdl, *Ergot Alkaloids: Chemistry, Biological Effects, Biotechnology*, Amsterdam, Elsevier, 1990, p. 92; Manley, "Central England Temperatures," p. 393; Mary Kilbourne Matossian, "Why the Quakers Quaked: The Influence of Climatic Change on Quaker Health, 1647–1659," *Quaker History*, 96 (2007), pp. 36-51.

[718] Matossian, "Why the Quakers Quaked," pp. 36-51.

Chapter 15. The Medical Evidence: Diseases Occurring with the Plague

The official cause of death published in the *Bills of Mortality* was determined by the "Searchers of the dead," usually elderly women who were employed by the parishes, whose ethics and diagnostic abilities have been questioned by some,[719] although these reports appear to have been as accurate as can be reasonably expected.[720] In the summer months of 1665 there was a marked increase in deaths recorded as "Feaver" "Spotted Feaver," "Aged," "Dropsie," "Childbed," "Convulsion," "Rickets," "Consumption," "Rising of the Lights," "Surfet," "Griping in the Guts," "Worms," "Teeth," and "Infants."[721] Kemp writes: "When the *Plague* first seizeth upon any particular person, before many have been infected, it is very hard to discern it, because it hath divers symptomes attending it that are common to other diseases, and there is no one perfect proper, infallible, and inseparable sign to distinguish it, and many excellent and learned Physicians have disputed and differed much about it ; but when it hath continued a while, and spread it self abroad among many, it is very

[719] Bell, *Plague in London*, pp. 17-20.

[720] Referring to previous Plagues in the seventeenth century, Graunt wrote in 1662: "To conclude, In many of these Cases the *Searchers* are able to report the Opinion of the *Physicians*, who was with the Patient." John Graunt, "*Natural* and *Political* Observations...upon the Bills of Mortality," *The Economic Writings of Sir William Petty*, C. H. Hull (ed.), Cambridge, Cambridge University Press, 1899, vol. II, p. 349.

[721] Graunt, *London's Dreadful Visitation*, pp. 24-40; Champion, *London's Dreaded Visitation*, p. 29; Twigg, "Plague in London," p. 13.

easie to be known."[722] According to Thomas Sydenham, in the spring of 1665 there appeared two separate types of diseases: inflammatory ailments, and a continual "Epidemic Fever":

> The foregoing Winter being extremely cold, and the Frost continuing without any intermission till Spring, it thaw'd suddenly at the end of *March*, in the year 1665, and *Inflammations of the Lungs, Pleurisies, Quinsies*, and such like inflammatory Diseases, made great slaughter on a sudden, and at the same time a continual Epidemic *Fever* appaear'd. It was very different from the nature of the *Continual Fevers* that reign'd in the foregoing Constitution...The Pain of the Head was more violent, and the Vomiting more severe...the *Looseness*...was now heighten'd.[723]

Sydenham described "Quinsy" as a suffocating throat disease that could kill in a few hours, which especially struck the young in spring and summer, causing fever, chills, swelling of the throat, nose bleed, and pulmonary hemorrhage producing the spitting up of blood.[724] Sydenham writes:

> This disease comes at any time of the year, but especially between spring and summer ; it chiefly attacks the young...It begins 1. with a chil[l]ness and shivering, 2. a fever succeeds, and 3. immediately after a pain and inflammation of the fauces, which, without speedy relief, hinder deglutition, and prevent breathing through the nose, whence suffocation is endangered from the inflammation and tumour of the uvula, tonsillæ, and larynx. This disease is extremely dangerous, and sometimes kills the patient in a few hours...[725]

John Huxham described this disease that occurred in the cold and wet years of 1751 and 1752, in which "Grain of all Kinds suffered greatly."

> The *Attack* of this Disease was very different in different Persons.—Sometimes a Rigor, with some Fullness and Soreness of the Throat, and painful Stiffness of the Neck...Sometimes alternate Chills and Heats, with some Degree of Head-ach, Giddiness, or Drowsiness...It seized others with much more feverish Symptoms, great Pain of the Head, Back, and Limbs, a vast Oppression on the Præcordia, and continual Sighing...it commonly began with Chills and Heats, Load and Pain of the Head, Soreness of Throat and Hoarsness, some Cough, Sickness at Stomach, frequent Vomiting and Purging, in Children especially,

[722] Kemp, *Treatise Of The Pestilence*, p. 29.

[723] Thomas Sydenham, *The Whole Works of that Excellent Practical Physician, Dr. Thomas Sydenham...*, J. Pechey (ed.), Seventh Edition, London, Printed for M. Wellington..., 1717, pp. 56-57.

[724] Thomas Sydenham, *The Works of Thomas Sydenham, M.D.*, Volumes I-II, Dr. Greenhill (trans.), London, Printed for the Sydenham Society, 1848–1850, vol. I, pp. 264-267.

[725] Sydenham, *Acute and Chronic Diseases*, pp. 235-236.

> which were sometimes very severe...The second, or third Day, every Symptom became much more aggravated...Restlessness and Anxiety greatly encreased...there was generally more or less of a Delirium, sometimes...perpetual Phrenzy, tho' others lay very stupid, but often starting and muttering to themselves...[726]

In 1758, James Johnstone reports that "Quinsy" was a disease that occurred predominantly among poor families living in wet, marshy lowlands, and children were the most severely affected by it.[727] Johnstone writes: "A dearth or scarcity of provisions, especially of the alimentary grains ; the frequent effect of intemperately wet seasons ; has generally, in every age, been accompanied with epidemic putrid sicknesses. These prevail most amongst the poorer sort of people ; many of whom, in these times, were almost half-starved."[728] According to Dr. Johnstone, women were more prone to this disease than men, which also caused sweating, fetid breath, fainting, skin eruptions, swellings of the hands, feet, neck, and face, and a swollen tumor on the neck.[729] John Fothergill writes: "it was sometimes observed to carry off whole Families together...Some had a violent Cough ; some were comatous ; others had a Delirium ; some died in a lethargic Stupor ; others bled to Death at the Nose ; whilst others again had none of these Symptoms, but were carried off suddenly by an instantaneous Suffocation...The Consequences of this Disease were often felt a long time after it had ceased."[730] There were 35 deaths attributed to "Quinsie" in the *Bills of Mortality* for the year 1665, clearly a name for *Fusarium* poisoning.[731]

The disease called "Pleurisy" also appears to have been caused by *Fusarium* poisoning. Sydenham writes:

> This disease, which is one of the most frequent, happens at any time, but chiefly between spring and summer...and frequently also attacks country people...It generally begins 1. with a chil[l]ness and shivering, which are followed 2. by heat, thirst, restlessness, and other well-known symptoms of a fever ; 3. in a few hours...the patient is seized with a violent pungent pain in one side, near the ribs, which

[726] John Huxham, *A Dissertation on the Malignant, Ulcerous Sore-Throat*, London, Printed for J. Hinton..., 1757, pp. 4-21.

[727] James Johnstone, *An Historical Dissertation Concerning the Malignant Epidemical Fever of 1756*, London, Printed for W. Johnston..., 1758, pp. 1, 7, 17.

[728] Ibid., p. 60.

[729] J. Johnstone, *A Treatise on the Malignant Angina: or, Putrid and Ulcerous Sore-Throat*, Worcester, Printed and Sold by E. Berrow..., 1779, pp. 27-40, 61-64.

[730] John Fothergill, *An Account of the Sore Throat Attended with Ulcers*, Fourth Edition, London, Printed for C. Davis..., 1754, pp. 11–18.

[731] Mary Kilbourne Matossian, "Mold poisoning: an unrecognized English health problem, 1550–1800," *Medical History*, 25 (1981), pp. 73-84.

> sometimes extends towards the shoulder blades, sometimes to the spine, and sometimes towards the breast ; 4. a frequent cough likewise afflicts the patient, and occasions great pain by shaking and distending the inflamed parts...5. the matter expectorated...[is] mixed and coloured with blood ; 6. in the mean time the fever keeps pace, and even grows more violent with the symptoms arising therefrom ; till at last...both the fever and its dreadful concomitants, as the cough, spitting of blood and pain, &c. abate by degrees...and consequently the fever and its concomitants remit not at all till they prove mortal... the patient cannot so much as cough, but having a great difficulty of breathing, is almost suffocated by the violence of the inflammation... though the original fever either goes off entirely, or at least abates, yet the danger is not over ; for an empyema and an hectic fever succeed, and the patient is destroyed by a consumption.[732]

There were 5,257 deaths in the *Bills of Mortality* attributed to "Ague and Feaver."[733] Nicholas Culpeper writes:

> An Ague, or Intermittent Tertian Feaver...is either Legitimate and Exquisite, or Illegitimate and bastard. A Legitimate or Exquisite Tertian Ague, is terminated in twelve hours...But a bastard Tertian hath fits that last above twelve hours. But if it exceed twenty four hours, it is termed *Tertiana extensa*, a stretched Tertian...Also Tertian Agues are Simple, or Double, or Triple. A Simple Tertian, is that whose Fits come every other day. A Double Tertian is that whose Fits come every day. And although herein it differ not from a Quotidian or every day Ague, yet they are known from the other by their proper Signs... Sometimes notwithstanding in a Double Tertian there are two fits in one day, the other day remaining free ; and this some latter Physitians do call two Tertians, and make it to differ from a double Tertian. Which Distinction notwithstanding is of smal[l] moment. A Triple Tertian is when there are three fits in the compass of two daies. This is a most rare and seldom seen sort of Feavers...The Signs to know an Exquisite Tertian by, are these : That this Feaver alwaies begins with great shaking Fits, whereas in a Quotidian Feaver or Ague, there is

[732] Sydenham, *Acute and Chronic Diseases*, pp. 217-218.

[733] Hunter and Gregory write: "[Samuel] Jeake describes his agues according to the standard terminology of the time, which classified them principally according to the cycles in which attacks returned. The commonest types were tertian and quartan. The former involved attacks of some twelve hours' duration, recurring on alternate days ('simple'), daily ('double'), or with three fits every two days ('triple'). A quartan, on the other hand, involved a visitation every fourth day if simple, on two succeeding days with a third day free if double, or daily if triple: the triple always emerged from the single or double. The rarer quotidian ague involved daily bouts, while Jeake also uses the concept of 'irregular' for agues which failed to fit into any of these categories. Apart from the pattern of attacks, the various agues were distinguished by different symptoms: quartan, for instance, was characterized by melancholy, while the onset of a tertian was marked by great shaking fits, as against a light shivering and coldness in a quotidian." Samuel Jeake, *An Astrological Diary of the Seventeenth Century: Samuel Jeake of Rye 1652–1699*, M. Hunter and A. Gregory (eds.), Oxford, Clarendon Press, 1988, pp. 51-52.

only a light shivering or coldness. After the cold shaking Fit, follows great Heat, sharp and biting, Intollerable Thirst, great and frequent breathing, want of Sleep, Head-ach[e], and som[e]times Ravings. After the shaking fit, sometimes there follows a vomiting...or a purging by Stool. The Urine is sometimes Yellow, Yellowish-Red, or Red. The Fits last not above twelve hours, and they are terminated by Sweat...In a bastard Tertian, all the foregoing Signs are more remiss than they are in a Exquisite one, but more intense than in a Quotidian Ague...the Fits come neerer to those of an Exquisite Tertian, or of a Quotidian ; but in respect of the vehemency of the Symptoms, and the length of the Fit itself. So that the Paroxysms of a bastard Tertian may be lengthened out to sixteen, eighteen, or more hours. Although they may be sometimes shorter...and be terminated within the space of eight, ten, or twelve hours. The Prognostick of this Disease, is taken out of *Hippocrates*, in Sect. 4. Aph. 59. *Exquisite or exact Tertian Agues last but for seven fits at most*. And in Aphor. 43 of the same Section, *All Intermitting Feavers are void of danger*. Which is to be understood only of such Tertians as are void of all malignity. For there are Malignant and Pestilent Tertians... [which] often kill the Patients. Furthermore, many things fall upon the Neck of a Tertian, which may breed danger...*Haly* writes, and common Experience shews, That if such as are sick of a Tertian Ague, have Ulcers, Scabs, or Pustules breaking out in their Lips, it is a token the Ague will leave them...A Quotidian Ague is so called, because its fits do return every day...This Feaver is most rarely seen, so that among six hundred Patients that have Agues which come every day, scarce one of them is troubled with a Quotidian or every day Ague...This Feaver is wont to be perpetually long...it may degenerate into a Cachexy, Dropsie, Lethargy, and other grievous Diseases...*Hippocrates* himself saith, do very often bring the Patient into Consumption...A Quartan Ague is that which hath its Fits returning every fourth day... Again, A Quartan Ague is either Single, Double, or Triple. A Single Quartan is when one Fit alone comes every fourth day. A Double is when two Fits happen upon two daies one immediately after the other, and the third day is free. A Triple Quartan is when the Fits come every day, as they do in a Quotidian, and in a Double Tertian...a Triple Quartan is distinguished from a Double Tertian, and a Quotidian...by the form of the Fits ; but also because it was first a Single, or a Double Quartan before it came to be a Triple Quartan. For very rarely, or never doth a Quartan Ague begin with a Triple ; but a Simple or Double Quartan degenerates into a Triple. As for what concerns the Prognostick ; this kind of Ague is wont to be longest of all others ; and that which begins in the fal[l] of the leaf continues al[l] Winter commonly, and goeth not away til[l] the Spring come. Yea and some Quartans continue a yeer or yeers. Summer Quartans are the shortest...and with these, for the patient to make black urine is a good token...This kind of Ague is wont to be very safe from danger, especially the legitimate, being accompanied with no grievous affection of any of the bowels. But the bastard Quartan is more dangerous, and if the Liver, Spleen, or any other part be grievously damnfied, it degenerates into a Dropsie. Aged persons above sixty years, being taken with a Quartan Ague, do for the most part dy [die] of it...An intermitting Quartan, being changed into a continual, is for the most part deadly...A Quartan Ague coming upon one that hath the falling sickness cures the same, according to *Hippocrates* in the 70 Aphorism of the 5 Section. *Those that have Quartan Agues are not much troubled with Convulsions. And if having first Convulsions, a Quartan Ague follows, they are freed from their Convulsions*...To bleed at the nose in a Quartan Ague, is a very bad sign...such bleeding is symptomatical, and if it continue wil[l] breed a Dropsie ; it must presently be stopped by opening the basilica vein, out of which the putrid blood may flow, because the pure blood comes from the Nose.[734]

"Agues" and "Fevers" were common in the marshlands of southeastern England,[735] and the inhabitants of these regions suffered enormously during the Plague. According to Sydenham, "Agues" occurred at harvest time in the spring and fall.[736] In 1772, William Grant attributed "Agues" and "Fevers" to "unwholesome food from the spoiled grain."[737] "Ague" was a name directly applied to convulsive ergot poisoning.[738] George Barger identifies the "Malignant Fever with the Cramp" as convulsive ergotism in an English translation of Daniel Sennert's *De Febribus*, published in 1658 under the title *Of Agues and Fevers*:

It seized upon men with a twitching and kind of benummedness in the hands and feet, sometimes on one side, sometimes on the other, and sometimes on both : Hence a Convulsion invaded men on a sudden when they were about their daylie employments, and first the fingers and toes were troubled, which Convulsion afterwards came to the arms, knees, shoulders, hips, and indeed the whole body, until the sick would lie down, and roul up their bodies round like a Ball, or else stretch out themselves straight at length : Terrible pains accompanied this evil, and great clamours and scrietchings did the sick make ; some vomited when it first took them. This disease sometimes continued some days or weeks in the limbs, before it seized the head, although

[734] Nicholas Culpeper, *The Practice of Physick*..., L. Riverius (trans.), London, Printed by *Peter Cole* in *Leaden-Hall*..., 1655, pp. 580-587.
[735] Dobson, *Contours of Death*, pp. 294-306.
[736] Sydenham, *Whole Works*, p. 38.
[737] William Grant, *Observations on the Nature and Cure of Fevers*, Second Edition, London, Printed for T. Cadell..., 1772, vol. I, pp. 9, 23; Creighton, *Epidemics in Britain*, vol. I, p. 505.
[738] Matossian, *Poisons of the Past*, p. 13.

fitting medicines were administered ; which if they were neglected, the head was then presently troubled, and some had Epilepsies, after which fits some lay as it were dead six or eight hours, others were troubled with drowsiness, others with giddiness, which continued till the fourth day, and beyond with some, which either blindness or deafness ensued, or the Palsie : When the fit left them, men were exceeding hungry contrary to nature ; afterwards for the most part a looseness followed, and in the most, the hands and feet swell'd or broke out with swellings full of waterish humours, but sweat never ensued. This disease was infectious, and the infection would continue in the body being taken once, six, seven, or twelve moneths. This disease had its original from pestilential thin humours first invading the brain and all the nerves ; but those malignant humours proceeded from bad diet when there was scarcity of provisions. This disease was grievous, dangerous, and hard to be cured, for such as were stricken with an Epilepsie, were scarce totally cured at all, but at intervals would have some fits, and such as were troubled with deliriums, became stupid. Others every yeer in the month of December and January, would be troubled with it.[739]

Barger adds the following:

This description by Sennertus...applies to severe cases only. Later authors...have distinguished a milder form, which might pass off in a few weeks without preventing the patients from following their ordinary occupations. It was characterized by a feeling of fatigue, heaviness in the head and limbs, giddiness, pressure and pain in the chest. This was sometimes accompanied by mild diarrhœa, with or without vomiting, and lasting for several weeks. A very common, early symptom, often persisting throughout the disease, was "the kind of benummedness" in the hands and feet...It gave the most common German name to the disease : Kriebelkrankheit.[740]

Barger also writes:

The next epidemic, in Hessen and Westphalia, during 1596 and 1597, was one of the most important in the history of the subject, for it led the Marburg medical faculty to publish their famous description and warning in the vernacular, for the benefit of the people [Marburg, 1597]. "Of an unusual, poisonous, infectious disease, hitherto unknown in these parts, which disease is called by the common people here in Hessen the tingling disease, the cramp or the spastic evil"...The exact cause of the disease remained as yet unknown and the Marburg

[739] Daniel Sennert, *Of Agues and Fevers. Their Differences, Signes, and Cures. Divided into four Books: Made English by* N. D. B. M. *late of* Trinity *Colledge in* Cambridge, London, Printed by *J. M.* for *Lodowick Lloyd*, at the Castle in Cornhil, 1658, pp. 114–115; Barger, *Ergot and Ergotism*, pp. 32-33.

[740] Barger, *Ergot and Ergotism*, p. 33.

> faculty merely attributed it to bad food in general. For a long time there were no further precise descriptions of convulsive ergotism. The disease was often regarded as a variety of scurvy ("Schorbock" or "Scharbock")...Drawitz [1647]...was the first to use the name Affectus scorbutico-spasmodicus or scharbockische Kriebelkrankheit ; he considered the disease due to bad food in times of scarcity. He still regarded it as infectious. The sufferers often seemed to be bewitched, or possessed by demons ; their cries could be heard four or five houses off...Sennertus, in his book *De Febribus*, speaks of Kriebelkrankheit as... "malignant fever with the cramp"...[741]

"Scurvy" was a name that was commonly applied to convulsive ergotism.[742] According to Sydenham: "A quartan [fever] now and then changes its face, and likewise produces abundance of morbid symptoms, as the scurvy, hard belly, a dropsy, &c."[743] Sydenham writes:

> [W]hen the patient....is severely afflicted during life with flying pains...they are commonly reckoned symptoms of the scurvy...most of those disorders we term scorbutic, are the effect of approaching ills, not yet formed into diseases...Neither are we ignorant that as many symptoms, resembling the scurvy, afflict gouty persons after the fit of the gout is over...this is to be understood not only of the gout, but also of a beginning dropsy...it is proverbially said, that where the scurvy ends the dropsy begins...And the same may be maintained of several other chronic diseases...And in reality...the name of scurvy (as it does at this day) will obtain universally, and comprize most diseases... there is another species of the rheumatism, which is near a-kin to the scurvy ; for it resembles it in its capital symptoms...and therefore I call it a scorbutic rheumatism. The pain sometimes affects one, and sometimes another part, but it does rarely occasion a swelling, as in the other species...sometimes it only attacks the internal parts, and causes sickness, which goes off again upon the return of the pain of the external parts...It chiefly attacks the female sex, and men of weak constitutions...[744]

There were 105 deaths attributed to "Scurvy" in the *Bills of Mortality*. Hodges writes:

> It being then granted, that this Plague first was brought from *Africa*, or *Asia*, to *Holland*, and from thence into *Britain*...producing different Degrees of Infection, and Series of Symptoms : But this

[741] Ibid., pp. 68-69.
[742] Ibid., pp. 31, 59, 65, 69; Bove, *Story of Ergot*, pp. 27, 154.
[743] Sydenham, *Acute and Chronic Diseases*, p. 77.
[744] Ibid., pp. 230-231.

> Variation would be more discernable in...the particular Diseases of each Country, and those which are as it were peculiar to them...in *Holland*, where the Scurvy extreamly reigns, and therefore, for Reasons before given, most liable to a pestilential Infection, it obtained only as a more aggravated *Scurvy*...[745]

Hodges compared the symptoms of the Plague with "Scurvy" and writes: "But the Affinity between a Pestilence and a *Scurvy* is not a slight, and a supposititious Conjecture, but strengthened and confirmed by a plain Union between them."[746] A rival of Hodges and George Starkey's friend, George Thomson concurred and thought that the symptoms of the Plague were identical to those of "Scurvy," and Thomson also thought that both diseases were contagious:

> It is undoubted that they are both Malignant, Poisonous, and Contagious, possessing and affecting the same parts alike...that is, the Stomack and Spleen...[causing] great oppression about those parts, suspirious respiration, Cardialgie, Nauseousness, Vomiting, extream pain in the Head, Vertigo, Swimming, Agrypnie or want of Sleep, Lethargical dullness, profound sleep, Lypothimy, Palpitation of the Heart, Fainting, Sudden debility of the parts, unreasonable fears, terrors, despair, and confusion, Dyspnæa, irregular and difficult breathing, a violent flux of the belly, various Tumours, Erysipela's, virulent Ulcers, Sugillations, Black and Blew stroaks, Spots red, black, and blew ; all which...are so co-incident with the present Pest and Scurvie, that they seem to differ but only *secundum magis & minus*."[747]

Barger writes: "In his *Medical Observations* [1581] Dodonæus attributes scurvy to bad food, particularly to bad rye, such as that imported in 1556 from Prussia into Brabant, 'when not a few began to suffer from scurvy ; in most the effects of the evil only showed themselves in the gums.'"[748] Sydenham's description of "Scurvy" clearly describes ergot poisoning: "It is accompanied with Heaviness of Body, Weariness...difficult Breathing...Rottenness or filthiness of the Gums and Teeth ; stinking Breath ; often bleeding at the Nose ; Difficulty and uneasiness in Walking. To these Signs add...spots on the Legs, which are black and blue, yellow, or of a lead or violet colour, the Legs

[745] Hodges, *Loimologia*, p. 64.
[746] Ibid., pp. 80-83.
[747] George Thomson, *Loimologia. Consolatory Advice, And some brief Observations Concerning the Present Pest*, London, Printed for *L. Chapman*, at his Shop in Exchange-ally, 1665, p. 6.
[748] Barger, *Ergot and Ergotism*, p. 65.

in the mean while being sometimes swol[le]n, and sometimes sunk, and the Face a colour between pale and tawny."[749]

Willis associated "Scurvy" with marshy areas and attributed the disease to an ill diet, characterized by fever, nausea, headache, sore gums, loose teeth, fetid breath, excessive salivation, difficulty breathing, coughing, a constriction and straightness of the breast accompanied by pain, gastrointestinal disturbances, continual diarrhea, lassitude, drowsiness, fainting, swooning, violent pains, pain in the loins, discolored urine, spots and ulcers on the skin, formication, swollen belly, giddiness, watching, twitching, convulsions, and gangrene.[750] "Dropsy," according to Willis, "is wont to happen upon the Scurvy."[751] There were 1,478 deaths in the *Bills of Mortality* attributed to "Dropsie and Timpany," the latter a form of "Dropsy."[752] Sydenham writes:

> [T]hose Symptoms which accompany *Agues* at their Declination, we must take notice, that very few, compared with the *Autumnal*, belong to *Agues* in the *Spring*...A *Dropsy* now and then concurring, is the chiefest, wherein the Legs swell first, and then the Belly also...this Disease seldom happens to young Men, unless it has been foolishly brought upon them by Purges repeated frequently in the course of the *Agues*...I have observ'd it is in vain to endeavour the Cure of such a *Dropsy* by purging Medicines, while the *Ague* continues...We must therefore wait till the *Ague* is cur'd...Infants are sometimes hectick after *Autumnal Fevers*, both Continual and Intermitting ; their Bellies are puffed up, swell'd and hard, and they have often a Cough and other Symptoms that Consumptive People are troubled with, and which plainly resemble the *Rickets*...I have cured a great many Children of the true *Rickets*...But, as I said before, great care must be taken that we do not begin to purge till the *Ague* is quite gone...It is worth noting, That when these *Autumnal Agues* have a long time molested children, there is no hope of recovery till the Region of the Belly, especially about the *Spleen*, begins to be harden'd and to swell, for the *Ague* goes gradually off as this Symptom comes on ; nor perhaps can you any other way better prognosticate the going off of the Disease in a short time, than by observing this Symptom, and the swelling of the Legs, which are sometimes seen in grown People. The swelling of the Belly, which comes upon Children after these *Agues*, in those Years the Constitution of the Air is Epidemically determin'd to propagate *Autumnal Agues*...

---

[749] Thomas Sydenham, *Dr. Sydenham's Practice of Physick. The Signs, Symptoms, Causes and Cures of Diseases...*, W. Salmon (trans.), London, Printed for *Sam. Smith*..., 1695, p. 165.

[750] Willis, *Practice Of Physick*, pp. 326-327.

[751] Ibid., p. 366.

[752] Sydenham, *Practice of Physick*, p. 100.

> which is worth noting : the true *Rickets* do not often happen, unless in those Years, wherein *Autumnal Agues* prevail.[753]

John Graunt wrote in 1662:

> My next Observation is, That of the *Rickets* we find no mention among the *Casualties*, until the Year 1634, and then but of 14 for that whole Year. Now the Question is, Whether that Disease did first appear about that time ; or whether a Disease, which had been long before, did then first receive its Name?...For in some years I find *Livergrown*, *Spleen*, and *Rickets*, put all together, by reason (as I conceive) of their likeness to each other...It is also to be observed, That the *Rickets* were never more numerous than now, and that they are still increasing ; for *Anno* 1649, there were but 190, next year 260, next after that 329, and so forwards, with some little starting backwards in some years, until the Year 1660, which produced the greatest of all.[754]

The *Bills of Mortality* attributed 557 deaths to "Rickets" in 1665. There were 4,808 deaths attributed to "Consumption and Tissick," in the *Bills of Mortality*, "Consumption" being the wasting away of the body. "Consumption" was also associated with fungal poisoning.[755] Sydenham included hair-loss as a symptom of "Consumption,"[756] and writes of this malady:

> There are many kinds of this Disease : The first of which arises from Cold taken in Winter time ; for a little before the Winter Solstice, a cold Season violently coming on, many are taken with a Cough...a mighty quantity of crude Flegm is heaped up...From thenceforth the Lungs being fill'd with *Pus* or Matter...And from hence comes a putrid Fever, whose Fits are excited and trouble the Sick about Evening, but go off the next morning with a plentiful Sweating, and diminishing the strength of the Sick. And to this heap of symptoms is added a *Diarrhœa*, or Flux of the Belly...And so in the approaching Summer the Sick at length submits to its fate, and even by that Disease to which the preceding Winter led the way, meets his Death.[757]

There were another 134 deaths attributed to "Collick and Winde" in the *Bills of Mortality*. "Hysteric fits" and "Colic" were diseases that Sydenham associated and called "Hysterical Colic," and these were all connected with convulsive ergot poisoning.[758] Willis identified

[753] Sydenham, *Whole Works*, pp. 51–53.
[754] Graunt, "Bills of Mortality," *Writings of Petty*, vol. II, pp. 357-358.
[755] Matossian, *Poisons of the Past*, p. 99.
[756] Thomas Sydenham, *The Entire Works of Dr. Thomas Sydenham*, J. Swan (trans.), Fifth Edition, London, Printed for F. Newbery..., 1769, p. 657.
[757] Sydenham, *Practice of Physick*, p. 169.
[758] Sydenham, *Works*, vol. I, pp. 189–199, vol. II, pp. 270-271; Matossian, *Poisons of the Past*, p. 65.

"Hysterick" as a convulsive disorder associated with women, characterized by gastrointestinal disturbances, suffocation of the throat, giddiness, laughing, weeping, idle talk, speechlessness, catalepsy, and convulsions.[759] Sydenham writes:

> Sometimes it attacks the head, and causes an apoplexy, which also terminates in a hemiplegia ; exactly resembling that kind of apoplexy, which proves fatal...Sometimes it causes terrible convulsions, much like the epilepsy...she talks wildly and unintelligibly, and beats her breast. This species of the disease...is commonly entitled the strangulation of the womb, or fits of the mother...Sometimes this disease attacks the external part of the head...and occasions violent pain...and it is likewise accompanied with very violent vomiting...Sometimes it seizes the vital parts, and causes so violent a palpitation of the heart...Sometimes it affects the lungs, causing an almost perpetual dry cough...and the patient's senses are also disordered. But this species of the hysteric cough is very rare...Sometimes attacking the parts beneath the scrobiculus cordis in a violent manner, it occasions extreme pain, like the iliac passion, and is attended with a copious vomiting...And frequently after the pain and vomiting have continued several days, and greatly debilitated the patient, the fit is at length terminated by an universal jaundice. Moreover, the patient is so highly terrified, as to despair of recovering...Sometimes this disease s[e]izes one of the kidneys, where by the violent pain it occasions, it entirely resembles a fit of the stone...so that it is hard to distinguish whether the symptoms arise from the stone or any hysteric disorder...The bladder also is occasionally affected...which not only causes pain, but a suppression of urine, as if there was a stone....Sometimes seizing the stomach, it causes continual vomiting....it sometimes attacks the external parts, and muscular flesh, sometimes causing pain, and sometimes a tumour in the FAUCES, shoulders, hands, thighs, and legs, in which kind the swelling which distends the legs is most remarkable...The teeth also... are subject to this disease....But the most frequent of all the tormenting symptoms of this disease is a pain of the back...the part they affect cannot bear the touch after they are gone off, but remain tender and painful, as if it had been severely beaten....all these symptoms are preceded by a remarkable coldness of the external parts, which seldom goes off before the fit ceases. And I have observed, that this coldness resembles that which is perceived in dead bodies...Whereto may be added, that most of the hysteric women that I have hitherto treated, have complained of a lowness, and (to use their expression) a sinking of the spirits...it is generally known that hysteric women sometimes laugh, and sometimes cry excessively, without the least apparent provocation...But their unhappiness does not only proceed from a

[759] Willis, *Practice Of Physick*, p. 297.

> great indisposition of body, for the mind is still more disordered ; it being the nature of the disease to be attended with an incurable despair ; so that they cannot bear with patience to be told that there are any hopes at all of their recovery, easily imagine that they are liable to all miseries that can befall mankind, and presaging the worst evils to themselves. Upon the least occasion also they indulge terror, anger, jealousy, distrust, and other hateful passions ; and abhor joy, and hope, and cheerfulnes[s]...[760]

There were 2,036 deaths attributed to "Convulsion and Mother" in the *Bills of Mortality*. "Rising of the Lights" was also "Mother,"[761] to which the *Bills of Mortality* attributed 397 deaths. In 1603, Edward Jorden composed an essay after recognizing that the women in England who were accused of witchcraft exhibited the symptoms of the disease named "Suffocation of the Mother," also called "Hysterica," which Matossian has determined to be caused by ergot poisoning.[762] Edward Jordan's description of the disease included the symptoms of formication, fever, body temperature disturbances, stomach pains, gastrointestinal disturbances, diarrhea, difficulty breathing, swelling of the throat, feeling of choking or suffocation, swelling of the body and feet, tumors, consumption, sighing, hiccough, sneezing, coughing, temporary loss of sight, hearing, taste, speech, and feeling, appetite loss or ravenous hunger, catalepsy, paralysis so that some were almost buried alive, coma, fainting, compulsive laughing, weeping, howling, tremors, convulsions, chorea (St. Vitus's dance), hallucinations, and frenzy.[763]

In 1584 Reginald Scot made the observation that witchcraft accusations in England were linked with the occurrence of "Apoplexies, Epilepsie, Convulsions, hot Fevers, Worms, &c." in children.[764] Barger writes: "a belief in witchcraft was still prevalent and many believed the sufferers from convulsive ergotism to be possessed by demons... Albrecht [1743] remarks that 'through ignorance of natural causes,' the common people were apt to 'ascribe the symptoms of this peculiar disease to the action of spells.'"[765] Included among Walter George

[760] Sydenham, *Acute and Chronic Diseases*, pp. 313-315.

[761] Graunt, "Bills of Mortality," *Writings of Petty*, vol. II, p. 352.

[762] Matossian, *Poisons of the Past*, pp. 64-65.

[763] Edward Jordan, *A Briefe Discourse of a Disease Called the Suffocation of the Mother*, London, Printed by *John Windet*..., 1603, pp. 5–18.

[764] Reginald Scot, *The Discovery of Witchcraft*, Third Edition, London, Printed for *Andrew Clark*..., 1665, p. 5.

[765] Barger, *Ergot and Ergotism*, p. 71; Leigh, "Account of strange Epileptic Fits," p. 1175. Willis, *Practice Of Physick*, p. 160.

Bell's collection of ominous signs foreshadowing the Great Plague, in March of that year three hapless women accused of witchcraft stood trial in Suffolk in East Anglia, and were convicted and hanged.[766]

Willis writes: "It happens that Infants and Children are so generally, and frequently troubled with Convulsive affects...for those kinds of Symptoms in Adult persons are denoted by other Names, and are wont to be refer'd to the Epilepsy ; Hysterick, Hypochondriak, or Colick passions, or also to the Scurvy ; but in Children...they are call'd Convulsive motions."[767] Culpeper writes:

> There is in *Galen*, and almost all Authors, a threefold Epilepsy... The first, as being chief, is called *Epilepsia* ; the second *Analepsia* ; the third, *Catalepsia* : But (by *Galens* leave) that division is superfluous... all ought to be called *Sympathicæ*, or Epilepsies by content...*Fernelius* saith, that because an Epilepsy is quickly dissolved, it should turn into a Palsey, as an Apoplexy doth...And it is false which *Fernelius* saith, That an Epilepsy never ends in a Palsey ; for we have seen a Palsey come after it. And sometimes Apoplexies at the first coming are turned into Convulsions...and also many Epileptick men die by an Apoplexy... Therefore al[l] signs of an eminent Epilepsy are to be propounded with this admonition, That al[l] signs do not meet in al[l] ; but some in one, some in another...unaccustomed disturbance of the mind and Body, threatneth an Epilepsy, heaviness of head, head-ach[e], vertigo or giddiness, or much sleep...troublesome[e] dreams, dullness of mind, or perplexity, forgetfulness, sorrow, fear, dread, sloth, graveness of actions, snatching and trembling of the parts, dul[l]ness of the senses, a down look, clouds and other things flying before the eyes, noise in the ears, a stink in the nostrils, a stiff tongue, and its inordinate Motion, yawning and [s]neezing...Anger, Beating or palpitation of the heart, strai[gh]tness of Breast, and alteration in Breathing...disdain of meat, or immoderate Appetite, Squeamishness, heart-burning...Much spittle, thin and crude Urine, often Nocturnal Pollutions...Paleness of Face, and swelling of the heart...a staring and thrusting forth of the eyes, gnashing of Teeth, a difficult breathing, as in those that are hanged ; the feed, dung, and urine are sent forth involuntarily, and about the end of the fit, he foameth at the Mouth and Nose, which happen only in a vehement Epilepsy, and the fit being ended, he forgets all things he then acted. Some of the Ancients make three kinds of Epilepsies : One which is like a deep sleep ; another which doth shake the body after divers motions ; a third which is made of both the former. The late Physitians deny the first kind, saying, That it is more like a Coma, or a Carus than an Epilepsy ; and these two Diseases cannot be

[766] Bell, *Plague in London*, p. 2.

[767] Willis, *Practice Of Physick*, p. 250.

> otherwise distinguished, but that in a Coma is a deep sleep without a Convulsion, and a Convulsion is a certain sign of an Epilepsie. But *Avicen* saith otherwise, namely, That an Epilepsy comes many times without an apparent Convulsion. And experience teacheth us, That many men in Epilepsies have fits like Coma : and it's known to be an Epilepsie, not a Coma, or a Carus by this ; The sleep in an Epilepsie cometh and goeth by fits, when in a Coma it comes all at once...the Cause of an Epilepsy is in the stomach : Disdain of meat, an inability to fast, loathing, vomiting, pain of the stomach, gnawing, pricking, and distention...often belching and breaking of wind, a swelling of the belly with rumbling and noise, sowr [sour] belchings, strai[gh]tness of the Midrif[f], and pain sometimes reaching to the back...An Hysterick fit, or the Mother, mixed with Convulsions...shews that it comes from the Womb...Lastly, The Signs of worms, shew that the disease come from them, as stinking sowr Breath, itching of the Nose, pain of the Belly, earthly Excrements, grating of the teeth, sleepiness, and the like, especially if sometimes worms are voided. But the extraordinary Causes, as Imposthumation, foulness of a Bone, stopping of urine, and the like, may be taken from their proper signs. As to the Prognostick, An Epilepsy is a Disease of long continuance, and very stubborn and deadly in Infants.[768]

According to Willis, those who were of "a Hypochondriack, Colick, Hysterick, and sometimes Asthmatick disposition" were predisposed to "Timpany," a hard "tumor" in the abdomen.[769] Sydenham says that the disease of women called "Hysterical" was in men named "Hypochondrial."

> [I]n the Head the Apoplexy, which ends in a Palsy of one half of the Body, comes presently after Child-bearing ; sometimes they are seized with Convulsions, that very much resemble the Epilepsy, and are commonly called the Suffocation of the Womb...The Paroxysm doth also counterfeit the Palpitation of the Heart, the Cough, the Cholick, the Illiack Passions, the Stone, and Suppression of Urine ; it is attended with prodigious Vomitings, and sometimes with a Diarrhæa ; outwardly in the musculous Flesh it causes sometimes Pains, and sometimes Swellings ; in the Legs it is like a Dropsy, nor...doth it leave the Teeth untouch'd ; the Back is often cruelly pained, and almost always the external Parts are so cold, that a dead body is not more ; the sick Persons break out ridiculously into excessive Laughter and Tears without any Cause, and are sometimes troubled with spitting to such a degree, as were enough to make one believe they had been anointed with Mercury. Hysterical

[768] Culpeper, *Practice of Physick*, pp. 29-30.
[769] Willis, *Practice Of Physick*, p. 160.

> Pains, whatsoever Part they affect, leave a Tenderness behind them that cannot endure to be touch'd, as if the Flesh had been beaten.[770]

"Palsy" was associated with convulsive ergot poisoning,[771] and there were 30 deaths attributed to "Palsie" in the *Bills of Mortality*. Culpeper writes:

> In the Disease called *Coma*...is a deep sleep ; but such an one as from which the Patient is raised, openeth his eyes, and answereth, but presently he is again in a deep sleep. In a *Lethargie*, the sleep is like that of *Coma*, but is joyned with a Feaver and Frenzy, or Dotage [Delirium]... In *Apoplexy* the sleep is most deep, and a total privation of sense and motion, except breathing : and so therefore the sick doth neither open his eyes, answer, nor feel when he is hurt ; as also he breatheth very difficultly...Men in Apoplexies die in seven daies, except a Feaver take them...but that Feaver must be a violent one...for if it be gentle, and only symptomatical...then the Feaver doth not diminish the Disease, but rather cause some symptomes of madness...The Disease called *Waking Coma*, or *Coma vigil*...is a disease in which the Patient lieth with his eyes shut as if he were asleep, when he is awake and distracted ; and if you touch him, he presently openeth his eyes, and looks strangely, and falls asleep again, which is hindered by divers strange imaginations and fancies. This Disease *Galen* placeth as a mean, between a Frenzy and a Lethargy, and calleth it *Typhomania*. The usual Cause of this Disease, is Choller mixed with Flegm, by which humors the Brain is made too moist, or it is swelled or inflamed ; from whence, either the Tumor called *Erisypelas œdematosum*, or *œdema Erisypelatosum*...they which have it, lie with their eyes shut, and seem to sleep, yet they cannot sleep, but toss and tumble, lift themselves up suddenly, strive to get out of the bed, and then fall again asleep...If the Dotage, or Delirium be strong, it produceth a Convulsion...A true *Coma vigil* is cured as a Frenzy and Lethargy ; and if it incline most to a Frenzy, then the Medicines proper for that are most to be used ; if to a Lethargy, then the Medicines proper for that. But a Coma that cometh by Sympathy, is cured by curing the Malignant Feaver from whence it cometh...the Palsey...often happen in Bastard Tertian Feavers, which are probable to be those which *Fernelius* saith turn into Palseys. Finally, This is very manifest from the Scorbutick Palsey, or that which is joyned with the Scurvy, which hath often a Convulsion accompanying it...[772]

There were 1,251 deaths in the *Bills of Mortality* attributed to "Surfet," or violent retching, and another 51 deaths attributed to

[770] Sydenham, *Entire Works*, pp. 6-7.
[771] Barger, *Ergot and Ergotism*, p. 32.
[772] Culpeper, *Practice of Physick*, pp. 9–19.

"Vomiting." There were also 1,288 deaths in the *Bills of Mortality* attributed to "Griping in the Guts." There were 625 deaths recorded as "Childbed" in the *Bills of Mortality*, considered a "Fever" after childbirth. Ergot poisoning can cause women to die in childbed.[773] The Plague also caused miscarriages,[774] and according to the *Bills of Mortality* there were 617 "Abortive and Stilborne" deaths in 1665. Ergot poisoning causes abortions and stillborn deaths.[775] Infants are prone to fatal convulsions if fed on pap made from ergot-infected bread, and can die after a single meal.[776] There were 1,258 deaths attributed to "Chrisomes and Infants" in the *Bills of Mortality*, and another 2,614 deaths attributed to "Teeth and Worms." "Teeth" was the death of a teething infant (i.e., 6 months to 3 years), and "Worms" or "Worm fitts" were convulsions associated with teething, fever, worms, and diarrhea.[777] Richard Kephale included "worms" as a common symptom of the Plague,[778] although Boghurst reports that death by worms was rare in the Plague in London, and he had only heard of one case.[779] Large numbers of round worms (*Ascaris*) have frequently been found in victims of convulsive ergot poisoning.[780]

---

[773] Bove, *Story of Ergot*, p. 160; Devitt *et al.*, "Ergot Poisoning," p. 441; Louis S. Goodman and Alfred Gilman (eds.), *The Pharmacological Basis of Therapeutics: A Textbook of Pharmacology, Toxicology, and Therapeutics for Physicians and Medical Students*, Fourth Edition, London, Macmillan Company, 1970, p. 898.

[774] Boghurst, *Loimographia*, p. 25.

[775] Barger, *Ergot and Ergotism*, p. 18; Goodman and Gilman, *Pharmacological Basis of Therapeutics*, p. 898; Matossian, *Poisons of the Past*, p. 9.

[776] William Moss, *An Essay on the Management, Nursing and Diseases of Children...*, Second Edition, London, Printed by and for C. Boult..., 1794, pp. 79-80.

[777] Graunt, "Bills of Mortality," *Writings of Petty*, vol. II, p. 348.

[778] Richard Kephale, *Medela Pestilentiæ: Wherein is contained several Theological Queries Concerning the Plague, With Approved Antidotes, Signes, and Symptoms...*, London, Printed by J. C. for *Samuel Speed...*, 1665, p. 81.

[779] Boghurst, *Loimographia*, p. 98.

[780] Barger, *Ergot and Ergotism*, pp. 23-26, 35.

Chapter 16. The Medical Evidence: Diseases Following the Plague

Symptoms of the Plague included delirium, convulsions, erratic and uncontrollable behavior, and loud shrieks.[781] Hodges writes: "All the Sick likewise quickly after Seizure grew delirious, running wildly about the Streets, if they were not confined by Force ; when some tired with Rambling, on Increase of their Distemper, would fall down, ignorant of their Condition...Many were seized with a *Vertigo*, which, without any Motion of external Objects, made them believe their Heads to turn round."[782] Hallucinations were one of the primary symptoms associated with the Plague. The poet George Withers remained in London during the Plagues of 1625 and 1665, and gave dozens of examples of the delirium in his *Britain's Remembrancer*, including a man who saw Death lurking in his room,

> [N]ow by the bed,
> He stands, now at the foot, now at the Head...
> He acted with a look so tragical
> That all bystanders might have thought his eyes
> Saw real objects, and no fantasies.[783]

According to Boghurst, fretfulness, fuming, and snarling were signs of impending death.[784] Vincent writes:

[781] Bell, *Plague in London*, p. 126.
[782] Hodges, *Loimologia*, p. 95.
[783] George Withers, *Britain's Remembrancer* (1628); in Nicholson, *Sources of Defoe's Journal*, p. 17.
[784] Boghurst, *Loimographia*, p. 98.

> Another, was of a man at the corner of the *Artillery-wall*, that as I judge through the diz[z]iness of his head with the disease, which seized upon him there, had dash't his face against the wall, and when I came by, he lay hanging with his bloody face over the rails, and bleeding upon the ground ; and as I came back he was removed under a Tree in *More-fields*, and lay upon his back, I went and spake to him ; he could make me no answer, but rat[t]led in the throat, and as I was informed, within half an hour dyed in the place. It would be endless to speak what we have seen and heard of some of their frensie, rising out of their beds, and leaping about their rooms ; others crying and roaring at their windows ; some coming forth almost naked, and running into the streets ; strange things have others spoken and done when the disease was upon them : but it was very sad to hear of one, who being sick alone, and it is like[ly] frantick, burnt himself in his bed.[785]

An anonymous author writes:

> For an Eighth Argument [against shutting up] I alledge the mischief and sad consequence that may arise from the high fits of Frenzy, that usually attend this and all other the like Distempers ; wherein the sick (if not restrained by main force of their Attendants) are ready to commit any violence, either upon themselves or other, whether Wife, Mother, or Child. A sad instance whereof we had this last week in *Fleet* Lane, where the Man of the House being sick, and having a great Swelling, but not without hope of being almost ripe for breaking, did in a strong fit rise out of his bed, in spite of all that his Wife (who attended him) could do to the contrary, got his Knife, and therewith most miserably cut his Wife, and had killed her, had she not wrapped up the sheet about her...[786]

A widespread epidemic of ergot poisoning occurred in Sweden, Russia, and elsewhere on the European continent in 1746 and 1747, in which "pain was most violent ; so that the fire in the limbs drove the victims hither and thither—some in their agony hurling themselves against walls or even into the water. Those grievously attacked generally died; those who survived became blind, dumb, or demented."[787] Similar hallucinations and suicidal behavior were also evident in the outbreak of ergot poisoning at Pont St. Esprit in 1951:

> Towards evening visual hallucinations appeared, recalling those of alcoholism. The particular themes were visions of animals and of flames.

[785] Vincent, *Gods Terrible Voice*, pp. 9–10.

[786] *The Shutting up Infected Houses as it is practised in England Soberly Debated. By way of Address from the poor souls that are Visited, to their Brethren that are Free*, London, s.n., 1665, p. 19.

[787] Clifford Allbutt and Humphry Davy Rolleston (eds.), *A System of Medicine*, London, Macmillan and Co., Limited, 1908, vol. II, pt. I, p. 887.

> All these visions were fleeting and variable. In many of the patients they were followed by a dreamy delirium. The delirium seemed to be systematized, with animal hallucinations and self-accusation, and it was sometimes mystical or macabre. In some cases terrifying visions were followed by fugues, and two patients even threw themselves out of windows. The delirium was of a confusional kind...Every attempt at restraint increased the agitation...The duration of these periods of delirium was very varied. They lasted several hours in some patients, in others they still persist.[788]

The Plague could recur in its victims, and those who survived the first attack would often perish from the second.[789] This is at odds with bubonic plague in which second attacks are not very common, and a considerable degree of resistance can be conferred from the first attack.[790] Allin wrote on 16 November: "God seemes now in divers familys to visit the 2nd time, after they have beene all well 6 or 8 weekes ; and fresh houses in divers places, besides some whole familys, swept away that have returned to ye City allready."[791] For reasons that are not understood, those who have been afflicted with ergot poisoning are more vulnerable to the toxins if they are again exposed.[792] Boghurst writes: "The evill Reliques and consequent diseases which the plague left in people's bodyes tormented many worse than the disease itself."[793] According to Boghurst, diarrhea, vomiting, and "gripping of the Guts" seized many people in the following summer.[794] Boghurst also noted that inflammatory ailments followed the epidemic:[795] "After the Plague, in

[788] Gabbai *et al.*, "Ergot Poisoning," p. 651.

[789] Hodges, "Letter from Hodges," pp. 20-21.

[790] Morris, "Plague in Britain," p. 213.

[791] Allin to Fryth, 16 November 1665; in Cooper, "Letters of Allin," p. 18.

[792] Matossian, *Poisons of the Past*, p. 12.

[793] Boghurst, *Loimographia*, p. 27.

[794] Ibid., p. 98.

[795] Samuel Jeake of Rye, the son of Samuel Jeake of the Rye circle that included friends of George Starkey, suffered from smallpox in 1666, and attacks of repeated "fits" from "Agues" that lasted until his death in 1699. Hunter and Gregory write: "It is worth noting that none of Jeake's siblings survived infancy...in 1666, he had an attack of smallpox...the colds started in his childhood—indeed his smallpox developed from a cold...while agues began in 1667 and recurred frequently, with particularly severe attacks in 1670–1 and 1678. Colds were characterized by headaches, hoarseness, blocked nose...In addition, they overlapped with ague, and Jeake sometimes conflated the two, both with one another and with the 'fevers' from which he also suffered. The agues were preceded by symptoms such as 'a great pain in my head'...they then took the form of successive more or less severe attacks, involving sweating, vomiting, giddiness, drowsiness, and shaking. There were often also accompanying pains and other corollaries...In addition Jeake complains of a miscellany of other ailments, such as swellings and twitches in different parts of his body, stomach pains, diarrhoea, dizziness, toothache—of which he had a particularly severe bout in 1671—and 'noise in the Ears' which began in the 1660s...and continued throughout his life. He was also concerned about his eyesight, which was affected by his smallpox in 1666...A further lifelong complaint was 'melancholy'...which Jeake seems to have suffered to an extreme

the two succeeding winters, the Small Pox was exceeding rife, and bastard Pleurisies, and strai[gh]tness of the bre[a]st and wheezing, with difficulty of breathing killed many."[796] Boghurst writes:

> The Plague is like the Serpent with two heades, for it kills at both ends...Many sorts of venomes doe the like, both minerall, vegetable and animall...The common diseases that the Plague carryes in its Tail are these. Extreeme paines in the backe, hipps, shoulders, heade, armes, great and painfull swellings on the shoulders, breast, leggs, thighs, losse of limbs, soare eyes, lamenesse, intolerable Itch, pimples, swelling of the yard. Alopecia, impostumes in the head, scurfes and inflammations, little red superficiall pimples in clusters like nettle stinging. Blindness, Feavers, Deafness, consumptions, forgetfullness, Toothach[e], excrescences...I have seen many of them this yeare, which have handled the Patient worse then the disease by far. Some have layne a quarter of a yeare in intollerable paine in the back and then dyed, others of as great and constant paine in their heads and then dyed, others soe miserably plagued with the Itch they have torne their skin off their flesh. Others become quite deafe, others ever lingering fall into consumptions, and are cured by Death. Their hair coming off was common, soe was the swelling of the yard towards the glans...I have seen some who have beene very laudably purged after the disease, yet fall into severall of these diseases.[797]

Hodges writes: "The Pestilence did not however stop for Want of Subjects to act upon, (as then commonly rumoured) but from the Nature of the Distemper, its Decrease was like its Beginning, moderate ; nor is it less to be wond[e]red at, that as at the Rise of the Contagion all other Distempers went into That, so now at its Declension, That degenerated into others, as *Inflammations, Head-achs, Quinseys, Dysenteries, Small-Pox, Measles, Fevers, and Hectics.*"[798]

Sydenham writes:

> In the year 1667, at the approach of the vernal equinox, the small pox, which during the immediately preceding pestilential constitution, appeared very rarely, or not at all, began to shew itself, and spreading more and more every day, became epidemic about autumn ; after which, its violence being abated by degrees, upon the coming on of

extent, complaining of its duration for ten years from 1667, reaching a 'most intolerable' climax in the winter of that year but continuing 'with Violence' till the summer of 1670...Later, perhaps the worst attack 'I ever were afflicted with' came in 1680, causing him prolonged sleeplessness." Hunter and Gregory (eds.), *Diary of Jeake*, pp. 51-52.

[796] Boghurst, *Loimographia*, p. 99.

[797] Ibid., pp. 93-94. Itching is a common symptom of ergot poisoning. Robertson and Ashby, "Ergot Poisoning," p. 303.

[798] Hodges, *Loimologia*, p. 26.

the winter, it decreased, but returned again the following spring, and prevailed, till it was checked, as before, by the subsequent winter...in August 1669, it totally disappeared, and was succeeded by an epidemic dysentery...At the same time the small pox first appeared, there arose a new kind of fever, not much unlike it, except in the irruption of the pustules...This fever, though it affected fewer persons by far than the small pox, did notwithstanding last as long ; but in the winter, when that abated, this prevailed, and when the small pox returned again in the spring, the fever went off...it totally disappeared, together with the small pox, in August 1669. These two epidemic diseases were accompanied by a third, especially the last summer, wherein this constitution prevailed, viz. a looseness...There are two kinds of this small pox...the distinct and confluent...The distinct kind begins 1. with a chil[l]ness and shivering, immediately followed by 2. extreme heat, 3. violent pain in the head and back, 4. vomiting, 5. and in grown persons a great tendency to sweat...6. pain in the parts immediately below the scrobiculus cordis, if they be pressed with the hand ; 7. sleepiness and stupor, especially in children, and sometimes convulsions...The distinct small pox comes out mostly on the fourth day...The eruption proceeds nearly in the following manner ; pale red pustules, as large as the head of a small pin, shew themselves here and there on the face first, or on the neck and breast, and afterwards on the whole body. During this stage of the disease, the throat is affected with soreness...For about the eighth day...the eye-lids are so filled and distended, as sometimes to make the patient blind...That kind of the small pox which we call the confluent, is attended with the same symptoms in common as the distinct, only they are all more violent ; the fever, anxiety, sickness, and vomiting, &c. being more severe...1. sometimes a sharp pain in the loins, resembling a fit of the stone ; 2. sometimes in the side, like a pleurisy ; 3. sometimes in the limbs, as in the rheumatism ; or lastly 4. in the stomach, attended with great sickness and vomiting...Sometimes this sort comes out like an erysipelas, and sometimes like the measles ; from which they cannot be distinguished, at least as to the outward appearance...The confluent small pox is attended with two other symptoms not less considerable than the eruptions, the swellings...1. a salivation or spitting in grown persons, and 2. a looseness in children...I have always observed when the disease proved very violent that the patient had a kind of fit towards evening, at which time especially, the more dangerous symptoms arose, and raged most severely...There are also other symptoms, which happen in any stage of the distemper, and which are equally common in the distinct and confluent kinds of small pox. For instance, a delirium sometimes seizes the patient...that he endeavours in a furious manner to get loose from those that confine him in bed. Sometimes the same cause produces a very different or contrary effect, as it seems, namely, a kind of coma, so that the patient dozes

> almost always, unless he be constantly roused. Sometimes also in this disease, as in the plague...purple spots appear in the spaces between the eruptions, which are generally forerunners of death...Sometimes small black spots, scarce so large as pins heads, and depressed in the middle, appear on the top of the eruption...sometimes, but not so frequently, a spitting of blood...Sometimes, also, especially in young person, there happens a total suppression of urine...there sometimes happens a troublesome swelling of the legs...the fever that prevailed during this variolous constitution...began and ended with the small pox...Now as this fever depended upon that epidemic constitution of the air, which at the same time produced the small pox ; so in effect, it seemed to be nearly of the same nature therewith...For they both attacked in the same manner, and were attended with the like pain and soreness of the parts below the pit of the stomach, there was the same colour of the tongue, and consistence of the urine, &c...Since therefore this fever did also prevail chiefly at the same time the small pox was more epidemic than I had ever known it here, no one can doubt their being of the same tribe. This, I certainly know, that all the practically indications were manifestly the same in both diseases, those excepted, with the eruption of the small pox, and the symptoms thence arising, afforded, which could not be expected in this fever, because it was not attended with an eruption...For these reason I must be allowed to call this a variolous fever...this fever, and of the small pox likewise, which so nearly resembles it...Before this fever went quite off, and particularly in the year 1668, a looseness became epidemic...I judge this looseness to be the same fever with the then reigning variolous fever, and that it only differed in form, and appeared under another symptom[799]...this disease, though naturally gentle, frequently proved mortal, as the bills of mortality of the current year sufficiently testified.[800]

Referring to the diseases prevalent in the following years, Sydenham continues:

> In the beginning of August, 1669, the cholera morbus,[801] the dry gripes, and likewise a dysentery that rarely appeared during the ten preceding years, began to rage...Between these gripes and the above-mentioned dysentery, which raged very universally, a new kind of fever arose, and attended both diseases...Now as this fever in

[799] Benjamin Rush writes: "See ! here again in the whole of this section an instance of the sagacity of our author, and another proof of the unity of disease, and of the influence of a belief in that unity, in leading to a successful practice."

[800] Sydenham, *Acute and Chronic Diseases*, pp. 111–120, 135–142.

[801] "The cholera morbus is easily known by the following signs : 1. immoderate vomiting, and discharge of vitiated humours by stool, with great difficulty and pain ; 2. violent pain and distension of the abdomen and intestines ; 3. heart-burn, thirst, quick pulse, heat and anxiety, and frequently a small and irregular pulse ; 4. great nausea, and sometimes colliquative sweats ; 5. contraction of the limbs ; 6. fainting ; 7. coldness of the extremities, and other like symptoms, which...often destroy the patient in twenty-four hours." Ibid., p. 146.

some measure resembled that which frequently attended the above-mentioned diseases, it must be distinguished from others, by the title of the dysenteric fever...Upon the approach of winter the dysentery vanished for a time, but the dysenteric fever raged more violently ; and a mild small pox also appeared in some places. In the beginning of the following year, namely, in January, the measles succeeded...This kind of the measles introduced a kind of small pox...the dysenteric fever and small pox raged the winter throughout. But about the beginning of February the following year, intermittent tertians arose, whence both diseases became less frequent...In the beginning of July the dysenteric fever again resumed the station it held in the preceding years ; and towards the decline of autumn, the dysentery returned a third time... but upon the approach of winter it vanished, and the dysenteric fever and small pox prevailed during the rest of that season. We observed above, that at the beginning of the two preceding years, two remarkably epidemic diseases raged...the measles at the beginning of 1670, and intermittent tertians at the beginning of 1671, and prevailed so considerably as to overpower the small pox...But in the beginning of 1672...this distemper [small pox] reigned alone, it of course proved very epidemic till July, when the dysenteric fever again prevailed, but soon gave place to the dysentery, which returned a fourth time in August... In the beginning of January, 1670, the measles appeared...1. It comes on with a chil[l]ness, shivering, and an inequality of heat and cold, which succeeded alternately during the first day ; 2. the second day these terminate in a perfect fever, attended with 3. vehement sickness ; 4. thirst ; 5. loss of appetite ; 6. the tongue white, but not dry ; 7. a slight cough ; 8. heaviness of the head and eyes, with continual drowsiness ; 9. an humour also generally distils from the nose and eyes, and this effusion of tears is a most certain sign of the approach of the measles ; whereto must be added, as a no less certain sign, 10. that though this disease mostly shews itself in the face, by a kind of eruption, yet, instead of these, large red spots, not rising above the surface of the skin, rather appear in the breast ; 11. the patient sneezes as if he had taken cold ; 12. the eyelids swell a little before the eruption ; 13. he vomits ; 14. but is more frequently affected with a looseness...The symptoms usually grow more violent till the fourth day, at which time generally little red spots, like flea-bites, begin to appear in the forehead and other parts of the face, which being increased in number and bigness, run together, and form large red spots in the face...These red spots are composed of small red pimples...these spots extend by degrees to the breast, belly, thighs, and legs ; but they affect the trunk and limbs with a redness only, without perceptibly rising above the skin...the measles in its nature nearly resembles the small pox...The measles, as we said before, introduced a different sort of small pox from that of the preceding constitution...About the beginning of July, 1673, there arose another kind of fever, which did not prove very epidemic...As the epidemic measles of 1670 introduced the black small pox above described, so the present kind, which appeared in the beginning of the current year 1673, being equally epidemic, was accompanied with a similar kind of small pox. For whereas the small pox of the preceding constitution...after the first two years gradually abated of their blackness, and also increased proportionally in size, till towards the end of the year 1673...it now returned with its former violence, and attended with very dangerous symptoms. This kind of small pox prevailed during the following autumn...but when cold weather came in, it abated, and soon gave place to the present epidemic fever. This fever, which had continued the whole year, made great devastation in the beginning of July, 1675, but at the approach of autumn it began to strike in upon the bowels, appearing sometimes with the symptoms of a dysentery, and at others with those of a diarrhœa ; though sometimes it was free from both, and rather seized the head, and caused a kind of stupor...This fever proceeded in this manner...till the end of October ; when the weather... changed suddenly to cold and moist, whence catarrhs and coughs became more frequent than I remember...the stationary fever of this constitution usually succeeded these coughs, and hence became more epidemic, and likewise varied some of its symptoms...now it chiefly seized the lungs and pleura, whence arose peripneumonic and pleuritic symptoms ; though it was still precisely the same fever than began in July, 1673, and continued without any alteration of its symptoms till the rise of these catarrhs. These catarrhs and coughs continued to the end of November, after which time they suddenly abated.[802] But the fever still remained the same as it was before the catarrhs appeared... Moreover upon their going off, a small pox, manifestly of the same kind with that of the preceding year, began to attack a few person here and there...Amongst the symptoms attending this fever, the principal one was a kind of coma, which rendered the patient stupid and delirious, so that he would doze sometimes for several weeks, and could not be awakened without loud noises, and then he only opened his eyes... [and] fell into a sleep again, which sometimes proved so very sound as to end in an entire loss of speech...Sometimes the patient did not sleep, but was rather silently delirious, though at times he talked wildly, as if in a passion ; but the fury never rose to so great a height, as is common

[802] "In 1675...towards the end of October...a cough became more frequent than I remember...for it scarce suffered any one to escape, of whatever age or constitution he were, and seized whole families at once...the cough assisted the constitution in producing the fever, so the fever on this account attacked the lungs and pleura, just as it had affected the head even the week preceding the cough ; which sudden alteration of the symptoms occasioned some...to esteem this fever an essential pleurisy or peripneumony...But nevertheless, as far as I could observe, the fever was the very same with that which prevailed to the day when this cough first appeared...And though the pungent pain of the side, the difficulty of breathing, the colour of the blood that was taken away, and the rest of the symptoms that are usual in a pleurisy, seemed to intimate that it was an essential pleurisy..." Ibid., p. 202.

> in a phrenzy in the small pox and other fevers...In autumn, 1675...this fever endeavoured to go off by a dysentery...As the epidemic measles, which appeared in the beginning of the year 1670, introduced the black small pox there described ; so that kind which arose in the beginning of 1674, and proved equally epidemic, introduced a sort of small pox, so extremely like the former, that it seemed to be the same revived...[803]

According to Sydenham, these diseases continued into the following decade:

> [T]he same sort of intermitting fevers....arose first in 1677, still prevails, viz. in 1681...These fevers...like all epidemics, chiefly raged in those seasons that conspired most with their nature....For instance, upon the coming in of winter they always gave way to the cough and peripneumonic fevers thereon depending, and likewise to the small pox ; but upon the return of the spring they re-appeared. So in the year 1680, when these intermittents had prevailed universally during the autumn, the small pox succeeded them in the winter and spread much ; but in 1681, the intermittents returned, though they did not spread so epidemically, the violence being abated, so that the small pox appeared along with them in a few places. But at the beginning of summer the small pox increased every day, and at length became epidemic, and killed abundance of persons...the spring and summer having been the driest seasons that any person living could remember, for the grass was burnt up in most places...whence the then reigning small pox was accompanied with a more considerable inflammation than ordinary, and the other symptoms thence arising were more violent. And this I conceive was the cause that purple spots frequently preceded the total eruption of pustules...And the disease proved so much more destructive, because the eruptions so readily ran together... It must, however, be owned, that the bloody urine and purple spots, which so certainly prognosticate death, do sometimes happen, when there is little sign of the appearance of the small pox, or only a very few eruptions coming out...[804]

According to Boghurst, "A cough, sneezing, and losse of limbes, which many Authors say troubled people in the Plague in other countryes, offended but little here, yet some lost their memory, haire, and senses, especially their hearing and sight, and some became meere Ideotts."[805] Victims of ergot poisoning can suffer hair-loss, permanent

[803] Ibid., pp. 143–145, 163–186, 195.
[804] Ibid., pp. 286-287.
[805] Boghurst, *Loimographia*, p. 28.

sight and hearing damage, permanent dullness and stupidity, as well as permanent insanity.[806]

Among the gentry in England, Bulstrode Whitelocke and his wife Mary were ill throughout 1664 and 1665, and in the following years they both frequently suffered from violent relapses of their "distemper."[807] During the year of the Great Plague, Whitelocke suffered from "Palsy" in March and "fluxe" and "Chollicke" between April and November,[808] and his son Carleton suffered from "Ague" in January of that same year.[809] In January 1671, Whitelocke suffered a severe relapse of his "distemper" that became aggravated by "dropsey" the following month,[810] which he recounted in the third person:

> Wh[itelocke] kept his bed, & was in a dying Condition...Dr Whistler left w^t^h Wh[itelocke's] wife a little viole w^t^h a very strong spirit in it, so that holding it att a little distance from her nose & though it was close stopped, yet she was not able to endure it butt thought she had fire in her head. The D[octo]^r^ told her that although he was not sent to, to bring such a thing with him, yett he knowing the constitution of Wh[itelocke] & that once formerly he had had a fitt of the Apoplexy, & fearing lest his want of blood & spirits might bring him into another fitt of the Apoplexy or into fainting fitts, he wished Wh[itelocke's] wife if he should fall into any such fitts then to put that viole w^t^h the Spirit, being open to his nose, w^c^h the D[octo]^r^ hoped might recover him again out of those fitts...About this time, in the evening, Wh[itelocke]... grew sick in his stomacke & his sight & speech began to fayle him, & stretching down his hands, his eyes sett, his throat rat[t]led, & he fell down out of his chayre in a sound.[811]

During a prolonged "Epidemic Fever," Whitelocke suffered several violent "swowning fitts" in the summer of 1674, and died the following year.[812] In the course of that summer Whitelocke's wife also suffered from "Chollicke" and "swowning,"[813] and his son Carleton was afflicted with "Scervey."[814] In December 1665, and again in March

[806] Barger, *Ergot and Ergotism*, pp. 29, 61, 64.

[807] Bulstrode Whitelocke, *The Diary of Bulstrode Whitelocke 1605–1675*, R. Spalding (ed.), Oxford, Oxford University Press, 1990, pp. 681-841.

[808] Ibid., pp. 690-699.

[809] Ibid., p. 689.

[810] Whitelocke wrote in February 1671: "they were a little fearfull bicause that legge was most swelled / & feared the dropsey humor falling into it, might endaunger it to gangrine." Ibid., p. 767.

[811] Ibid., pp. 764-766.

[812] Ibid., pp. 827-828, 841; Creighton, *Epidemics in Britain*, vol. II, p. 56.

[813] *Diary of Whitelocke*, pp. 825-826.

[814] Ibid., p. 829.

1667, it was determined by "Mrs Goddard & others, who said that his wife[']s distemper was the St Anthonies fyre,"[815] or ergot poisoning.

Barger provides a description of the permanent effects of convulsive ergotism:

> In severer cases, the whole body was attacked by general convulsions, often so suddenly "that some at the table dropped knife or spoon and sank to the floor, and other fell down in the fields while ploughing"...If not confined to bed, the sufferers "tumbled about as if drunk"...The loud cries of the sufferers are often referred to in a graphic manner. A cold sweat covered the whole body and the spasm of the abdominal muscles caused violent retching. Occasionally the disease first showed itself by convulsions, two or three days after eating the poisonous bread...In the intervals between the convulsions many patients suffered little discomfort...Taube and Hensler considered that the convulsive attacks were aggravated by the enormous number of round worms (*Ascaris*) frequently infesting their patients...Several descriptions mention severe insomnia. In extreme cases the patients would lie for six or eight hours as if dead ; in the 1597 epidemic some narrowly escaped being buried alive (Marburg). In such cases there followed a pronounced anæsthesia of the skin, the lower limbs became paralysed, and the arms subject to violent jerky movements ; epileptiform convulsions, delirium, imbecility, and loss of speech were apt to occur in such patients, who became unconscious and generally died on the third day after the onset of the first symptoms. In severe but non-fatal cases, the disease might last for six to eight weeks, and convalescence took several months. Convalescents apparently remained very sensitive to ergot, for Hussa recorded the deaths, due to a single meal of dumplings, in February, of two patients who had more or less recovered from an attack in the previous August. Relapses were frequent...These relapses were accompanied by epilepsy, hemiplegia, and paraplegia (von Leyden). Among the after-effects of a severe attack may be mentioned : general weakness, trembling of the limbs, gastric pains, chronic giddiness, permanent contractures of the hands and feet, anæsthesia of finger and toes, impairment of hearing and of sight, and various mental derangements...The effects on the mind consisted of dullness and stupidity, even in the less severe cases... the more general disturbances in severe cases was dementia...Minor nervous defects, spasms and a dull intellect may persist for a long time in the adult, and serious relapses occurred years after the first attack... Psychoses due to ergot have been especially studied by Gurewitsch [1911] and von Bechterew [1892]. A graphic early description of a

[815] Whitelocke wrote in March 1667: "Wh[itelocke's] wife was very ill. She would have no D[octo]r sent for. She kept her bed, her illnes[s] was thought to be St Anthonies fire[.] Wh[itelocke] wrote to Dr Whistler for his advise about it. She recovered pretty well." Ibid., pp. 699, 716.

> patient with delusional insanity, seven months after the harvest, was already given by Hoffmeyer [1742] ; the constant movements of the hands and feet were only interrupted by tetanic convulsions. The "looseness" mentioned in the English description quoted above, followed frequently, but not invariably, after severe convulsions ; this severe diarrhœa often persisted for months and was apt to prove fatal... The "swellings full of waterish humours" were...commonly observed by the Marburg physicians and by Scrinc...The belief that the disease was infectious is already contained in the title of the Marburg account, and was maintained for a long time ; it no doubt originated in the circumstance that all members of a family, living on the same diet, were often taken ill at the same time.[816]

Hodges divided the symptoms of the Plague into two classes; the first are symptoms that belong to ergot poisoning, and the second occur with *Fusarium* poisoning:

> I think it necessary to premise, that a Pestilence puts on sometimes one, and at another, another Appearance, and sometimes even contrary ones...The Symptoms of the first Class are Horror, Vomiting, Delirium, Dizziness, Head-ach, and Stupefaction. Of the second, a Fever, Watching, Palpitation of the Heart, Bleeding at Nose, and a great Heat about the *Precordia*. The Signs more peculiar to a Pestilence are those Pulstules which the common People call *Blains*, Buboes, Carbuncles, Spots, and those Marks called *Tokens*...[817]

Kemp writes of the Plague:

> It sometimes begins with a cold shivering like an Ague, sometimes continues with a mild warmth like a Hecktick Fever or a Diary, and encreaseth with violent heat like a Burning Fever. It corrupteth the *Blood* and all the humours, it afflicteth the *Head* with pain, the *Brain* with giddiness, the *Nerves* with Convulsions, the *Eyes* with dimness, making them look as if they had wept, and depriving them of their lively splendor, it makes the *Countenance* look ghastly, troubling the *Ears* with noise and deafness ; it infecteth the *Breath* with stinking, the *Voice* with hoarseness, the *Throat* with soreness, the *Mouth* with drought, and the *Tongue* with thirst ; the *Stomach* with worms and want of appetite, with hickhop, nauseousness, retching, and vomiting ; the *Bowels* with looseness and the bloody Flux, the *Sides* with stitches, the *Back* with pains, the *Lungs* with flegme, the *Skin* with fainty and stinking Sweats, Spots, Blains, Botches, Sores, and Carbuncles, the *Pulse* with weakness, the *Heart* with sounding and faintness. It makes feeble like the *Palsie*, it causeth sleepiness like the *Lethargy*, watchfulness and madness like

[816] Barger, *Ergot and Ergotism*, pp. 34-37.
[817] Hodges, *Loimologia*, pp. 86-87.

> a *Phrensie*, and sudden death like the *Apoplexy*. And these symptoms happen not alike to all, but differ and vary...[818]

The symptoms of the Plague varied from person to person and changed from week to week;[819] the disease also killed some slowly and others almost immediately.[820] Most victims died in five or six days,[821] while some died after ten or twenty days.[822] Kemp writes: "And, though this grievous *Sickness*, most commonly comes in state, attended with a Fever, and strengthened with other maladies, yet it is not alwayes so, for sometimes it comes stealing into the heart, whereby many have died suddenly, without the sense of fore-going pain or preceding distemper."[823]

Boghurst writes: "I have heard some say 'How much am I bound to God who takes me away by such an easy death' ; and they commonly say they are not sicke when Death is just at hand."[824] On 17 June Pepys was traveling by coach along Holborn, and the driver abruptly stopped the vehicle and informed his passenger that he was "suddenly stroke very sick and almost blind, he could not see."[825] According to Boghurst, the prognosis for this unfortunate man was discouraging: "When the disease seizeth people violently at first as with convulsions, swooning fits, blindness, vehement headache, stupefaction of the senses, such

[818] Kemp, *Treatise Of The Pestilence*, pp. 30-31.

[819] Tillison to Sancroft, 14 September 1665; in Ellis, *Original Letters*, pp. 36-37.

[820] Hodges, *Loimologia*, p. 50.

[821] "In the medical literature we can see that fear was so closely connected with the plague that it was regarded as one of its causes. Consequently, the advice to maintain a calm mind was regarded as one of the more important preventative measures that could be taken...This notion was later mainly developed within the Paracelsian tradition, but was integrated in all kinds of plague writings, independent of the general theory of the disease. The connection between fear and plague was also expressed in the proverb : 'worse than the plague is the fear of the plague'... In medieval Swedish the word for 'sudden death' (*brådöda*) became a name for the plague. The feature of the disease which provoked most fear was its rapid course. In the first description of plague in Sweden, a letter from King Magnus to the Diocese of Linköping in 1349, it was emphasized that people were dying suddenly without being able to prepare themselves. That also remained the most frightening aspect of plague after the Reformation. This feature of the disease was also always emphasized in the plague tracts as a reason for preparing oneself in time with both spiritual and natural preservatives...There was in particular a fear of being so unbalanced at the hour of death as to talk disdainfully about God and to persist in sin. Since the plague often led to severe disorders of the mind and to frenzy, this made the disease even more frightening. This was a point which perhaps was more important in Lutheran societies, where it was emphasized that forgiveness of sins and the sacraments should not be granted if the receiver had not been able to confess with full reason and knowledge of the meaning of the act." Per-Gunnar Ottosson, "Fear of the Plague and the Burial of Plague Victims in Sweden 1710–1711," *Maladies Et Société*, pp. 377-378.

[822] Boghurst, *Loimographia*, p. 27.

[823] Kemp, *Treatise Of The Pestilence*, p. 31.

[824] Boghurst, *Loimographia*, p. 27.

[825] *Diary of Pepys*, vol. VI, p. 131.

seldome live about two days."[826] According to Boghurst, when entire families died of the Plague, they "were commonly taken all alike, affected alike, proceeded in their sickness alike, lay the like tyme and dyed alike."[827]

[826] Boghurst, *Loimographia*, p. 98.
[827] Ibid.

Chapter 17. The Medical Evidence: Symptoms of the Plague

William Boghurst, the London apothecary who kept shop near the parish of St. Giles-in-the-Fields, recorded a description of the Plague that is considered to be the best account of the epidemic in existence. Boghurst believed that all other diseases turned into the Plague, and there were few diseases those months other than Plague.[828] Kemp agreed: "And as in the times of great Infection all Diseases turn to the *Plague*, so the *Plague* discovers the symptoms of all those Diseases whereof it had its beginning and original."[829] Hodges also believed that all other diseases turned into the Plague.[830] Boghurst listed forty-six primary indicators of the disease in his chapters, "Evil Signes or Presagers of the Plague," and "Further Comments on the Evil Signs of the Plague."[831] The following is the complete list of the "Evil Signs" of the Great Plague observed and recorded by William Boghurst.[832]

1. *Of the Tokens* — These were spots on the skin that were black, purple, and red in color. Ergot poisoning causes dark purple spots on the skin.[833] *Fusarium* poisoning also causes purple spots on the skin.[834]

---

[828] Ibid., p. 26.
[829] Kemp, *Treatise Of The Pestilence*, p. 31.
[830] Hodges, "Letter from Hodges," p. 15.
[831] Boghurst, *Loimographia*, pp. 22-23, 31-44.
[832] Unless noted, the following symptoms of *Fusarium* and ergot poisoning are quoted from Matossian, *Poisons of the Past*, pp. 11–12, 16.
[833] Leigh, "Account of strange Epileptic Fits," p. 1174.
[834] Moreau, *Moulds, Toxins and Food*, pp. 225-227.

2. *Stopping of the Stomach* — This was commonly accompanied by a great fever.[835] Graunt writes: "There seems also to be another new Disease, called by our Bills *The stopping of the Stomach*, first mentioned in the Year 1636...I conjectured that this *stopping of the Stomach* might be the *Mother*, forasmuch as I have heard of many troubled with *Mother fits*...I was somewhat taken off from thinking this *stopping of the Stomach* to be the *Mother*, because I ghessed rather the *Rising of the Lights* might be it. For I remembered that some Women, troubled with the *Mother-fits*, did complain of *a choaking in their Throats*."[836] There were 332 deaths in the *Bills of Mortality* attributed to "Stopping of the Stomack."

3. *An Hiccough* — Boghurst said that there were not many who suffered this symptom, "yet in those that had it it proved, as it doth commonly in all other diseases, a forerunner of present death." Hiccough was recorded as a symptom of the convulsive disorder named "Suffocation of the Mother," or ergot poisoning.

4. *Continual vomiting* — Violent retching is common with ergot and *Fusarium* poisoning.

5. *Buboes* — These were soft, white swellings that appeared *only* on the neck, armpit or groin. The flea-borne bubonic plague causes hard, red swellings that appear anywhere on the body. Boghurst said that these buboes "were filled with wind or humour, or both, and fell flat commonly just before they dyed." Boghurst distinguished between buboes, tokens, carbuncles, and blains. Carbuncles broke out on almost any part of the body, which started as a hard red swelling that turned black and hard, sometimes even with the skin and sometimes slightly raised, which could be as large as a man's head. Hodges said that the carbuncles contained pustules.[837] Boghurst said that he never saw a carbuncle turn into a bubo. Blains were also called blisters, and these were smaller skin eruptions that were either black and blue or whitish in color, which sometimes contained pustules, and appeared singularly or in clusters.[838] *Fusarium* poisoning causes miliary eruptions, and dark red skin eruptions with pustules, and is thought to cause swelling of the lymphatic glands in the neck, armpit, and groin. Sydenham identified these suppurations of the glands in the Plague with "Erysipelas," or ergot poisoning, which caused

[835] Boghurst, *Loimographia*, p. 50.
[836] Graunt, "Bills of Mortality," *Writings of Petty*, vol. II, pp. 358-359.
[837] Hodges, *Loimologia*, p. 119.
[838] Boghurst, *Loimographia*, pp. 49-53.

identical buboes on the neck, armpit, and groin.[839] Ergot poisoning also causes swelling and blistering of the skin, which also occurred in the Plague.[840]

6. *Suddain Looseness* — "Dysentery" was also called the "Plague in the Guts."[841] Severe diarrhea is a common symptom of ergot poisoning. Boghurst said that the stool was black, which is typical of *Fusarium* poisoning: "They that in the Plague fall into a looseness within 2 or 3 dayes, their excrements are black as Inke, and they generally all dye."[842]

7. *Short, difficult thick breathing* — Boghurst described this as "Shortness and difficulty of breathing as if they were choaked." The feeling of choking or suffocating is common with ergot and *Fusarium* poisoning. Boghurst included the oppression of the chest and stomach as a symptom of the Plague.[843] The oppression of the chest is a symptom of ergot poisoning.[844]

8. *Stopping of urine on a suddain* — This was accompanied by pain across the bottom of the belly and the groin. Ergot poisoning causes suppression of the urine.[845]

9. *Inward heat and outward cold* — Cold skin and fever are symptoms of ergot poisoning, and chills and fever are common with *Fusarium* poisoning.

10. *A lisping, faultering voice* — Muscular paralysis and loss of speech are symptoms of ergot poisoning.

11. *Continual sleep and drowsyness* — Boghurst said that sleeping at the beginning of the disease was a bad sign. Ergot poisoning causes drowsiness and unconsciousness, and *Fusarium* poisoning causes listlessness and stupor.

12. *Great Thirst continuing* — Boghurst writes: "This drinking much commonly kept their vomiting afoot, and soe did them mischiefe ; it also swelled their belly and lungs...Those that are poisoned are commonly very thirsty, and if they drink much it swells and kills them." Unquenchable thirst is a common symptom of ergot poisoning.

[839] Sydenham, *Whole Works*, pp. 62-63.
[840] Boghurst, *Loimographia*, p. 95.
[841] Graunt, "Bills of Mortality," *Writings of Petty*, vol. II, p. 369.
[842] Boghurst, *Loimographia*, p. 99.
[843] Ibid., pp. 21-22.
[844] Barger, *Ergot and Ergotism*, p. 33.
[845] Leigh, "Account of strange Epileptic Fits," p. 1174.

13. *Want of eruptions or sores* — According to Boghurst, the absence of skin eruptions was a bad sign.

14. *Sleeping with the eyes half open, half shutt* — To Boghurst this "showed a cessation of strength and stupefaction of sense." Ergot poisoning causes weakness and confusion, and *Fusarium* poisoning causes delirium and disorientation.

15. *Concerning the Pulse* — Boghurst observed two different pulses. One was an uneven, faltering, stopping pulse, and the other a weak, trembling pulse that would disappear and had to be found again. *Fusarium* poisoning causes a weak, fluttering pulse. Ergot poisoning causes constriction of the arteries and veins, and produces a weak, imperceptible pulse.[846] Kephale included "*Cardialga*, commonly called heart-ache," as a symptom of the Plague.[847] Ergot poisoning can cause cardiac arrest.

16. *Trembling hands, lips, shaking of the head* — Ergot poisoning causes tremors and spasms.

17. *Swelling of the Neck* — Boghurst writes: "sometymes the whole neck swelled round about, sometymes but one side, but alwais very hard and much distended, soe that it quickly troubled them in swallowing, and at last in breathing." *Fusarium* poisoning causes inflammation of the tonsils and uvula, which produces the difficulty in swallowing and breathing, and also causes the lymph nodes and cervical glands of the neck to swell, which can lead to strangulation.[848]

18. *Staggering* — This was common with people who had a fever or dizziness. Feelings of intoxication and dizziness occur with ergot poisoning, which also causes ataxia and a staggering gait.[849] *Fusarium* poisoning also causes dizziness and inebriation; the outbreaks in Russia have been named "staggering grains."[850]

19. *Sick Fits, Swooning, etc.* — Boghurst also called this "Sicke faint fitts." Convulsive ergotism was also called "Fits," "Hysteria," "Frenzy," and "Vapours."[851]

20. *Distraction, Raving, etc.* — Boghurst also called this "Distraction or idle talke, raving phrensy." Delirium, disorientation, and delusions

[846] Gabbai *et al.*, "Ergot Poisoning," p. 651.
[847] Kephale, *Medela Pestilentiæ*, p. 80.
[848] Moreau, *Moulds, Toxins and Food*, pp. 225-227.
[849] James Robertson and Hugh T. Ashby, "Ergot Poisoning among Rye Bread Consumers," *British Medical Journal*, 1 (1928), pp. 302-303.
[850] Joffe, *Fusarium Species*, pp. 5-6, 287.
[851] Matossian, *Poisons of the Past*, p. 13.

are common symptoms of ergot and *Fusarium* poisoning. "Phrensy" was also a name given to fungal poisoning.[852]

21. *Miscarriage in a woman with child* — Ergot poisoning causes abortions.

22. *A settled pain in the back and sides, etc.* — *Fusarium* poisoning causes pains in the back, sides, joints, and elsewhere.

23. *Cold Sweat* — Cold sweating is a symptom of ergot poisoning.[853]

24. *Sweating too freely* — According to Kephale, this sweat produced a terrible odor.[854] *Fusarium* poisoning causes profuse sweating. Ergot poisoning also causes sweating with a fetid odor.[855]

25. *Swelling and suddain falling of the belly* — Boghurst writes: "The swelling of the belly I judge was the cause either by the venome of the disease or the much vomiting or windiness and looseness. I read one author who sayeth that the whole body used to swell, but I saw not one person in such a condition, yet it is very likely such a thing may bee in some places or tymes, for many sorts of poysons will swell the whole body." Kephale included swelling of the legs, feet, and testicles with extreme pain as common symptoms of the Plague.[856] Ergot poisoning causes the extremities to become painfully swollen, and leads to "Dropsy."[857]

26. *Blindness and Headach* — *Fusarium* poisoning causes headache, and ergot poisoning causes blindness and headache.

27. *Much Belching and windiness* — Both ergot and *Fusarium* poisoning cause gastrointestinal disturbances.

28. *Sores decreasing or drying up, etc.* — Boghurst said that this was caused by a number of things, including fever, relapse, purging, bleeding, and vomiting.

29. *Urine* — Boghurst said that bloody or black urine was a bad sign. *Fusarium* poisoning causes bloody urine, and ergot poisoning also causes discolored and bloody urine.[858]

30. *Melancholy, Frightfull Dreames* — Boghurst writes: "some would dream they were among Graves and Tombs in Churchyards, others that they tumbled down from some high place and fell amongst

[852] Ibid., p. 56.
[853] Gabbai *et al.*, "Ergot Poisoning," pp. 650-651.
[854] Kephale, *Medela Pestilentiæ*, p. 81.
[855] Gabbai *et al.*, "Ergot Poisoning," p. 651.
[856] Kephale, *Medela Pestilentiæ*, p. 81.
[857] Barger, *Ergot and Ergotism*, p. 30.
[858] Leigh, "Account of strange Epileptic Fits," p. 1174; Gabbai *et al.*, "Ergot Poisoning," p. 650.

coffins, others cryed out in their dreams that they were all on fire, and these were harbingers of destruction." Culpeper writes:

> *Melancholly* is a Doting or *Delirium* without a Feaver with fear and sadness. It is distinguished from a Phrenzy by want of Feaver ; and from Madness, by Fear and Sadness, because that comes with Fury and Boldness. We say this Disease hath no Feaver : namely, of its own nature of it self : but a Feaver may Accidentally be joyned with it... but this Feaver wil[l] not be essentially in it, as in a Phrenzy, where a Feaver is essential to the Dissease. But we may doubt how Fear and Sadness may be said to be of the essence of Melancholly : when we perceive that in many Melancholick people there is much laugher and appearance of joy. For some laugh, some sing, some think themselves to be very rich Kings and Monarchs. We Answer, That there are divers degrees of Melancholy...Also great variety of Doting arisith from the various disposition of the Melanchollick humor : Hence it is that some think themselves to be Kings, Princes, Prophets : Others, that they are made of Glass, or Potters-Clay ; or that they are barely [barley] Corns ready to be devoured by the Hens : Some think they are melting Wax, and dare not approach the Fire : Others, That they are Dogs, Cats, Wolves...[859]

31. *Restlessness* — Boghurst writes: "Their legs would never lye still, but alwaies tossing them up and down, and flinging their arms carelessly out of the bed, and many tymes their whole body was constantly tossed up and down from one side of the bed to the other with a certain raging." Kephale included restlessness and insomnia as symptoms of the Plague.[860] Ergot poisoning causes insomnia, restlessness, spasms, and writhing.[861]

32. *Sighing* — Boghurst said that this was involuntary. Sighing was recorded as a symptom of "Suffocation of the Mother," or ergot poisoning, and "Malignant Angina," or *Fusarium* poisoning.

33. *Contraction of the Jaws* — Ergot poisoning causes permanent constrictures and muscular paralysis. *Fusarium* poisoning causes the lymph nodes and cervical glands of the neck to swell, making it difficult to open the mouth.[862]

34. *A Laughing Countenance, a faint forced smile* — Kephale included loss of memory and foolish behavior as symptoms of the Plague.[863]

---

[859] Culpeper, *Practice of Physick*, pp. 48-49.

[860] Kephale, *Medela Pestilentiæ*, pp. 80-81.

[861] Gabbai *et al.*, "Ergot Poisoning," pp. 650-651.

[862] Moreau, *Moulds, Toxins and Food*, pp. 225-227.

[863] Kephale, *Medela Pestilentiæ*, p. 81.

Ergot poisoning causes confusion, and *Fusarium* poisoning causes delirium and disorientation.

35. *A livid Countenance and sharp countenance* — Boghurst described the complexion as pale and bluish black. Kephale included jaundice as a symptom of the Plague.[864] Jaundice is a symptom of *Fusarium* and ergot poisoning.[865]

36. *Being Deaf and Dumb and unwilling to speak* — Boghurst writes: "Anguish of mind, dejection of spirits, fear of death, make some unwilling to speak or to bee spoke too...But the reservation of speech which I mean here is a kind of stupidity of soul and spirit." Ergot poisoning causes deafness, speechlessness, and stupidity.[866]

37. *Hoarseness* — According to Kephale, the tongue and mouth were inflamed and furred.[867] *Fusarium* poisoning causes hoarseness, swelling and necrosis in the oral cavity, and painful inflammation and swelling of the tonsils and uvula causing loss of the voice.[868]

38. *Fumbling with the Bedclothes* — Boghurst writes: "This signe is rather prophetical then predictive, and because it is common to almost all diseases it needs the less discourse about it ; it shewed itself also as in other diseases just before death."

39. *Vomit pouring out of the side of the Mouth* — Boghurst writes: "commonly they had a certain stupidity of spirits and dissolution of strength who vomited thus, for they never stirred or moved their body or head when they vomited." Vomiting and stupor are common symptoms of ergot and *Fusarium* poisoning.

40. *Cramps* — These were in the arms, legs, and elsewhere. Ergot poisoning causes muscle pain and cramps, and was known as the "Cramp Disease,"[869] and "Malignant Fever with the Cramp."[870]

41. *Stiff, soar neck* — Ergot poisoning causes wryneck and neck pain.[871]

42. *White soft tumours or Buboes* — Boghurst writes: "Children and old people were free from it, but from sixteen to forty odd they suffered

[864] Ibid.

[865] Leigh, "Account of strange Epileptic Fits," p. 1174; Barger, *Ergot and Ergotism*, p. 21; Moreau, *Moulds, Toxins and Food*, pp. 225-227.

[866] Leigh, "Account of strange Epileptic Fits," p. 1174; Barger, *Ergot and Ergotism*, pp. 32-33, 35-36.

[867] Kephale, *Medela Pestilentiæ*, p. 81.

[868] Moreau, *Moulds, Toxins and Food*, pp. 225-227.

[869] Bove, *Story of Ergot*, pp. 153–154.

[870] Barger, *Ergot and Ergotism*, p. 69.

[871] Gabbai *et al.*, "Ergot Poisoning," p. 650.

under it," adding, "I judge these to bee the right Buboes." These were found only on the neck, armpit, or groin.

43. *Being overloaden with the disease* — Boghurst defined this as "haveing riseings, carbuncles, and many blains on the body all at once."

44. *Great swelling under the Chyn* — Boghurst said that this "sometymes hindred their speaking & swallowing and choakt them." *Fusarium* poisoning causes swelling in the neck glands and oral cavity, which can lead to strangulation.

45. *Continuall Cold, aguish, difficult to sweat and raise blisters* — Boghurst said that this included shivering. *Fusarium* poisoning causes chills, and ergot poisoning causes cold skin.

46. *Bleeding at the nose, etc.* — *Fusarium* poisoning causes bleeding from the nose, mouth, vagina, and bloody urine and stool. Ergot poisoning causes bloody urine, and premature menstruation and hemorrhage.[872] Hodges considered vaginal bleeding always fatal.[873]

[872] Ibid., p. 651.
[873] Hodges, *Loimologia*, p. 143.

Chapter 18. The Medical Evidence: Fever and the Plague

Thomas Sydenham writes: "*Continual Fever* and *Agues*...were almost the only Epidemick Diseases that reign'd during the Constitution of the Years 61, 62, 63, 64 ; but how many Years they reign'd before, I cannot say ; this I certainly know, that from the Year 64, to the Year 67, they very rarely appear'd at all in *London*."[874] According to Boghurst, the Plague was "like an Ague at the beginning and the Lues Gallica at the latter end."[875] Sydenham compared the Plague to the "continual Fever" called "Erysipelas," a word many dictionaries and encyclopedias define as ergot poisoning.[876] Sydenham provides a detailed description of the "Erysipelatous Fever":

> This disease affects every part of the body, but especially the face, and it happens at all times of the year, but chiefly at the close of summer, at which time it frequently attacks the patient whilst he is abroad. 1. The face swells of a sudden, with great pain and redness, and 2. abundance of small pimples appear, which, upon the increase of the inflammation, often rise up into small blisters, and spread considerably over the forehead and head, the eyes in the mean time being quite closed by the largeness of the tumour. The country people call it a blast, or blight ; and in reality it differs little from those symptoms which accompany the wounds made by stings of bees, or wasps, excepting only that there are pustules. And these are the signs of the common and most remarkable species of the erysipelas. But whatever part is affected by this disease, and at whatever time of the year it

[874] Sydenham, *Whole Works*, p. 56.

[875] Boghurst, *Loimographia*, p. 97.

[876] Barger, *Ergot and Ergotism*, p. 53; E. A. Cameron and E. B. French, "St. Anthony's Fire Rekindled: Gangrene due to Therapeutic Dose of Ergotamine," *British Medical Journal*, 2 (1960), p. 28.

> comes, a chil[l]ness and shivering...generally attend this inflammation, with a thirst, restlessness, and other signs of a fever. As the fever in the beginning occasioned the pain, swelling, and other symptoms (which increasing daily sometimes terminate in a gangrene) so in the course of the disease these symptoms greatly conduce to the increase of the fever, till both are taken off by proper remedies. There is another species of this disease,[877] though it happens less frequently. This attacks at any time of the year...It begins with a slight fever, which is immediately succeeded by an eruption of pustules almost over the whole body, resembling those occasioned by the stinging of nettles, and sometimes they rise up into blisters, and soon after disappear, and lie concealed under the skin, where they cause an intolerable itching...[878]

The skin condition called "Erysipelas" was also known as St. Anthony's blush.[879] In the context provided by Sydenham, the *disease* "Erysipelas" was considered a mild form of the Plague, which like "Scurvy" was the result of consuming grain infected by ergot and *Fusarium*, a disease which Sydenham also says produced the buboes that formed in the lymph glands, and elsewhere he included gangrene as a symptom of this disease:[880]

> [T]he *Plague* is a peculiar *Fever* of its own kind...that Inflammation which is call'd an *Erysipelas* is much like the *Plague* ; for it is reckon'd by the best Physicians a continual *Fever*...whereon a Tumour, or rather red broad Spots, dispersed (for a very visible Tumour does not often appear) arise ; which they call an *Erysipelas*...there is sometimes a pain in the Glandules of the Arm-pit, or Groin, as in the *Plague* ; and it begins also almost like it with a Shaking and Shivering, and a febrile Heat following ; so that they who have not had this Disease before, think they are seiz'd with the *Plague*...But the *Plague* far exceeds an *Erysipelas*... If any one be dissatisfy'd with my Opinion, *viz.* That this Disease is occasion'd by an Inflammation, let him consider, that not only the presence of a *Fever*, but also many other things favour this Opinion : for instance, the...Appearance of a *Carbuncle*...the *Buboes*, which are as much inclin'd to an Inflammation as other Swellings of any other sort, and they end in *Abscesses*, as most Inflammations are wont to do : and also the Season of the Year wherein an *Epidemick Plague* breaks out, most commonly seems to confirm the same ; for at the same time, *viz.*

[877] "There is also another kind of eruption, that generally appears in the breast, being a broad spot with yellow scales, which scarce rises higher than the skin..." Sydenham, *Acute and Chronic Diseases*, p. 467.

[878] Ibid., pp. 232-233.

[879] Steven Shapin, "The Philosopher and the Chicken: On the Dietetics of Disembodied Knowledge," *Science Incarnate: Historical Embodiments of Natural Knowledge*, C. Lawrence and S. Shapin (eds.), Chicago, University of Chicago Press, 1998, p. 27.

[880] Sydenham, *Works*, vol. I, p. 260.

> betwixt Spring and Summer, *Pleurisies*, *Quinsies*, and other Diseases... are wont to be Epidemical ; and I never knew them more common than they were some Weeks before the beginning of the *London Plague*...[881]

Among others, Sydenham considered the Plague to be an extreme form of Erysipelas, that is, St. Anthony's Fire, and *Ignis Sacer* ("Holy Fire"), all terms that exclusively denote ergot poisoning:

> The plague usually begins with chil[l]ness and shivering, like the fit of an intermittent ; soon after, a violent vomiting, a painful oppression at the breast, and a burning fever, accompanied with its common symptoms succeed, and continue till the disease proves mortal, or the kindly eruption of a bubo, or parotis, discharges the morbific matter, and cures the patient. Sometimes the disease, though rarely, is not preceded by any perceptible fever, and proves suddenly mortal ; the purple spots, which denote immediate death, coming out, even whilst the persons are abroad about their business...In my opinion, the inflammation which the Latins call ignis sacer, and we St. Anthony's fire, or an erysipelas, is a good deal like the plague. For the skil[l]ful physicians esteem it a continued fever...where a tumour, or rather (for frequently there is no very remarkable tumour) a large red spreading spot, usually called a rose, arises ; but the fever is critically terminated in a day or two by this tumour, or eruption, and is sometimes accompanied with a pain in the glands of the arm-pit, or groin, as in the plague.[882] Moreover, the erysipelas begins much in the same manner as the plague, viz. with a shivering, followed by a feverish heat ; so that such as never had this disease before judge it to be the plague...To this may be added, that some authors suspect there is a kind of malignity joined with this disease...But the plague is much more violent than an erysipelas...[883]

According to Culpeper:

> All Authors (very neer) who have writ[ten] of Feavers, do distinguish a pestilential Feaver from a malignant...so that by the name of a pestilential Feaver they doe understand the true Pestilence or Plague ; and by a malignant Feaver, they mean that which is commonly called the spotted Feaver, or such a Feaver which...more live than die of it : whereas the true essence of the Plague consists in this [:] that more

[881] Sydenham, *Whole Works*, pp. 61-63.
[882] Commenting on this passage, Dr. Benjamin Rush, Professor of Medicine at the University of Pennsylvania, writes in 1815: "The unity of fever is clearly hinted at in this account of the sameness of the plague and the erysipelas. Dr. Chisholm says, the erysipelas preceded the yellow fever in one of the West India islands. The former was probably a modification of the latter."
[883] Sydenham, *Acute and Chronic Diseases*, pp. 95-97.

> die thereof than recover...these Feavers differ one from another only according to the greater or lesser degree of malignity...[884]

Culpeper continues:

> So, many Feavers, which at first were not pestilentiall...they turn in the end to malignant and Pestilentiall Feavers...And hence arose that common distinction of pestilential Feavers among Physitians, into a Pestilential feaver simply and properly so called and into a Malignant Feaver ; calling that a pestilentiall Feaver properly so termed wherein is the true Plague...The symptomes are Head-ach[e], Watchings, Raveings, Dead sleepes, Thirst, Stomach-Sickness, and Vomiting, want of Appetite, Swooning, Fainting, Hiccoughing, Unquietness, Loos[e]ness, Sweats and such like, which are common also to other kind of Feavers. But there is one Symptom proper and peculiar to a pestilential Feaver, which doth not happen in other Feavers ; *viz.* Purple Specks, or Spots on the whol[e] Body, but especially the Loyns, the breast and back, like unto Flea-bitings for the most part ; which the Italian Physitians name *Peticulæ* or *Petechiæ* ; the these Feavers which have these Symptoms, are commonly called *Purpuratæ* or *Petechialis*, Purple or Spotted Feavers. For these Purple Spots do not appear in all Pestilential Feavers ; but when they appear, they are a most certain Sign of pestilential Feaver. Now we called them Purple Spots, because they are for the most part of a Purple colour. Yet they are many times of a violet colour, Green, blewish, or black, and then they are far worse, and do signifie greater Malignity. And although these Spots are for the most part like Flea-bitings : yet they appear som[e]what greater : So as to represent those black and blew marks which remain after whipping, and then they are worse. And sometimes they are very large, and possess whol[e] Members and a great part of the body ; *viz.* the Arms, Thighs, and back, and then the parts appear tainted with redness...As to those Diseases which are joyned to a Pestilential Feaver, we may affirm what hath been said of the Symptoms ; *viz.* that many deadly Diseases are joyned with these Feavers ; namely Phrensies, Squinizes, Pleurisies, Inflam[m]ations of the Lungs, Inflam[m]ations of the Liver, bloody Fluxes, and very many more. But the chief Diseases which shew themselves in a Pestilential Feaver, are two, *viz.* a Pestilent *Bubo*, and a Carbuncle, which declare the venomous quality to be in the highest degree, and are not found but in the true Pestilence, and are wont commonly to accompany the same : So that the common People call them by the very name of the Pestilence...we must note, that there is no true, proper and Pathognomonick sign of these Feavers[,] viz. Such an one as wherever that signe is, there is the pestilence, and where that sign is not, there is no pestilence ; no not the Bubo or swelling in the Groyn nor the

[884] Culpeper, *Practice of Physick*, p. 611.

Carbuncle ; seeing that many have them not though they have the plague, and many have Buboes and Carbuncles that have no malignitie in them : neither are those purple spots any such pathognomonick sign, although a malignant Feaver is from them termed the spotted Feaver, forasmuch as many have a malignant Feaver without any such spots ; & those spots doe sometimes appear on women that want, their courses, and in some Children...without any Feaver...Likewise the signs of this Feaver are *Cardialgia*, Heart burning or pain at the mouth of the stomach...Sometimes great thirst...and sometimes want of thirst, with a vehement Feaver, and dryness of the Tongue...Great want of appetite, which make many abhor al[l] kinds of meat...Stomach sickness and vomiting...A frequent and inordinate shivering, which comes divers times in a day...The pulse...is small, unequal, frequent and very weak...Wearyness of the whol[e] bodie, Heaviness...Paines of the Head, Watchings and Raveings...the eyebrowes are cut asunder as it were with pain, and sometimes other parts as the shoulder-blades, the sides, the back, &c...In some patients drousie and sleepie dispositions happen...al[l] evil dispositions of urines, doe happen in this disease.... Chollerick fluxes of the belly...An abundance of Worms...Frequent sweates...[Redness] of the eyes is often seen in pestilential Feavers... Finally, purple spots like Flea-bitings, called by late Physitians *Peticulos* or *Petechiæ*, are the proper and peculiar Signs of a malignant Feaver. For they are found in no other kind of Feaver...Som[e]times they are seen in al[l] parts of the body, but most frequently in the Loynes, bre[a]st and Neck. Now the Diseases which come, upon a pestilential feaver for the most part are, som[e]thing coming out like Pox called *Exanthemata* ; Pushes and Ulcers of the Mouth ; Carbuncles ; Risings in the Groyn and behind the Ears. The Exanthemata aforesaid, differ from the purple spots, because in the spots there is only the color changed, but here is a certain rising in these *Exanthemata* to an head. Sometimes they are like warts, and sometimes less, resembling millet feed. Sometimes they are red...som[e]times white...yellow...purple...blewish or black...Some dry away, others come to matter, others grow to be ulcerous. To these may be referred pushes appearing in the Mouth which have al[l] the differences of the for[e]said *Exanthemata*, and are som[e]times so malignant, that the sick can hardly endure to eat and drink. From Children they often Cause Death because they wil[l] not endure the pain of eating and drinking. These pushes do som[e]times degenerate into Ulcers, which breed very great trouble to the patients, hindering the motion of their tongue and especially their swallowing. But sometimes Ulcers are bred in the Mouth immediately without any pustules or pushes foregoing which the *Greeks* cal[l] *Aphthe* : of which there are many sorts[885]...Carbuncles and Buboes, are wont to rise in

[885] According to Sydenham, dysentery accompanied by "aphthæ generally foreshew imminent death." Sydenham, *Acute and Chronic Diseases*, p. 150.

divers parts of the body, but especially where the Glandules are... behind the Ears...the Arm-pits...the Groyns ; whence swellings under the Ears called *Parotides*, & those in the Groyn called *Bubones*, do arise. Which kinds of tumors do chiefly appear in a true pestilential feaver, especially those in the Groyn, which therefore the common people call the Pest or Plague...Raving is very common in this Feaver...But a persevering dotage [delirium], is pernicious, because it's a token that it degenerates into a true Phrenzy. The contractions and hoppings of the Members, which do often happen in this Feaver, are Convulsive motions and very pernicious, and that the more if joyned with raveing... Trembling motions of the hands and Tongue are wont to be deadly... Deafness...Sneezing...Frequent Heart-burnings or Hiccoughings, do portend danger...avers[e]ness to meat...Suppression of al[l] evacuations...A Fat and oyly Urine, black or livid, with a black or blewish settling, doth certainly betoken death. Very much Urine being made, and no abatement of the Feaver thereupon is dangerous...A Loos[e]ness is very frequent in this Disease...Worms do very often vex those that have these Feavers...When the malignant and venomous Quality bears sway in these Feavers, that they come neer the Nature of the true Pest : in such Patients there are commonly risings behind their Ears, and Carbuncles...That Feaver which is commonly attended by the Measles and small Pox, may justly be reckoned among Malignant and pestilential Feavers...and kills very many children, to whom it commonly happens. What is the difference between the Measles and small Pox, authors are not yet well agreed. But custom hath obtained that those same larger pustules or Whelks like unto Warts (from whence they have their name) should be called in latin *Variolae*, in English the small Pox ; but those little Pustules and as it were asperities of the Skin with a deep redness like St. Anthonies fire or the rose, which are discussed within five or seven daies without suppuration, are called in latin *Morbilli* and in English Measles. There is also another kind of pust[u]les common to Children, like unto the small Pox in respect of the fashion and size ; but herein differs, in that the small Pox begins with redness and inflam[m]ation ; but these are white....which within three daies break and dry up and are wont to cause no danger, and commonly break forth without a Feaver.[886]

Sennert identifies "Measles" and "Small Pox" as Erysipelas:

First, there happen Feavers wherein pushes or eminent tubercles break forth, and sometimes certain spots shew themselves : the Greeks call them *Exanthemata*, and *Ecthamata*[,] the Latines *Papulas* and *Pustulas*, and at this day they are called the Measles and Small Pox... To the Poxes or Measles certain small red tubercles do belong, which invade with heat, and a cough and other symptomes...Moreover there

[886] Ibid., pp. 612-638.

are other breakings out which seem to be referr'd to Poxes...they are called Purples, and Eruthemata, yet some call the red spots or Patechii, purples : They are red, as it were fiery spots...coming out over all the body, as it were certain small Erysipelaes at the beginning of the sickness...In the progress of the disease it spreads over all the body, as if it were on fire, or as if one were sick of a universal Erysipelas...[887]

Among others, Samuel Pepys survived the Plague, and on 14 September he reviewed his proximity to the disease, thought by many physicians to be infectious:

[M]y meeting dead corps[e]'s of the plague, carried to be buried close to me at noonday through the City in Fanchurch-street—to see a person sick of the sores carried close by me by Grace-church in a hackney-coach—my finding the Angell tavern at the lower end of Tower-hill shut up; and more then that, the alehouse at the Tower-stairs; and more then that, that the person was then dying of the plague when I was last there...to hear that poor Payne my water[man] hath buried a child and is dying himself—to hear that [a labourer I sent but the other day to Dagenhams to know how they did there is dead of the plague; and that] one of my own watermen, that carried me daily, fell sick as soon as he had landed me on Friday morning last, when I had been all night upon the water...is now dead of the plague...both my servants, W Hewers and Tom Edwards, have lost their fathers, both in St. Sepulcher's parish, of the plague this week...[888]

Symon Patrick told a similar story:

I set myself to consider the great goodness of God to me since this plague, and how many dangers I had been in, by people coming to speak to me out of infected houses, and by my going to those houses to give them money...One thing I cannot but remember, that preaching a funeral sermon at Battersea, I was desired to let a gentleman come back to London in a coach which I had hired to wait upon me. The gentleman proved an apothecary, who entertained me all the way home with a relation of the many persons he had visited, who had the plague ; how they were affected, with the nature of their swellings and sores. But, blessed be God, I was not in the least affrighted, but let him go on, without any conceit that he might infect me. My poor clerk, a very honest man, found his house infected, and acquainted me with it. I was so pitiful as to bid him come out of the house himself, and attend his business, and I should not be afraid of him. He did so, and his wife and child died of the plague ; but he was preserved...[889]

[887] Sennert, *Agues and Fevers*, pp. 97-98.
[888] *Diary of Pepys*, vol. VI, p. 225.
[889] *Autobiography of Patrick*, pp. 54-55.

William Boghurst writes:

> Wherefore I commonly drest forty soares in a day, held their pulse sweating in the bed half a quarter of an hour together...[I] held them up in their bedds to keepe them from strangling and choking half an houre together, commonly suffered their breathing in my face severall tymes when they were dying, eate and dranke with them, especially those that had soares, sate downe by their bedd sides and upon their bedds discoursing with them an houre together if I had tyme, and stay[e]d by them to see the manner of their death, and closed up their mouth and eyes (for they dyed with their mouth and eyes very much open and stareing) ; then if people had noe body to helpe them (for helpe was scarce at such a tyme and place) I helpt to lay them forth out of the bedd and afterwards into the coffin, and last of all accompanying them to the grave.[890]

Consequently, the highly competent Boghurst determined that the Plague was not infectious:

> Great doubting and disputing there is in the world whether the plague bee infectious or catching or not...I have as much reason almost as any to think it is not infectious, having passed through a multitude of continuall dangers *cum summo vitæ periculo*, being employed all day till ten a clocke at night, out of one house into another, dressing soares and being allwaies in their breath and sweate, without catching the disease of any through God's protection, and soe did many Nurses who were in the like danger.[891]

[890] Boghurst, *Loimographia*, pp. 30-31.
[891] Ibid., pp. 96-97.

Chapter 19. The Death of George Starkey: A Victim of Plague

The American alchemist and physician George Starkey was described by his friend and colleague George Thomson as a man with "a sweet flexible nature, and a hearty desire to do good for Mankinde, he taught many Philosophical Arcana's : for which they rewarded him with persecution, calumnies, and base detracting language."[892] Starkey's remarkable life was overshadowed by his perpetual fight against poverty, which resulted in his imprisonment,[893] and appears to have contributed to his change of residency at least eight times in the last ten years of his life,[894] during which time his wife Susanna died.[895] In the rigid class society of Britain, Starkey's greatest enemy was his own poverty, and it was this poverty that would lead to his early demise.

Prior to Starkey's death, Thomson, who received his M.D. from the University of Leiden,[896] performed a post-mortem examination of a young Plague victim that had been enclosed in a coffin for a few days, after which Thomson fell gravely ill.[897] The disease became apparent to Thomson after he had finished the examination,[898] which began

[892] Thomson, *Loimotomia*, p. 102.

[893] Hartlib to Boyle, 28 February 1653/54; in Boyle, *Works*, vol. VI, pp. 79-80.

[894] Newman, *Gehennical Fire*, pp. 82-83, 247.

[895] [Starkey], *Dignity of Kingship*, p. xix n. 7.

[896] Charles Webster, "The Helmontian George Thomson and William Harvey: The Revival and Application of Splenectomy to Physiological Research," *Medical History*, 15 (1971), pp. 154–167.

[897] Thomson, *Loimotomia*, pp. 70-71.

[898] Although it has been thought that those present contracted the disease from the post-mortem examination, this is highly unlikely either for fungal poisoning or bubonic plague, since the

with a stiffness and numbness in his hand.[899] He then continued his medical duties,[900] and the next morning, stricken and despondent, Thomson sent for "that Excellent expert Chymist and legitimate Physician, Dr. *Starkey*," who had cured him of his previous bout with the disease.[901] According to Thomson, he twice survived the Plague in 1665 through Starkey's "Chymical" preparations.[902] Mortally afflicted by the disease, Starkey arrived to care for Thomson, who recounted Starkey's condition that fateful day:

> [H]e was stricken, and had as much need of mine, as I of his succour...However this worthy Gentleman came to me, whose very Aspect exhilarated and solaced my drooping spirits ; of whom when I had taken something, I was wonderfully composed for some time. This brave man, (that did, I dare maintain it, more good than all the *Galenists* in *England* put together) was that night after he had been with me, forced to yeeld himself prisoner to that insolent Conqueror... and so I was left destitute of two of my dearest friends in my saddest Condition ; For Honest and learned Dr. *Dey*, of whom I shall speak further hereafter, could not afford me his Assistance, being much afflicted and oppressed (after a former deliverance) with the same venemous Sickness.[903]

Thomson writes:

> [T]wo of my most esteemed Consorts, Dr. *Joseph Dey*, and Dr. *George Starkey*, two Pillars of *Chymical Physick*, were both reposed in their Graves, before I knew of their Deaths...They are gone, and at rest free from Persecution, Slanders and Obloquies of their Enemies...'Tis not for my own so much, though great, as for the publick interest, that I heartily lament the Translation of these two brave Souls into a better place, whose Praises all Candid and Ingenuous men will, I dare say, Celebrate ; and none but Sordid, Envious, Ignorant Spirits, will ever detract from them...for 'tis the nature of this sly Venome to strike so unawares at the brain...that a man hath not free liberty of his Reason, to act what may be convenient for the securing his own life, as it was too apparent in these Physicians of singular parts. One of whom, *i.e.* Dr. *Dey* I visited in the Evening, a little after I had Anatomized the body... At my departure from him (that night I was surprized by the Pest) he seemed chearfull, having felt about a fortnight before the smart of

pathogen Y. *pestis* would be easily destroyed by the conditions in the coffin. Shrewsbury, *History of Plague*, p. 471.

[899] Thomson, *Loimotomia*, pp. 77-78.

[900] Ibid., p. 79.

[901] Ibid., p. 83.

[902] Ibid., p. 3.

[903] Ibid., p. 84.

> this, piercing stroak, of which he was healed, and so had likewise upon his second encounter, had he not been too vent[u]rous to save others, rather than his own life : such was the true Charity of this Worthy Honest Gentleman...leaving the way of Riches, Honour, Pleasure and Ease his Colleagues trod in (having been admitted one of the College of *London* about twenty years) whom he deserted...[Dr. *Starkey*] was then infected when he came to me, having his Imagination dislocated, yeelded himself Prisoner to this Cruel Enemy that very night, being wounded in his Groin, a Bubo appearing there, which I conceive, if rightly ordered, might have been a meanes to have saved him, had he not poured in an unreasonable quantity of Small beer. After which understanding what he had done, he told those then present, That all the Medicines that he had in possession were of no force to do him any good...I have reason to beleeve, that had Dr. *Starkey* made use of his own Noble Chymical preparations in the beginning, and followed them close, as I did, it had been no difficult thing for himself to have escaped, that had an extraordinary Gift bestowed on him of Curing others in a far worse condition then he was reported to be. But for my part I am perswaded, that being very sensible of the impiety, hypocrisie, dishonesty, the imposture, subtile Frauds, disrespect of Real worth, odious Ingratitude, and other notorious Crimes of the Times, he was willing to resign himself to Death, so that he was not much sollicitous to live.[904]

Drinking alcohol while afflicted with the Plague was thought by Boghurst to be disastrous, causing madness and frenzy.[905] Following Starkey, Thomson recommended opium for the Plague, "menaged with that Method and Discretion, as that practically learned Dr. *Starkey* hath taught the world, ought to be highly esteemed for the quieting and mitigating the rage of the *Archeus* [i.e., Spirit]."[906] Willis writes of the "Epidemick Fever" in 1658: "let Opiates be used in this Fever with great Caution ; for the Frenzy appeas'd by them is oftentimes chang'd into a Lethargy, or a deep Stupor."[907] Boghurst writes: "I much dislike opium, wherefore in my Antidote I omitted it, and many in the Parish, haveing taken Mathews his Pill (soe called) were destroyed by it."[908]

Referring to the post-mortem examination and the deaths of his colleagues, Thomson writes:

904 Ibid., pp. 96–101.
905 Boghurst, *Loimographia*, pp. 25, 28, 56-57.
906 Thomson, *Loimotomia*, p. 153.
907 Willis, *Practice Of Physick*, p. 672.
908 Boghurst, *Loimographia*, p. 91.

[T]here was a confident Report (the broachers of which it is no difficulty to conjecture) that Dr. *Dey*, and Dr. *Starkey*, were both present at the Dissection of this Pestilent Body : It is an absolute untruth...either of those Gentlemen would willingly have joyned with me in this Anatomy, had not the Opportunity offered to me occurred so unexpectedly, that I could not conveniently gain any leisure to send to them. Moreover they were both seized upon by this Truculent Disease, before I entered upon this Dissection : so that Dr. *Dey* was not capable to assist me therein, being infirm ; and Dr. *Starkey* went to and fro with this mortal Arrow sticking in his side unfelt : and withall, so great was his employment, and medicinal negotiation at that time, that it was both hard to finde him out, and likewise to divert him from those engagements of visiting his Patients he had taken upon him...Mr. *Pick* my Patient then recovering...gave order that his Servant should attend me, who was the only man that stood by and looked on...[909]

Following Thomson's post-mortem of Mr. Pick's servant that was witnessed only by his other servant, John Tillison wrote to William Sancroft on 14 September:

Dr. Burnett, Dr. Glover, and one or two more of the College of Physicians, with Dr. O'Dowd, which was licensed by my Lord's Grace of Canterbury, some surgeons, apothecaries, and Johnson the chemist, died all very suddenly. Some say (but God forbid that I should report it for truth) that these, in a consultation together, if not all, yet the greatest part of them, attempted to open a dead corpse which was full of the tokens ; and being in hand with the dissected body, some fell down dead immediately, and others did not outlive the next day at noon.[910]

In a letter to Philip Fryth also dated at London, 14 September 1665, Starkey's close friend John Allin repeated a similar rumor that circulated in the city following Starkey's death and the examination.

Our friend Dr. Starkey is dead of this visitation [the plague], wth about 6 more of them chymicall practitioners, who in an insulting way over the Galenists, and in a sorte over this visitation sicknes[s], which is more a judgment then a disease, because they could not resist it by their Galenical medicines, wch they were too confident yt their chymical medicines could doe, they would give money for the most infected body they could heare of to dissect, which yey had, and opened to search for the seate of this disease, &c. ; upon ye opening whereof a stinch ascended from the body, and infected them every

[909] Thomson, *Loimotomia*, pp. 103–105.

[910] Tillison to Sancroft, 14 September 1665; in Ellis, *Original Letters*, p. 37.

> one, and it is said they are all dead since, the most of them distractedly madd, whereof G. Starkey is one.[911]

Besides George Starkey and Joseph Dey, those who were rumored to have died as a result of the post-mortem examination were Alexander Burnett, a member of the College of Physicians; John Glover, a Harvard graduate of 1650 who also belonged to the College;[912] and Thomas O'Dowde, a Chemical Physician who opposed the College.[913] Pepys recorded the death of Burnett the previous night on the morning of 25 August and does not mention the post-mortem examination.[914] According to his daughter Mary, O'Dowde and his wife also died in late August, his death occurring on a Wednesday of the "week before the sickness was at the Highest pitch," and she also makes no reference to Thomson's examination.[915] According to Sibley, John Glover survived the Plague and died around 1668.[916] There appears to be sufficient reason to doubt the presence of these physicians at the post-mortem examination, as Thomson denies they were present, and the accounts that place others with him are clearly based on false rumors. The deaths of Starkey and Dey were clearly unrelated to Thomson's examination, and the deaths of O'Dowde and Burnett were also unconnected to this event.

Tillison's previous letter to Sancroft is dated 23 August, the day before the death of Burnett and the week before or the day of O'Dowde's death.[917] There were two facts among the numerous false rumors recorded in Allin's and Tillison's letters of 14 September: Thomson's post-mortem examination, and the death of Starkey following the examination. John Allin had previously written to Fryth on 7 September and did not mention Starkey or the examination.[918] Allin

[911] Allin to Fryth, 14 September 1665; in Cooper, "Letters of Allin," p. 10; Sibley, *Graduates of Harvard*, vol. I, pp. 134–135.

[912] S. D. Clippingdale, "A Medical Roll of Honour. Physicians and Surgeons who Remained in London during the Great Plague," *British Medical Journal*, 1 (1909), p. 352.

[913] Sir Henry Thomas, "The Society of Chymical Physitians: An Echo of the Great Plague of London, 1665," *Science Medicine and History: Essays on the Evolution of Scientific Thought and Medical Practice written in honour of Charles Singer*, E. A. Underwood (ed.), London, Oxford University Press, 1953, vol. II, pp. 56-71; Harold J. Cook, "The Society of Chemical Physicians, the New Philosophy, and the Restoration Court," *Bulletin of the History of Medicine*, 61 (1987), pp. 61-77.

[914] *Diary of Pepys*, vol. VI, pp. 124, 203; Pepys to Lady Carteret, 4 September 1665; in *Memoirs of Pepys*, p. 597.

[915] Mary Tyre, *Medicatrix, or the Woman-Physician: Vindicating Thomas O'Dowde, a Chemical Physician...*, London, Printed by *T. R. & N. T...*, 1675, pp. 47, 55-59.

[916] Sibley, *Graduates of Harvard*, vol. I, pp. 208-211.

[917] Tillison to Sancroft, 23 August 1665, Harleian MSS. 3785, fol. 49; in Ellis, *Original Letters*, pp. 33-35.

[918] Allin to Fryth, 7 September 1665; in Cooper, "Letters of Allin," pp. 9–10.

obviously became aware of the fate of his close friend and Thomson's examination on the week ending 14 September, which he reported to Fryth the very same day that Tillison recorded another story that circulated in the city following the post-mortem. This strongly suggests that Starkey's death occurred that same week, as it seems unlikely that separate accounts of the two unrelated and history-making events of Thomson's examination and Starkey's death could circulate in London for nearly two weeks or longer without either being communicated in writing. The *Bills of Mortality* were compiled on Tuesday, and according to Thomson, Starkey attended to him on Saturday and died that night, most likely the night of 9 September 1665.[919] Further support for this later date is found in a seventeenth century handwritten note inscribed in a copy of Morhof's defense of alchemy to Langelott, once owned by Starkey's brother-in-law William Stoughton, which states that Starkey died "towards the end of the raging epidemic."[920]

When the Plague struck in 1665, only a handful of the elite College of Physicians remained in London to save the honor of the Galenists.[921] Out of their ranks, Wharton, Brooke, Hodges, Witherley, Barwick, Allen, Glisson, Baber, Paget, Glover, Deantry, Davies, Harrison, and Wyberd survived, while according to Hodges, "eight or nine fell in this Work," yet the only existing accounts are of the deaths of Burnett recorded by Pepys, and Conyers by Hodges.[922] The death toll was much higher among the less advantaged physicians who practiced in the poor suburbs outside of the city walls. According to Allin, over one hundred and forty physicians, apothecaries, and surgeons were deceased by mid-September.[923] Pepys wrote on 16 October: "they tell me that in Westminster there is never a physitian, and but one apothecary left, all being dead."[924]

[919] Thomson, *Loimotomia*, pp. 79, 100.

[920] Boston Athenaeum, $HJZ.L26, [Alv]; in Newman, *Gehennical Fire*, p. 52.

[921] In 1647 there were fifty-two members of the College of Physicians, fifty-four in 1650, fifty-eighth *circa* 1654, and one hundred and forty-three members in 1669. It appears that at least three-quarters of the College of Physicians fled London in 1665. Clippingdale, "Medical Roll of Honour," pp. 351-53; Cook, *Old Medical Regime*, p. 275.

[922] Hodges, *Loimologia*, p. 15; Clippingdale, "Medical Roll of Honour," pp. 351-353; Clark, *College of Physicians*, vol. I, p. 322; *Annales Collegii Medicorum*, vol. IV, 88a-88b; in Cook, *Old Medical Regime*, p. 156. According to Clark, Conyers survived the Plague, which contradicts the account given by Hodges.

[923] Allin to Fryth, 14 September 1665; in Cooper, "Letters of Allin," p. 10.

[924] *Diary of Pepys*, vol. VI, p. 268.

With Starkey's death the Chemical Physicians were without their greatest champion, and the Galenists of the College of Physicians seized the opportunity to silence their opponents. Hodges writes: "nothing was otherwise wanting to aggravate the common Destruction ; and to which nothing more contributed than the Practice of Chymists and Quacks."[925] Hodges wrote in May the following year:

> [C]oncerning our pretended Chymists, I can only make you this return ; that the People are convinced of their Designs, their most admired Preparations proving altogether unsuccessful, and their Contrivances being chiefly bent upon more secret Ways...These scandalous Opposers of the *College* are now for ever silenced, since that so many Members of the most honourable Society have ventured their Lives in such hot Service ; their Memory will doubtless survive Time, who died in the Discharge of their Duty, and their Reputation flourish, who (by God's Providence) escaped.[926]

Hodges also took the opportunity to denigrate the Chemical Physicians as unlearned "*Empericks* with their noble *Arcana*," and "I very much wonder that the honorable Mr. *Boile* should so much favor the *practise* of *Empericks*," whom Hodges classed with "*Midwives, Barbers, Old Women, Empericks, and the rest of that illiterate crew*," adding that "I hope they have no reason to take it ill, if I remind them of the several *Callings* in which they were *educated*, and ought still with *care* and *industry* to have *exercised:* The most eminent of our *Empericks* are HEEL-MAKERS, GUN-SMITHS, TAYLORS, WEAVERS, COBLERS, COACHMEN, BOOKBINDERS, and infinite more of the like quality, beside a great number of the *other* SEX."[927] George Thomson's response to the Galenists was that their "Bleeding, cutting holes in the Skin, Blisterings, and such like butchering Tortures" were as dangerous as the Plague, and challenged Hodges to compete in a medical trial similar to the one proposed by Starkey years earlier.[928] The Galenists of the College of Physicians, who were also known as the Methodists and Dogmatists, made numerous medical recommendations during the Plague, most of which were worthless at best, and some directly

[925] Hodges, *Loimologia*, p. 21.

[926] Hodges, "Letter from Hodges," p. 35.

[927] Nathaniel Hodges, *Vindiciæ Medicinæ & Medicorum: Or An Apology For the Profession and Professors of Physick*, London, Printed by *J. F.* for *Henry Broom*, 1666, pp. 18, 42-[43], 47.

[928] Thomson, *Galeno-pale*, p. 8; Thomson, "The Authors Apology against the Calumnies of the *Galenists*," *Loimotomia*, pp. 188–189.

contributed to the destruction of the population,[929] including their leading the flight of the gentry from the city.[930] Kemp classed the College of Physicians with other quacks who published, "observations which they have met with in the cure of Diseases...and yet not one Medicine found out to preserve the Doctor."[931]

In 1667, Thomas Sprat wrote: "The mortality of this *Pestilence* exceeded all others of later Ages," and declares the Plague to be the "greatest *Terror* of mankind."[932] Sprat writes:

> Why may we not believe, that in all the vast compass of Natural virtues of things yet conceal'd, there is still reserv'd an *Antidote*, that shall be equal to this *poyson*? If in such cases we only accuse the *Anger* of *Providence*, or the *Cruelty* of *Nature* : we lay the blame, where it is not justly to be laid. It ought rather to be attributed to the *negligence* of men themselves, that such *difficult Cures* are without the bounds of their *reasons power*.[933]

[929] For more on the history of Galen and the Galenists fleeing epidemic disease, see Patrick Wallis, "Plagues, Morality and the Place of Medicine in Early Modern England," *English Historical Review*, 121 (2006), pp. 1-24.

[930] "Necessary Directions For The Prevention and Cure Of The Plague in 1665. With divers Remedies of small Charge, by the College of Physicians," *Collection Of Very Valuable and Scarce Pieces*, pp. 36-52; Bell, *Plague in London*, p. 237; Thomas, "Society of Chymical Physitians," p. 58.

[931] Kemp, *Treatise Of The Pestilence*, p. 3.

[932] Thomas Sprat, *The History of the Royal-Society of London...*, London, Printed by *T. R.* for *J. Martyn...*, 1667, p. 123.

[933] Ibid.

Chapter 20. George Starkey's Secret Remedy

The Reverend John Allin was George Starkey's Harvard roommate and lifelong friend who graduated Harvard in 1643 and soon afterwards returned to England.[934] According to Cooper, in 1653 Allin became vicar of Rye, in Sussex, "and continued vicar till December 1662, when he was ejected under the Bartholomew Act. On leaving Rye he came to London and studied physic."[935] Throughout the Plague, Allin frequently mentioned in his letters a mysterious plant which he "described under the general term of 'Materia prima'...It generally appears between the vernal and autumnal equinoxes after rain, in dry, parched, and sandy soils."[936] Quoting Cooper, Sibley writes:

> Allin dabbled in alchemy, and attached a high value to the *Materia prima*. "It was to be gathered with great mystery, and preserved with much care, for the purposes of distillation; and he intended, in September, 1665, to set up 'divers chemical stills and one furnace for the main worke.' He was a disciple of Paracelsus, who says that 'the saline spirit unites with the earthly principle, which always exists in the liquids, but in a state of *materia prima*.'" The plant was formerly known by the "name of *cœlifolium*, as the popular belief that it fell from heaven in the night. Paracelsus gave to it the name of *nostock* or *cerefolium*...The alchemists took it to contain the universal spirit, and an extract to be the solvent of gold." Being reduced to a powder, it was

[934] "Harvard College Records, College Book I," *Publications of the Colonial Society of Massachusetts*, 15 (1925), p. 9.
[935] Cooper, "Letters of Allin," pp. 1-2; Sibley, *Graduates of Harvard*, vol. I, p. 93.
[936] Cooper, "Letters of Allin," p. 4.

said to cure ulcers, however "obstinate and rebellious they may be": hence possibly its use in the plague.[937]

Allin wrote to Fryth on 22 September 1665:

> Friend get a piece of angell gold, if you can of Eliz. coine (yt is ye best), wch is phylosophicall gold, and keepe it allways in yor mouth when you walke out or any sicke persons comes to you : you will find strange effects of it for good in freedome of breathing, &c. as I have done ; if you lye wth it in your mouth wthout yor teeth, as I doe, viz. in one side betweene your cheke and gumms, and so turning it sometimes on one side, sometimes on ye other.[938]

It remains an open question whether Allin discovered a remedy for the Plague, although according to reliable accounts, his friend George Starkey possessed the only cure in London. Sibley writes:

> In a manuscript lecture on Sir George Downing by Charles Wentworth Upham, it is stated that Stirk "rendered himself famous, by his professional skill, during the dreadful plague in London in 1665. His extraordinary knowledge of chemistry led him to the discovery of a remedy which, if properly applied, was always found effectual. He was the only physician in the city who could cure the plague. As may be well supposed, he was in such constant demand that his constitution became debilitated by fatigue and exposure, and at length the disease fastened upon him. His remedy was required to be administered at a particular stage of the malady when the patient had passed into a delirium. As he felt himself approaching that state, he gave the most minute directions to his attendants in reference to the mode of administering his medicine. When the delirium had passed off he made inquiries as to the treatment he had received, and found that an irremediable and fatal error had been committed. He had scarcely time to declare that he was a dying man. His remedy died with him."[939]

According to members of the Hartlib circle, upon his arrival in London Starkey was successfully curing "Agues,"[940] "Feaver," "Falling

[937] Ibid., pp. 3-4; Sibley, *Graduates of Harvard*, vol. I, p. 94.

[938] Allin to Fryth, 22 September 1665; in Cooper, "Letters of Allin," p. 15; Sibley, *Graduates of Harvard*, vol. I, p. 95.

[939] Quoting Upham, Sibley writes, "For the circumstances in reference to this connection with the plague of London, and his tragical and sudden death, I am indebted to the late venerable and learned Doctor Edward Augustus Holyoke, of Salem. He related them to me when in his one hundred and first year. They had been brought to his knowledge by tradition, which, when it reached him, however, was so recent as to have a very high degree of authority." Sibley, *Graduates of Harvard*, vol. I, p. 134.

[940] Hartlib, *Ephemerides*, [September 1653], MM-MM4; in Wilkinson, "Hartlib Papers II," p. 103.

Sicknes[s]," "Dropsy,"[941] "Rickets,"[942] and "Palsy and other incurable diseases."[943] In his challenge to the Galenists in *Natures Explication*, Starkey had promised unlimited success in curing "Palsie, Epilepsie, Gout, Agues, Kings-evill, *&c*...[and] all acute diseases, as Feavers, Fluxes, and Pleuresies, Calentures, Small-pox, and Measles, at the utmost in four daies."[944] Starkey also claimed that he could cure "*all Feavers continual, Fluxes, and Pleuresies, in four daies ; in Agues (not Hyemal quartanes) in four fits, in Hecticks and Chronical diseases in thirty (at most forty) daies.*"[945] Starkey writes:

> And here is the goodness of the most High, that no man can truly boast himself to be a real son of Art, but he hath at command Medicines to cure the most common and truculent diseases, as for instance, Feavers, Pleurisies, Fluxes of all sort, Agues of all sort, small Pox and Measles which are indeed but a branch of Feavers, Calentures, also which belong to the same head, the Jaundies, Head-aches, Tooth-aches, with all running pains, Hypochondriacal Colicks, affections of the Mother, and obstructions of all sorts causing indigestion, Palpitations, Syncopes, Convulsions, Vertigoes, *&c.*[946]

Starkey also writes:

> Yet can I confidently affirm and make good, that I yearly cure more Feavers, Agues, and Pleurisies, then any one in the *Galenical* way have in nigh twice the time ; but my cures are too contemptible for the rich, Counsel and Medicine in almost two thirds of my cures scarce exceeding, sometimes not amounting to a crown...For many hundreds know and can testifie for me, that besides my own cures, many both in City and Countrey practice by my medicines, to the cure and relief of some thousands yearly, mine own practice in some years reaching to nigh two hundred Agues...with many more Feavers, Pleurisies, Fluxes, and vomitings, of all which scarce five in a year not perfectly cured, and those only such who hearing of the sudden effect of my medicines, send for some of them, and without observing the difference of season of the year, expect the same speed in cure with others, and not finding

[941] Ibid., 11 December 1650, K-L7; in Ibid., p. 87.

[942] Hartlib to Boyle, 28 February 1653/54; in Boyle, *Works*, vol. VI, p. 80; Wilkinson, "George Starkey," p. 130 n. 58.

[943] Hartlib, *Ephemerides*, [early 1650], E-F1, [summer 1650], G-H3; in Wilkinson, "Hartlib Papers II," pp. 86-87.

[944] Starkey calls "small Pox" a "very malignant sort of feaver," and writes, "*under continual Feavers I comprehend Calentures, small Pox, Measles,* &c. *which are of that head.*" Starkey described "Pleurisie" as a "most dangerous Feaver," accompanied by pain, cough, spitting of blood, and suffocation. Starkey, *Natures Explication*, pp. A7^r, b6^r, 245, 258, 266.

[945] Ibid., p. A7^r.

[946] Ibid., p. 185.

> the cure perfect (although notably abated) are discouraged, and leave off, whose error herein is not to be charged upon the medicine.[947]

In *Natures Explication*, Starkey called the Plague a "Feaver," for which he also possessed an antidote: "The saddest of all Feavers, the Pestilence (called by a general name, the Plague among us) as being reputed, and not without cause, the saddest of temporal plagues : that I shall passe over at present in silence, as never having (to my knowledge) experience in that disease, though of Feavers commonly known by the name of pestilential Feavers, and judged to be a degree of the Pestilence it self, I have known and cured many."[948] In his *Pyrotechny*, Starkey claimed to possess several remedies for "Feavers," which differed in their preparation: "I know many *specifick remedies* for *feavers*, which I have oft made, and used, yet when I find a medicine of no more difficulty of preparation ; and far larger extent in vertue when prepared, I wave the making of others, and content my self with the one."[949] As Boyle observed, Starkey and other alchemists typically abused the terms they employed, "that as they will now and then give divers things one name ; so they will oftentimes give one thing many names."[950] Starkey's homeopathic cures for "Agues" and "Feavers" appears to have been different preparations of the same medicine hidden under various names, which he revealed in *Natures Explication*:

> And first for the *Arcanum Corallinum*, which is *Paracelsus Diaceltatesson*, and is Mercury precipitated by mean of the Liquor Alchahest...a most certain cure for all Feavers, Agues, Pleuresies, *&c.* yea the Hectique it perfectly restores, as also Dropsies, with all Ulcers inward and outward, and the venereal distemper, with the Gout *&c*...the Horizontal gold, which is the same essentially with the Mercury corallated, cures all the forementioned distempers...and the same doth the *Ladanum* or sweet Oyle of Mercury (which is *Helmont* and *Paracelsus* true *Ladanum* without *Opium*)...the *Ladanum* of Mercury curing universally all diseases...The same may be said of an Antimonial Panacæa, which I know, and is a certain cure for Agues, Feavers, and Pleuresie...But moreover I shall give the studious Reader to understand, that in many vegetable Simples under the mask of virulency, great and noble virtues are hidden, which are kept by the poisonous appearance from rash hands, as the apples of the *Hesperides* were feigned to be kept by a watchful Dragon ; or as the passage to the Tree of life, was guarded

[947] Ibid., pp. 232-233.
[948] Ibid., pp. 253-254.
[949] Starkey, *Pyrotechny*, p. 167.
[950] Boyle, *Sceptical Chymist*; in Boyle, *Works*, vol. I, p. 520.

> by a flaming sword in the hand of Cherubims. Thus is Hellebore under the churlish vomitive poyson caused with convulsion both of stomach and nerves, is hidden a most noble remedy against Hypochondriack melancholy, the Gout, Epilepsie, Convulsions, and quartain or third day Ague...[951]

In *Pyrotechny*, Starkey asserts that the hidden virtues of "Hellebore" are discovered "through the discipline of the fire, which is our so much commended *Pyrotechny*."[952] In Philalethes's *George Ripley's Vision*, Starkey writes that "Mercury" drawn from the "Chalybs" was "the same which the fair *Medea* did prepare, and pour upon the two Serpents which did keep the Golden Ap[p]les, which grew in the hidden Garden of the Virgins *Hesperides*."[953] In Philalethes's *Transmutation of Metals*, Starkey calls the "Vegetable *Saturnia*" the "*Double Sword* in the hand of the *Cherub*, that defends the way of the Tree of *Life*,"[954] the *Arborem Vitæ* that in *Pyrotechny* Starkey identifies as the source of the hidden Manna.[955] In Philalethes's *Secrets Reveal'd*, Starkey calls "our Water" a product of "Fire," "the Liquor of the Vegetable *Saturnia*," and the "bond of ☿ [Mercury]," and declares the "Chalybs" of the Magi to be the "Vegetable *Saturnia*," also named "our Fiery *Dragon*...our *Arsenick*, our *Air*, our ☽ [Silver], our *Magnet*, our *Chalybs* or *Steel*."[956] In *Pyrotechny*, the "Philosopher by Fire" identifies van Helmont's herbal preparation named *Circulatum majus* as the Alkahest,[957] which he compares to "the Medicine of the *Magi*, their *Aurum potabile* ; attained by means of their Stone."[958] In Philalethes's *Secrets Reveal'd*, Starkey calls the "*Oyle*," "*Honey*," "*true Fire*," "*burning Wine*...*Royal Mineral*, and *Triumphant Vegetable Saturnia*,"[959] the "Medicine Universal, both for prolonging Life, and Curing of all Diseases,"[960] which is "the Secret *Golden-Fleece*...our Gold-making POWDER (which we call our *Stone*)...called *Our Gold*."[961]

In *Pyrotechny*, Starkey calls "Horizontal Gold" a cure for "*Leprosie, Gout, Palsey, Epilepsie, Cancers, Fistulaes, Wolves, Scorbute* [Scurvy],

[951] Starkey, *Natures Explication*, pp. 275-282. Paracelsus also recommends "Hellebore" for these ailments. Bruce T. Moran, "The Herbarius of Paracelsus," *Pharmacy in History*, 35 (1993), pp. 99–127.

[952] Starkey, *Pyrotechny*, p. 102.

[953] Philalethes, "Sir George Ripley's Vision," *Ripley Reviv'd*, pp. 6-7.

[954] Philalethes, "Ars Metallorum Metamorphoseωs," p. 82.

[955] Starkey, *Pyrotechny*, p. A4r.

[956] Philalethes, *Secrets Reveal'd*, pp. 5-6.

[957] Starkey, *Pyrotechny*, p. 94.

[958] Ibid., p. 49.

[959] Philalethes, "Ars Metallorum Metamorphoseωs," p. 82.

[960] Philalethes, *Secrets Reveal'd*, p. 119.

[961] Ibid., pp. 1-2.

Kingsevil, venereal disease...all *Fevers*, and *Agues*...also any sort of *Consumption*."[962] Starkey records an obscure procedure for obtaining the "Liquour" of *Gold*, and then claims that exceeding this medicine in virtue was his preparation of "Oil of *Venus*," the *Ens Veneris* made of calcined *Vitriol*, "the true *Nepenthe* of Philosophers," which like the "sweet Oyle of Mercury" was "the true *Ladanum* without *Opium*."[963] Starkey writes:

> This is the highest preparation of *Gold* that can be made by means of this Liquour, being its fift[h] Essence, and is of power to cure the most deplorable diseases to which the nature of Man is subject. But the magistery of *Gold*, which is the first preparation of it, by means of this Liquour, is a most eminent Medicine against all *Malignant Fevers*, the *Pestilence, Palsies, Plagues, &c.* Most excellent also is the fift[h] essence of *Silver*, and *Silver* potable, made by the same way and process ; but the sweet Oil of *Venus* doth exceed in Virtue both the one and the other... Nature hath not a more soverain remedie, for most (not to say all) diseases : This is the true *Nepenthe* of Philosophers.[964]

Robert Boyle told Samuel Hartlib in 1653 that Starkey's *Ens Veneris* was an excellent remedy for "Agues," "Fevers," and other such diseases, and was a true "Medicina Pauperum because for 5 Sh[illings] so much may be prepared with it as may serve a 100 Poore People."[965] The *Ens Veneris* ranked among Starkey's great *Arcana*, or "Secret" drugs,[966] which he prepared for Boyle of calcined "Copper" sublimated with "sal armoniack."[967] Concealing his secrets "in the language of the *Magi*,"[968] Starkey writes in his *Liquor Alchahest*: "For as there is a *Sal Armoniack* Vulgar, which scarce any Fool but knows ; so is there also a *Sal Armoniack* of Philosophers, which only true elect Sons of Learning know : In the circulation of which, is the perfection of the hope of all true adept Brothers of Art, so far as concerns this fire of Hell, which is Fire and yet Water."[969]

---

[962] Starkey, *Pyrotechny*, p. 47.
[963] [Starkey], *Pill Vindicated*, p. 7.
[964] Starkey, *Pyrotechny*, pp. 32-33.
[965] Hartlib, *Ephemerides*, [September 1653], MM-MM4; in Wilkinson, "Hartlib Papers II," p. 103.
[966] Starkey, *Natures Explication*, pp. 216, 239; Sloane Collection, British Library, MS 3750; in Newman, *Gehennical Fire*, p. 300 n. 88.
[967] Boyle, *Usefulnesse of Experimental Philosophy*; in Boyle, *Works*, vol. II, pp. 135, 215-216.
[968] Starkey, *Natures Explication*, p. 239.
[969] [Starkey], *Liquor Alchahest*, p. 35.

In Philalethes's *Ripley Reviv'd*, Starkey writes: "our work is nothing else but an uncessant boyling of thy Compound,"[970] to which he adds: "Thus is verified the saying of the Philosopher, that our Stone retaineth life, and is perfected...by continual boyling and subliming."[971] In *George Ripley's Vision*, Starkey again quotes "the saying of the Philosopher : Boyl, boyl, and again boyl."[972] Starkey's experiments involving the decoction of medicinal plants was known to the Hartlib circle,[973] and in *Secrets Reveal'd* Starkey writes: "if this one Secret were but openly discovered, Fools themselves would deride the Art ; for that being known, nothing remains, but the Work of Women and the Play of Children, and that is Decoction."[974]

In his *Pill Vindicated*, an undated public advertisement for his medicines published around 1663, Starkey writes:[975]

> I may say confidently that I was the first that made this [*Ens Veneris*] in *England* (that is known) in the year 1[6]52 I prepared it, for the Honourable *Robert Boyl* Esq; one of the Royal Society, who hath wrote of its excellency, as his extant Treatise thereof can testifie. Of this *Helmont* truly saith, nothing works more powerfully on the radical humidity then it, it is most excellent in feminine disease, especially menstrual stoppages, rickets of children, lingering fevers, obstructions of the spleen and mesaraicks, but especially for women after childbed, taking away after pains to admiration, helping their cleansings so kindly, and without violence, and is admirable, and hind[e]ring all feverish symptomes to which many women after trav[a]el are subject, to the incredible case and comfort of them, but this I must assure those that are concerned, that what I made in 16[5]1 for the Honourable *Robert Boyl* Esq; and is by him commended, is so inferiour a preparation to that as is made by me now, that the former deserves not the same name with this latter, which is yellow as the purest gold, and in taste rich of the Venus, so that it exceeds the others beyond all comparison, and nearer approaches the Element of fire of Venus...This is the *Nepenthes verum* of the Phylosophers, the true *Ladanum* without *Opium*, lengthening the life by Gods permission, and conquering powerfully monstrous tragical maladies, such as the Epilepsie, Apoplexy, Palsie, Coma, Vertigo, madness[,] Hypocondriack melancholy, and uterine frensies, with several other like diseases to the stupefaction of nature, of this

[970] Philalethes, "An Exposition Upon Sir George Ripley's Epistle to King Edward IV," *Ripley Reviv'd*, p. 43.

[971] Philalethes, "An Exposition Upon the First Six Gates of Sir George Ripley's Compound of Alchymie," Ibid., p. 301.

[972] Philalethes, "George Ripley's Vision," Ibid., p. 5.

[973] Hartlib, *Ephemerides*, [April 1651], E-E2; in Wilkinson, "Hartlib Papers II," p. 94.

[974] Philalethes, *Secrets Reveal'd*, p. 89.

[975] Newman, *Gehennical Fire*, pp. 193–194.

> and Mercury, is made the Horizontal Gold that is truly and perfectly universal. In the *Belgick* Edition of *Helmonts* works under the [t]itle of preparation of medicines, mention is made of a middle Diaceltatesson, which is Antimonial...and it is excellent in Chronick diseases...The other Diaceltatesson is my true Corallatum, curing radically the Gout, Lues, and all ulcers in the throat, lungs, bladder, and kidneys, as also it roots out all Agues ; and hecktical fevers...In Histerical distempers there is great variety...causing all sorts of diseases, as palpitations, vertigoes, Epileptic palsies, with other distempers...in these cases the Salt of Steel (I mean the magisterial Salt of it)...These with several other specifick abstersives, as spirit of Salt, spirit of Sulphur vive, the essential sulphurous spirit of vitriol of Venus, Antimony, and other bodies, with the Magistery and spirit of Saturn, are truly prepared by *George Starkey*...[976]

In two letters of Starkey to Samuel Hartlib, Starkey mentions his "Wine of Corne" and *Aqua vitae*, the "Water of life," both made with grain and honey, the sugars contained in the honey producing alcohol, the "*Sal Armoniack* of Philosophers." Starkey writes:

> [Honey] is good to eat, both pleasant and wholesome, in Chirurgery and Medicine of excellent force, and inriched with a rare Quintessence. But besides, *by help of it and grain, may be made most excellent Wine, nothing inferiour to the richest Canary or Greek wines, and by the mixture of it with the Juyce of fruits, the best French or Rhenish Wines may be paralell'd, if not surpassed. Nor will any of the Specifick Odour, either of the Honey, or of the Corn, after a threefold fermentation remain.* It also will yeild a most excellent *Aqua vitæ*...[977]

Starkey responded to Hartlib's admonishments of this letter by openly revealing his method of preparing by distillation his Hippocratic medicines from poisonous plants, particularly of grain,[978] the homeopathic cure for grain poisoning:[979]

> [T]o my confident Assertion of the fecibility of *Aqua vitæ* out of grain unmalted, &c. and the producing of Wine out of fruit and also

[976] [Starkey], *Pill Vindicated*, pp. 6-7.

[977] Starkey to Hartlib, n.d.; in [Samuel Hartlib], *The Reformed Common-Wealth of Bees. Presented in severall Letters and Observations to Sammuel Hartlib Esq...*, London, Printed for *Giles Calvert* at the *Black-Spread-Eagle* at the West-end of Pauls, 1655, p. 30.

[978] Ergot was included in the *Book of Poisons*, a work ascribed to the highly influential and enigmatic Jabir, an eighth century Arabic figure considered the greatest alchemist of his time. The tenth century Persian physician Muwaffak was also familiar with ergot, which he considered to be a powerful poison. Bove, *Story of Ergot*, pp. 141–142.

[979] Poisonous grain has long been recognized as the cause of fatal epidemics in history, although confusion has arisen from the circumstances that these outbreaks were caused by various fungal parasites, and were often a combination of mycotoxins. Elinor Lieber, "Galen on Contaminated Cereals as a Cause of Epidemics," *Bulletin of the History of Medicine*, 44 (1970), pp. 332-345.

grain, equal to Spanish and French Wines, by the meanes of Honey, I conceive, that what was written is sufficiently full and plain. For to write a Receipt is a thing...not so convenient...a Receipt is every mans meat, and to such who lesse understand Nature, what Receipt can be full enough...I could not possibly throw the Receipts into the mouth of every one that could but gape...My work is to hint to the ingenuous what may be done, and let it be sufficient that (*fide bona*) I deliver what is really true in Nature, and adde the onely Meanes, which is by reiterate fermentation...First as to the *Aqua vitæ*, let Pease be taken and steeped in as much water as will cover them, till they swell and Corn, and be so ordered as Barley is for the Malting, onely with this difference, that for this work if they sprout twice as much as Barley doth in making Malt it is the better: these Pease thus sprouted if beaten small, which is easily done they being so tender, put into a vessel, and stopt with a Bung and a Rag as usually, these will ferment, and after two, or three, or four moneths, if distilled, will really perform what I promised. The Water that soaked them, it is good to save, either for the soaking of fresh, or for putting on them, being beaten, which else require some quantity of water to be added to them, but not much, and the like may be done in all other Grain, which the addition of refuse Honey will advance (as to quantity of Spirit) exceedingly. Thus may a Spirit of *Aqua vitæ* be made out of any green growing thing, of which the leaves being fermented, will yeild a small quantity of such a Spirit. So Roots, Berries and Seeds, which are not oyly, yea and those which are oyly, whose fatnesse is essential, that is, which may be distilled over an Alembick with water, will afford some more, some lesse of *Aqua vitæ*. Let me adde, that the Spirit which is made out of Grain not dryed into Malt, is more pleasant than the other.[980]

[980] Starkey to Hartlib, n.d.; in [Hartlib], *Common-Wealth of Bees*, pp. 32-35.

CHAPTER 21. EPILOGUE

Following an extremely severe winter throughout Europe that caused wheat prices in England to double, and then triple the following year, a lethal "Epidemic Fever" in London grew in magnitude in the years 1709 and 1710, during which time a number of scholars at Wadham College perished from the disease while other colleges remained unaffected.[981] The mortal "Fever" continued and soon caused many in London to fear that it was a precursor of the Plague that had been raging in Asia Minor, Eastern Europe, the Baltic, and Scandinavia.[982] The rural communities were severely affected by this epidemic, which in six months in 1711 killed 25,000 in Copenhagen and 40,000 in Stockholm.[983] In 1713, the Royal Society published John Gottwald's description of this Plague that struck Warsaw in 1707 and Danzig in 1709,[984] as well as John Chamberlayne's brief account of the disease in Copenhagen in 1711.[985] In Danzig the epidemic demonstrated the usual "Plague curve," a gradual increase and decline in the weekly

[981] Creighton, *Epidemics in Britain*, vol. II, pp. 54-59.

[982] F. H. W. Sheppard (ed.), *Survey of London: The Parish of St. James, Westminster*, London, Athlone Press, 1963, vol. XXXI, p. 196; Richard West, *The Life & Strange Surprising Adventures of Daniel Defoe*, London, HarperCollins, 1997, pp. 266-270.

[983] Charles F. Mullett, *The Bubonic Plague and England*, Lexington, University of Kentucky Press, 1956, pp. 263-265; David Kirby, *Northern Europe in the Early Modern Period: The Baltic World 1492–1772*, London, Longman Group UK Limited, 1990, p. 352.

[984] John Christoph. Gottwald, "An Abridgement of a Book intitl'd, A Description of the Plague, which happened in the Royal City of *Dantzick*, in the Year *1709*," *Philosophical Transactions*, 28 (1713), pp. 101–144.

[985] John Chamberlayne, "Remarks upon the *Plague* at *Copenhagen* in the Year *1711*," *Philosophical Transactions*, 28 (1713), pp. 279-281.

mortality rates that peaked in September, striking the working poor the hardest, while among the nearly 25,000 casualties, "few of the People of Condition and Quality died," and not a single physician perished.[986] Gottwald believed the agent of the disease to be a poison, but he was unable to identify it.[987] In his story-bound periodical in the August 1712 issue of *Review*, Daniel Defoe called the "Fever" in London "the new and unaccountable distemper" that would "be very mortal, and...contagious," and printed the Bill of Mortality for the destructive week ending 19 September 1665.[988] Notwithstanding Defoe's fears, England remained unaffected by the Plague, which has not since returned to the great city of London, although nearly three and a half centuries after the "Great Visitation" disagreement remains over the nature of the pathogen and Dr. George Starkey's secret remedy.

[986] Gottwald, "Description of the Plague," p. 107.
[987] Ibid., pp. 106, 110, 113.
[988] West, *Life of Defoe*, pp. 266-270.

Chapter 22. Conclusion

Like all diseases, there are physical parameters that exist within the presumably immutable laws of nature that determine the existence, survival, and movement of bubonic plague in a human population. These include a suitable candidate for the initial infection of the disease, the temperature and time factors that are needed for flea eggs to hatch, the need for a great number of fleas, the necessary rodent epizootic, and then the predictable slow movement of the disease geographically from an identifiable location. Any reasonable person who takes into consideration the hard facts that are present in the historical record will conclude that all of these parameters contradict the identification of bubonic plague as the agent of "Plague" in the seventeenth century. The medical literature and literary evidence associated with the "Plague" is unambiguous, and it is consistent with ergot and *Fusarium* poisoning, which certainly infected the rye crop of southeastern England.

Of all the cereals, rye is the most susceptible to ergot infection, the ergot fungus that produces the highest level of alkaloids, and rye also provides the best substrate for the development of the extremely poisonous *Fusarium* fungus, a potential hyperparasite of ergot. Ergot poisoning, usually through the rye crop, was known on the European Continent as the "Peasant's Disease" because it usually afflicted the poor peasantry, and it was also named *Ignis Plaga*, or "Fire Plague," and *Plaga Invisibilis*, or "Invisible Plague."

The English poor of the seventeenth century subsisted on rye, a cheaper grain compared to wheat, and maslin, a mix of rye and wheat, while the gentry generally used the more expensive wheat as their cereal staple. While few who lived comfortably were affected by the "Great Plague" in 1665, the poor suffered enormously during the epidemic, so much so that it became known as the "Poor's Plague," which can be logically attributed to the consumption of poisonous rye by the English poor. Also, it was children and healthy adults who were most severely affected by the "Plague," and children and healthy adults are most susceptible to ergot poisoning, although the "Plague" in 1665 affected almost everyone except the wealthy (who ate white bread).

The death rates associated with the "Plague" reflected the distribution of wealth in London, which does not square with the epidemiology of bubonic plague. The intensity of the disease in the poor out-parishes of London is completely inconsistent with what is known of rat biology, which selects the wealthier parts of a city where warmth is greater and the food is more plentiful, not to mention the relative absence of a truly viable animal candidate for the initial infection of the disease. However, considering the diet of the English poor, these death rates are consistent with the consumption of rye infected with ergot or *Fusarium*, or both. The infection rates within the households of London and the identical symptoms shared by the victims within a household are also inconsistent with the rat/flea theory, yet these rates and shared symptoms do perfectly match the prediction of the ergot/*Fusarium* theory. The misguided argument that the poor were filthy and this attracted rats will forever remain an offensive reminder of the self-deluded elitism that exists in modern medical society, since it is clear that the poor suffered inordinately during the "Plague" partly because the College of Physicians led the flight of the gentry from London. This act of cowardice caused the economy of London to shut down, forcing many of the unemployed poor into begging, and causing the population of limited means to rely even more on the cheaper cereal staple available, that is, they were eating less maslin and more poisonous rye.

Furthermore, the epizootic of cattle that preceded the "Great Plague" in 1665, as well as other "Plagues" in Britain that followed the introduction of rye by the Anglo-Saxon invaders, could in no conceivable way be related to an outbreak of bubonic plague. An epizootic of cattle is entirely consistent with outbreaks of ergot poisoning among humans, and there were those who observed an increase in mushrooms and fungus in the years 1664 and 1665. Also, the drought and extremely cold weather conditions of winter and early spring in 1665 were entirely unsuitable for the development of bubonic plague, a tropical disease, but these are ideal weather conditions for the growth of ergot and *Fusarium* on the rye crop, and the "Plague" started in April and peaked during the fall harvest, as one would expect with ergot and *Fusarium* poisoning. It is also difficult to see how, without exception, the resurgence of the "Plague" following rain can be squared with the epidemiology of bubonic plague, but rain is associated with the growth of fungus, including ergot and *Fusarium.* Moreover, the extremely high death rates in 1665, some rural towns losing half their population, contradicts an identification of the "Plague" with bubonic plague, since between 1896 and 1917 bubonic plague killed only 0.135 per cent of the population *per annum* in India, and the death rates in other countries were comparable.

The geographical distribution of the "Great Plague" in London is also at odds with the epidemiology of bubonic plague, as bubonic plague moves slowly and regularly from an identifiable location, and it is a fact that the numbers of "Plague" deaths recorded by the parishes week by week does not conform to this epidemiological model. While some of the wealthier parishes within the walled city were completely unaffected or scarcely affected by the "Plague," in some of the parishes outside the city wall, in the poorest neighborhoods, there were thousands who died of the "Plague" that year. Also, within the most severely affected regions in southeastern England there were townships that were completely unaffected by the epidemic, while neighboring townships a few miles away lost half their population. This geographical distribution is inconsistent with the epidemiology of bubonic plague. In contrast, the distribution of the disease in London and rural England can readily be explained by the known parameters of the growth of ergot or *Fusarium* on rye, or both.

Finally, the medical evidence that is conspicuously absent from other papers devoted to this subject is probably the most damaging of all the evidence against the bubonic plague theory. The various diseases, including the "Plague," that preceded the "Great Plague" in 1665, the higher death rates in diseases that occurred with the "Plague," and the diseases that followed the "Plague" do not conform to the epidemiology of bubonic plague, but all can be identified as caused by ergot or *Fusarium* poisoning. Among other symptoms of the "Plague," hallucinations, suicidal and homicidal behavior, convulsions, dementia, permanent blindness, permanent deafness, hair loss, permanent dullness and stupidity, and permanent insanity are inconsistent with bubonic plague, and are identical to the symptoms of ergot poisoning. The wide range of symptoms associated with the "Plague" is also inconsistent with the symptoms of bubonic plague. This includes the soft and white buboes of the "Plague" that only appeared on the neck, armpit, or groin, compared to the hard and red buboes of the bubonic plague that appear anywhere on the body. Every symptom of the "Plague" in 1665 that was observed and recorded by numerous authors fits the epidemiological model of either ergot or *Fusarium* poisoning, or both. Furthermore, many physicians thought that the "Plague" was a "Fever" and an extreme form of "Scurvy," which at the time ergot poisoning was thought to be a form of "Scurvy," as well as an extreme form of "Erysipelas," St. Anthony's Fire, and *Ignis Sacer* ("Holy Fire"), all terms that denote ergot poisoning. The only alternative explanation of the literature that is possible for the advocates of the bubonic plague theory is to suggest that "Scurvy," "Erysipelas," St. Anthony's Fire, and *Ignis Sacer* were all various forms of the bubonic plague, and despite the modern day spread of *Yersinia pestis* on all the continents, it is a stable organism and there is still only one serotype.

The young George Starkey tragically perished in the "Plague" of 1665, departing this world on 9 or 10 September of that year during the worst month of the epidemic, in which thousands of others also lost their lives, and the good doctor was likely buried with some of these victims in one of London's infamous "Plague Pits." In some respects Starkey's life was not only tumultuous, it was also tragic. The Hartlibians, including Robert Boyle, subjected Starkey to factionalism, manipulation, and appropriation, and later in life Starkey was also publicly slandered by his enemies, of which he apparently had more than a few. Because of this and debt Starkey was in and out of prison, his poverty also reflected in his constant change in residences, and in the end he died a widower. Yet Starkey was also the famous Philalethes and was a Promethean figure that had befriended some of the most influential individuals of the period in both the Old and New Worlds, and he had written several highly influential books and manuscripts that were later published, and he also successfully created a number of effective medicines that alleviated the suffering of many and saved countless lives.

Starkey had claimed the ability to cure all of the diseases associated with the "Plague," and even the "Plague" itself, and his Galenic opponents in the London College of Physicians never dared to take up Starkey's challenge to test his medicines against their absurd procedures and dangerous pharmacopoeia, with which they undoubtedly injured or killed more people than they helped. The shameful attacks and shady political maneuvers that the members of the College of Physicians made upon Starkey and his colleagues before and after Starkey's death constituted a grave injustice in the history of medicine, especially considering that it is now glaringly apparent that Starkey was a superior chemist and physician compared to all of those who opposed his Chemical Medicines.

George Starkey was an Anti-Galenic, Anti-Aristotelian, Neoplatonic philosopher and Hippocratic physician, and therefore his alchemical writings and medical writings reflect this tradition. Starkey explicitly admitted in his published books that he used simples (i.e., powdered matter of a single plant) in his practice, and he was also known to use distillations of plants, and Boyle even recorded Starkey's recipes for producing alcohol, including distilled liquor from "Corne," a generic word for cereal during this period. One of Starkey's medical preparations for the diseases of his day, which he shared with Samuel Hartlib, and specifically his cure for the "Plague," his *Aqua vitae*, or "Water of life," was a Hippocratic, homeopathic remedy that was produced by fermenting and distilling the poisonous rye (which others were also doing at that time, but they were producing whiskey from the plant), and presumably in a diluted form the remedy was apparently administered at a certain time in the course of the disease. It is absolutely clear from the historical record that Dr. George Starkey possessed the ability to cure those afflicted with the dreaded "Plague" as well as those afflicted with the other diseases of his day. The "American Philosopher" was the only physician in London during the

"Great Plague" in 1665 who was in possession of an effective remedy, for which he has never been properly recognized.

Bibliography

Allbutt, Clifford, and Rolleston, Humphry Davy (eds.), *A System of Medicine*, Volume II, Part I, London, Macmillan and Co., Limited, 1908.

Anderson, Robert G. W., "The Archaeology of Chemistry," *Instruments and Experimentation in the History of Chemistry*, F. L. Holmes and T. H. Levere (eds.), Cambridge, MIT Press, 2000.

Applebaum, Wilbur (ed.), *Encyclopedia of the Scientific Revolution from Copernicus to Newton*, New York, Garland Publishing, Inc., 2000.

Ashley, William, "The Place of Rye in the History of English Food," *Economic Journal*, 31 (1921), pp. 285-308.

Ashley, William, *The Bread of our Forefathers: An Enquiry in Economic History*, Oxford, Clarendon Press, 1928.

Ashmole, Elias, "Annotations and Discourses, Upon Some part of the preceding Worke," *Threatrum Chemicum Britannicum*, E. Ashmole (ed.), London, Printed by *J. Grismond* for Nath: Brooke, at the Angel in *Cornhill*, 1652.

Aut Helmont, Aut Asinus: Or, St. George Untrust; Being a Full Answer to his Smart Scourge, London, Printed for *R. Lowndes*, at the White Lion in *S. Pauls* Churchyard, 1665.

Barger, George, *Ergot and Ergotism: A Monograph Based on the Dohme Lectures Delivered in Johns Hopkins University, Baltimore*, London, Gurney and Jackson, 1931.

Barger, George, "The Alkaloids of Ergot," *Handbuch der Experimentellen Pharmakologie*, Volume VI, W. Heubner and J. Schüller (eds.), Berlin, Verlag von Julius Springer, 1938.

Baxter, Richard, *Reliquiæ Baxterianæ...*, M. Sylvester (ed.), London, Printed for *T. Parkhurst...*, 1696.

Bell, John, *Londons Remembrancer...*, London, Printed and are to be sold by *E. Cotes...*, 1665.

Bell, Walter George, *The Great Plague in London in 1665*, Second Edition, London, Bodley Head, 1951.

Bell, Walter George, *The Great Plague: Grim Tragedy Told in Church's Records of 1665*, London, Published by the Guild of St. Bride, 1958.

Beretta, Marco, *The Enlightenment of Matter: The Definition of Chemistry from Agricola to Lavoisier*, USA, Science History Publications, 1993.

Biraben, Jean-Noël, *Les hommes et la peste en France et dans les pays européens et méditerranéens, Tome I*, Paris, Mouton, 1975.

Birch, Thomas, *The History of the Royal Society of London*, Volume I, London, Printed for A. Miller in the Strand, 1756.

Birch, Thomas (ed.), *The Works of the Honourable Robert Boyle*, Volumes I-VI, London, Printed for J. and F. Rivington [etc.], 1772.

Black, Robert C., *The Younger John Winthrop*, New York, Columbia University Press, 1966.

Boghurst, William, *Loimographia, An Account of the Great Plague in London in the Year 1665*, J. F. Payne (ed.), London, Shaw and Sons, 1894.

Bonser, Wilfrid, "Epidemics During the Anglo-Saxon Period," *Journal of the British Archaeological Association*, Third Series, 9 (1944), pp. 48-71.

Borrichius, Olaus, "Conspectus Scriptorum Chemicorum Celebriorum," *Bibliotheca Chemica Curiosa*, Volumes I-II, J. J. Manget (ed.), Genevæ, Sumpt...De Tournes, 1702.

[Borrichius, Olaus], *Olai Borrichii Itinerarium 1660-1665*, Volume III, H. D. Schepelern (ed.), Copenhagen, Danish Society of Language and Literature, 1983.

Bove, Frank James, *The Story of Ergot: For Physicians, Pharmacists, Nurses, Biochemists, Biologists and Others Interested in the Life Sciences*, Basel, S. Karger, 1970.

Boyle, Robert, *Some Motives and Incentives To the Love of God [Seraphic Love]*, London, Printed for *Henry Herringman...*, 1659.

Boyle, Robert, *The Sceptical Chymist...*, London, Printed by F. Cadwell for F. Crooke, 1661.

Boyle, Robert, *Some Considerations touching the Usefulnesse of Experimental Naturall Philosophy*, Oxford, Printed by Hen: Hall..., 1663.

Boyle, Robert, *The Origine of Formes and Qualities...*, Oxford, Printed by H. Hall..., 1666.

Boyle, Robert, "An Account of the two Sorts of the *Helmontian Laudanum*," *Philosophical Transactions*, 9 (1674), pp. 147-149.

Boyle, Robert, *Suspicions about some Hidden Qualities of the Air...*, London, Printed by *W. G.* and are to be Sold by *M. Pitt*...,1674.

[Boyle, Robert], "Of the Incalescence of *Quicksilver* with *Gold*, generously imparted by *B. R.*," *Philosophical Transactions*, 10 (1675-1675/76), pp. 515-533.

Boyle, Robert, *The Producibleness of Chymical Principles*, Oxford, Printed by Henry Hall..., 1680.

Boyle, Robert, *A Free Enquiry Into the Vulgarly Receiv'd Notion of Nature...*, London, Printed by *H. Clark*..., 1685/6.

Boyle, Robert, "Medicinal Experiments," *Some Receipts of Medicines, For the most part Parable and Simple. Sent to a Friend in America*, London, s.n., 1688.

Boyle, Robert, *The General History of the Air*, London, Printed for *Awnsham* and *John Churchill*..., 1692.

[Boyle, Robert], *Letters and Papers of Robert Boyle: From the Archives of the Royal Society*, M. Hunter *et al.* (eds.), Bethesda, Md., University Publications of America, 1990, [microform].

Bradley, Leslie, "Some Medical Aspects of Plague," *The Plague Reconsidered: A new look at its origins and effects in 16th and 17th Century England*, Matlock, Eng., Local Population Studies, 1977.

Breger, Herbert, "*Elias Artista*—A Precursor of the Messiah in Natural Science," *Nineteen Eighty-Four: Science Between Utopia and Dystopia*, E. Mendelsohn and H. Nowotny (eds.), Dordrecht, D. Reidel Publishing Company, 1984.

Brock, C. Helen, "The Influence of Europe on Colonial Massachusetts Medicine," *Publications of the Colonial Society of Massachusetts*, 57 (1980), pp. 101-143.

Bronson, Henry, "Medical History and Biography," *Papers of the New Haven Colony Historical Society*, 2 (1877), pp. 239-388.

Browne, Alice, "J. B. van Helmont's Attack on Aristotle," *Annals of Science*, 36 (1979), pp. 575-591.

Cameron, E. A., and French, E. B., "St. Anthony's Fire Rekindled: Gangrene due to Therapeutic Dose of Ergotamine," *British Medical Journal*, 2 (1960), p. 28.

Cantor, G. N., and Hodge, M. J. S., "Introduction: Major themes in the development of ether theories from the ancients to 1900," *Conceptions of ether: Studies in the history of ether theories 1740-1900*, G. N. Cantor and M. J. S. Hodge (eds.), Cambridge, Cambridge University Press, 1981.

Cantor, G. N., "The theological significance of ethers," *Conceptions of ether: Studies in the history of ether theories 1740-1900*, G. N. Cantor and M. J. S. Hodge (eds.), Cambridge, Cambridge University Press, 1981.

Capp, B. S., *The Fifth Monarchy Men: A Study in Seventeenth-century English Millenarianism*, London, Faber and Faber, 1972.

Capp, Bernard, *Astrology and the Popular Press: English Almanacs 1500-1800*, London, Faber & Faber, 1979.

Cardilucius, Johann Hiskias, *Magnalia Medico-Chymica Continuata*, Nuremberg, Wolffgang Moritz Endters, 1680.

Chamberlayne, John, "Remarks upon the *Plague* at *Copenhagen* in the Year *1711*," *Philosophical Transactions*, 28 (1713), pp. 279-281.

Champion, J. A. I., *London's Dreaded Visitation: The Social Geography of the Great Plague in 1665*, London, Centre for Metropolitan History, University of London, 1995.

Clark, Sir George, *A History of the Royal College of Physicians of London*, Volume I, Oxford, Clarendon Press, 1964.

Clark, William, "The Scientific Revolution in the German Nations," *The Scientific Revolution in National Context*, R. Porter and M. Teich (eds.), Cambridge, Cambridge University Press, 1992.

Clericuzio, Antonio, "Robert Boyle and the English Helmontians," *Alchemy Revisited: Proceedings of the International Conference on the History of Alchemy at the University of Groningen 17-19 April 1989*, Z. R. W. M. von Martels (ed.), Leiden, E. J. Brill, 1990.

Clericuzio, Antonio, "From van Helmont to Boyle: A study of the transmission of helmontian chemical and medical theories in seventeenth-century england," *British Journal for the History of Science*, 26 (1993), pp. 303-334.

Clericuzio, Antonio, "The Internal Laboratory. The Chemical Reinterpretation of Medical Spirits in England (1650-1680)," *Alchemy and Chemistry in the 16th and 17th Centuries*, P. Rattansi and A. Clericuzio (eds.), Dordrecht, Kluwer Academic Publisher, 1994.

Clericuzio, Antonio, *Elements, Principles and Corpuscles: A Study of Atomism and Chemistry in the Seventeenth Century*, Dordrecht, Kluwer Academic Publishers, 2000.

Clippingdale, S. D., "A Medical Roll of Honour. Physicians and Surgeons who Remained in London during the Great Plague," *British Medical Journal*, 1 (1909), pp. 251-253.

Cogan, Thomas, *The Haven Of Health*, London, Printed by Anne Griffin..., 1636.

Cohausen, Johann Heinrich, *Hermippus Redivivus...*, E. Goldsmid (ed.), Edinburgh, Privately Printed, 1885.

Cohen, I. Bernard, "The Beginning of Chemical Instruction in America: A Brief Account of the Teaching of Chemistry at Harvard Prior to 1800," *Chymia*, 3 (1950), pp. 17-44.

Cohen, I. Bernard, "Ethan Allen Hitchcock: Soldier-Humanitarian-Scholar, Discoverer of the 'True Subject' of the Hermetic Art," *Proceedings of the American Antiquarian Society*, 61 (1951), pp. 29-136.

Cohen, I. Bernard, "The *Principia*, Universal Gravitation, and the 'Newtonian Style', Relation to the Newtonian Revolution in Science: *Notes on the Occasion of the 250th Anniversary of Newton's Death*," *Contemporary Newtonian Research*, Z. Bechler (ed.), Dordrecht, D. Reidel Publishing Company, 1982.

Cohen, I. Bernard, "Newton's Third Law and Universal Gravitation," *Journal of the History of Ideas*, 48 (1987), pp. 571-593.

Cohen, I. Bernard, and Westfall, Richard S. (eds.), *Newton: Texts Backgrounds Commentaries*, New York, W. W. Norton & Company, Inc., 1995.

Coleman, D. C., *The Economy of England, 1450-1750*, London, Oxford University Press, 1977.

Collins, E. J. T., "Dietary Change and Cereal Consumption in Britain in the Nineteenth Century," *Agricultural History Review*, 23 (1975), pp. 97-115.

Comer, James, "Distilled Beverages," *The Cambridge World History of Food*, Volume I, K. F. Kiple and K. C. Ornelas (eds.), Cambridge, Cambridge University Press, 2000.

Cook, Harold J., *The Decline of the Old Medical Regime in Stuart London*, Ithaca, Cornell University Press, 1986.

Cook, Harold J., "The Society of Chemical Physicians, the New Philosophy, and the Restoration Court," *Bulletin of the History of Medicine*, 61 (1987), pp. 61-77.

Cooper, William Durrant, "Notices of the Last Great Plague, 1665-6; from the Letters of John Allin to Philip Fryth and Samuel Jeake," *Archaeologia*, 37 (1857), pp. 1-22.

Cowell, E. B., and Thomas, F. W. (trans.), *The Harsa-carita of Bāna*, Delhi, Motilal Banarsidass, 1961.

Creighton, Charles, *A History of Epidemics in Britain*, Volumes I-II, Cambridge, Cambridge University Press, 1891.

Culpeper, Nicholas, *The Practice of Physick...*, L. Riverius (trans.), London, Printed by *Peter Cole* in *Leaden-Hall...*, 1655.

D'Mello, J. P. F., *et al.*, "Fusarium Mycotoxins," *Handbook of Plant and Fungal Toxicants*, J. P. F. D'Mello (ed.), Boca Raton, Fl., CRC Press, 1997.

Debus, Allen G., *The Chemical Dream of the Renaissance*, Cambridge, W. Heffer & Sons Ltd., 1968.

Debus, Allen G., "The Chemical Philosophers: Chemical Medicine from Paracelsus to Van Helmont," *History of Science*, 12 (1974), pp. 235-259.

Debus, Allen G., *The Chemical Philosophy: Paracelsian Science and Medicine in the Sixteenth and Seventeenth Centuries*, Volumes I-II, New York, Science History Publications, 1977.

Debus, Allen G., *Man and Nature in the Renaissance*, Cambridge, Cambridge University Press, 1978.

Debus, Allen G., *Paracelsus, Five Hundred Years: Three American Exhibits*, Bethesda, Md., Published by the Friends of the National Library of Medicine..., 1993.

Debus, Allen G., "Paracelsianism and the Diffusion of the Chemical Philosophy in Early Modern Europe," *Paracelsus: The Man and his Reputation, his Ideas and their Transformation*, O. P. Grell (ed.), Leiden, Brill, 1998.

Debus, Allen G., "Paracelsus and the Delayed Scientific Revolution in Spain: A Legacy of Philip II," *Reading the Book of Nature: The Other Side of the Scientific Revolution*, A. G. Debus and M. T. Walton (eds.), Kirksville, Mo., Truman State University, 1998.

Debus, Allen G., "Chemists, Physicians, and Changing Perspectives on the Scientific Revolution," *Isis*, 89 (1998), pp. 66-81.

Debus, Allen G., *Chemistry and Medical Debate: van Helmont to Boerhaave*, USA, Science History Publications, 2001.

Devitt, Roy E., *et al.*, "Ergot Poisoning," *Journal of the Irish Medical Association*, 63 (1970), pp. 441-445.

Dexter, Franklin Bowditch, *Biographical Sketches of the Graduates of Yale College...*, Volume II, New York, Henry Holt and Company, 1896.

Dexter, Franklin Bowditch (ed.), *The Literary Diary of Ezra Stiles*, Volumes I-III, New York, Charles Scribner's Sons, 1901.

Dexter, Franklin Bowditch (ed.), *Documentary History of Yale University: Under the Original Charter of the Collegiate School of Connecticut 1701-1745*, New Haven, Yale University Press, 1916.

"Diary of John Quincy Adams," *Proceedings of the Massachusetts Historical Society*, Second Series, 16 (1903), pp. 291-464.

Dictionary of American Biography, Volume XX, D. Malone (ed.), New York, Charles Scribner's Sons, 1943.

Dirac, P. A. M., "Is there an Æther?" *Nature*, 168 (1951), pp. 906-907.

Dobbs, Betty Jo Teeter, *The Foundations of Newton's Alchemy or "The Hunting of the Greene Lyon"*, Cambridge, Cambridge University Press, 1975.

Dobbs, Betty Jo Teeter, "Newton's Copy of *Secrets Reveal'd* and the Regimens of the Work," *Ambix*, 26 (1979), pp. 145-169.

Dobbs, Betty Jo Teeter, "Newton's Alchemy and His Theory of Matter," *Isis*, 73 (1982), pp. 511-528.

Dobbs, Betty Jo Teeter, "Newton's Rejection of the Mechanical Aether: Empirical Difficulties and Guiding Assumptions," *Scrutinizing Science: Empirical Studies of Scientific Change*, A. Donovan, L. Laudan, and R. Laudan (eds.), Dordrecht, Kluwer Academic Publishers, 1988.

Dobbs, Betty Jo Teeter, "Newton's Alchemy and his 'Active Principle' of Gravitation," *Newton's Scientific and Philosophical Legacy*, P. B. Scheurer and G. Debrock (eds.), Dordrecht, Kluwer Academic Publishers, 1988.

Dobbs, Betty Jo Teeter, "From the Secrecy of Alchemy to the Openness of Chemistry," *Solomon's House Revisited: The Organization and Institutionalization of Science*, T. Frängsmyr (ed.), USA, Science History Publications, 1990.

Dobbs, Betty Jo Teeter, *The Janus Faces of Genius: The role of alchemy in Newton's thought*, Cambridge, Cambridge University Press, 1991.

Dobbs, Betty Jo Teeter, "Stoic and Epicurean doctrines in Newton's system of the world," *Atoms, pneuma, and tranquillity: Epicurean and Stoic Themes in European Thought*, M. J. Osler (ed.), Cambridge, Cambridge University Press, 1991.

Dobbs, Betty Jo Teeter, "Gravity and Alchemy," *The Scientific Enterprise; The Bar-Hillel Colloquium: Studies in History, Philosophy, and Sociology of Science*, E. Ullmann-Margalit (ed.), Dordrecht, Kluwer Academic Publishers, 1992.

Dobbs, Betty Jo Teeter, "'The Unity of Truth': An Integrated View of Newton's Work," *Action and Reaction: Proceedings of a Symposium to Commemorate the Tercentenary of Newton's Principia*, P. Theerman and A. F. Seeff (eds.), Newark, University of Delaware Press, 1993.

Dobbs, Betty Jo Teeter, and Jacob, Margaret C., *Newton and the Culture of Newtonianism*, Atlantic Highlands, N.J., Humanities Press, 1995.

Dobson, Mary J., *Contours of Death and Disease in Early Modern England*, Cambridge, Cambridge University Press, 1997.

Doolittle, I. G., "The Plague in Colchester—1579-1666," *Transactions of the Essex Archeological Society*, Third Series, 4 (1972), pp. 134-145.

Duncan, C. J., and Scott, S., "What caused the Black Death?" *Postgraduate Medical Journal*, 81 (2005), pp. 315-320.

Dunn, Richard S., *Puritans and Yankees: The Winthrop Dynasty of New England 1630-1717*, Princeton, Princeton University Press, 1962.

Dussauce, H., *A General Treatise of the Manufacture of Soap, Theoretical and Practical*, Philadelphia, Henry Carey Baird, 1869.

Eadie, Mervyn J., "Convulsive ergotism: epidemics of the seratonin syndrome?" *The Lancet: Neurology*, 2 (2003), pp. 429-434.

Echard, Laurence, *The History of England. From the First Entrance of Julius Cæsar and the Romans, to the Conclusion of the Reign of King James the Second, and the Establishment of King William and Queen Mary upon the Throne, in the Year 1688*, Third Edition, London, Printed for Jacob Tonson..., 1720.

Einstein, Albert, "Ether and relativity," *Sidelights on Relativity*, G. B. Jeffrey and W. Perrett (trans.), London, W. Perrett, 1922.

Ellis, Henry (ed.), *Original Letters, Illustrative of English History...*, Volume VI, Fourth Series, London, Harding and Lepard, 1827.

Evans, R. J. W., *Rudolf II and His World: A Study in Intellectual History 1576-1612*, Oxford, Clarendon Press, 1973.

Evelyn, John, *The Diary of John Evelyn*, E. S. de Beer (ed.), London, Oxford University Press, 1959.

Faber, J. A., "The Decline of the Baltic Grain-Trade in the Second Half of the 17th Century," *Acta Historiae Neerlandica*, 1 (1966), pp. 108-131.

Fairley, T., "Notes on the History of Distilled Spirits, Especially Whiskey and Brandy," *Analyst*, 30 (1905), pp. 293-306.

Fajardo, J. E., *et al.*, "Retention of Ergot Alkaloids in Wheat During Processing," *Cereal Chemistry*, 72 (1995), pp. 291-298.

Faustius, Johann Michael (ed.), *Philaletha illustratus, sive Introitus apertus...*, Francofurti ad Moenum, Sumpt...Andreae, 1706.

Ferguson, John, *Bibliographica Chemica: A Catalogue of the Alchemical, Chemical and Pharmaceutical Books in the Collection of the late James Young of Kelly and Durris*, Volume II, Glasgow, J. Maclehose and Sons, 1906.

Figala, Karin, "Zwei Londoner Alchemisten um 1700: Sir Isaac Newton und Cleidophorus Mystagogus," *Physis*, 18 (1976), pp. 245-273.

Figala, Karin, "Newton as Alchemist," *History of Science*, 15 (1977), pp. 102-137.

Figala, Karin, and Petzold, Ulrich, "Alchemy in the Newtonian circle: personal acquaintances and the problem of the late phase of Isaac Newton's alchemy," *Renaissance and Revolution: Humanists, scholars, craftsmen, and natural philosophers in early modern Europe*, J. V. Field and F. A. J. L. James (eds.), Cambridge, Cambridge University Press, 1993.

Figala, Karin, "Newton's alchemy," *The Cambridge Companion to Newton*, I. B. Cohen and G. E. Smith (eds.), Cambridge, Cambridge University Press, 2002.

Figlio, Karl, "Psychoanalysis and the scientific mind: Robert Boyle," *British Journal for the History of Science*, 32 (1999), pp. 299-314.

Forbes, R. J., *Short History of the Art of Distillation: From the Beginnings up to the Death of Cellier Blumenthal*, Leiden, E. J. Brill, 1948.

Forbes, R. J., *Studies in Ancient Technology*, Volume III, Leiden, E. J. Brill, 1955.

Force, James E., "The Virgin, the Dynamo, and Newton's Prophetic History," *Millenarianism and Messianism in Early Modern European Culture*, Volume III, J. E. Force and R. H. Popkin (eds.), Dordrecht, Kluwer Academic Publishers, 2001.

Fothergill, John, *An Account of the Sore Throat Attended with Ulcers*, Fourth Edition, London, Printed for C. Davis..., 1754.

Fraser, Antonia, *Cromwell, The Lord Protector*, New York, Alfred A. Knopf, 1973.

Friend, D. Johannis, "Epistola D. Johannis Friend ad Editorem missa, de Spasmi Rarioris Historia," *Philosophical Transactions*, 22 (1701), pp. 799-804.

Fulton, John F., "Boyle and Sydenham," *Journal of the History of Medicine and Allied Sciences*, 11 (1956), pp. 351-352.

Fulton, John F., *A Bibliography of the Honourable Robert Boyle; Fellow of the Royal Society*, Second Edition, Oxford, Clarendon Press, 1961.

Gabbai, and Lisbonne, and Pourquier, "Ergot Poisoning at Pont St. Esprit," *British Medical Journal*, 2 (1951), pp. 650-651.

Gadbury, John, *Collectio Geniturarum...*, London, Printed by James Cottrel, 1662.

Garencieres, Theophilus, *A Mite Cast into the Treasury of the Famous City of London...*, London, Printed by *Tho. Ratcliffe...*, 1666.

Garrod, H. W. (ed.), *Keats: Poetical Works*, Oxford, Oxford University Press, 1956.

Geoghegan, D., "A Licence of Henry VI to Practise Alchemy," *Ambix*, 6 (1957), pp. 10-17.

Gifford, Jr., George E., "Botanic Remedies in Colonial Massachusetts, 1620-1820," *Publications of the Colonial Society of Massachusetts*, 57 (1980), pp. 263-288.

Goodman, Louis S., and Gilman, Alfred (eds.), *The Pharmacological Basis of Therapeutics: A Textbook of Pharmacology, Toxicology, and Therapeutics for Physicians and Medical Students*, Fourth Edition, London, Macmillan Company, 1970.

Gottfried, Robert S., *Epidemic Disease in Fifteenth Century England: The Medical Response and the Demographic Consequences*, New Brunswick, Rutgers University Press, 1978.

Gottfried, Robert S., "Plague, Public Health and Medicine in Late Medieval England," *Maladies Et Société (XIIe-XVIIIe siècles)*, N. Bulst and R. Delort (eds.), Paris, Éditions Du Centre National De La Recherche Scientifique, 1989.

Gottwald, John Christoph., "An Abridgement of a Book intitl'd, A Description of the Plague, which happened in the Royal City of *Dantzick*, in the Year *1709*," *Philosophical Transactions*, 28 (1713), pp. 101-144.

Grant, Edward (ed.), *A Source Book in Medieval Science*, Cambridge, Harvard University Press, 1974.

Grant, William, *Observations on the Nature and Cure of Fevers*, Volume I, Second Edition, London, Printed for T. Cadell..., 1772.

Gras, Norman Scott Brien, *The Evolution of the English Corn Market: From the Twelfth to the Eighteenth Century*, Cambridge, Harvard University Press, 1915.

Graunt, John, *London's Dreadful Visitation: Or A Collection of All the Bills of Mortality for this Present Year...*, London, Printed and are to be sold by *E. Cotes...*, 1665.

Graunt, John, "Reflections on the Weekly Bills of Mortality...," *A Collection of Very Valuable and Scarce Pieces Relating to the Last Plague in the Year 1665*, Second Edition, London, Printed for *J. Roberts...*, 1721.

Graunt, John, "*Natural* and *Political* Observations...upon the Bills of Mortality," *The Economic Writings of Sir William Petty*, Volume II, C. H. Hull (ed.), Cambridge, Cambridge University Press, 1899.

Greenwood, Isaac, "Rev. Richard Blinman of Marshfield, Gloucester and New London," *New-England Historical and Genealogical Register*, 54 (1900), pp. 39-44.

Grell, Ole Peter, "The Acceptable Face of Paracelsianism...," *Paracelsus: The Man and his Reputation, his Ideas and their Transformation*, O. P. Grell (ed.), Leiden, Brill, 1998.

Gyford, Janet, *Witham 1500-1700: Making a Living*, Witham, Eng., Privately Printed, 1996.

Hall, A. Rupert, and Hall, Marie Boas (eds. and trans.), *Unpublished Scientific Papers of Isaac Newton*, Cambridge, Cambridge University Press, 1962.

Hall, A. Rupert, and Hall, Marie Boas (eds.), *The Correspondence of Henry Oldenburg*, Volume II, Madison, University of Wisconsin Press, 1965.

Hall, A. Rupert, *Isaac Newton: Adventurer in Thought*, Oxford, Blackwell Publishers, 1992.

[Hall], Marie Boas, *Robert Boyle and Seventeenth Century Chemistry*, Cambridge, Cambridge University Press, 1958.

Hall, Marie Boas, "Newton's Voyage in the Strange Seas of Alchemy," *Reason, Experiment, and Mysticism in the Scientific Revolution*, M. L. R. Bonelli and W. R. Shea (eds.), New York, Science History Publications, 1975.

Harrison, John, and Laslett, Peter, *The Library of John Locke*, Oxford, Oxford University Press, 1965.

Harrison, John, *The Library of Isaac Newton*, Cambridge, Cambridge University Press, 1978.

Harrison, William, *The Description of England*, Ithaca, Cornell University Press, 1968.

[Hartlib, Samuel], *The Reformed Common-Wealth of Bees. Presented in severall Letters and Observations to Sammuel Hartlib Esq...*, London, Printed for *Giles Calvert* at the *Black-Spread-Eagle* at the West-end of Pauls, 1655.

"Harvard College Records, College Book I," *Publications of the Colonial Society of Massachusetts*, 15 (1925), pp. 3-168.

Haskins, Charles Homer, *Studies in the History of Mediaeval Science*, Second Edition, Cambridge, Harvard University Press, 1927.

Hecker, I. F. C., *The Epidemics of the Middle Ages: No. II. The Dancing Mania*, B. G. Babington (trans.), Philadelphia, Haswell, Barrington, and Haswell, 1837.

Heimann, P. M., "Ether and imponderables," *Conceptions of ether: Studies in the history of ether theories 1740-1900*, G. N. Cantor and M. J. S. Hodge (eds.), Cambridge, Cambridge University Press, 1981.

Helmont, Jean Baptiste van, *Ortus Medicinæ...*, Amsterodami, Apud Ludovicum Elzevirium, 1648.

Helmont, Jean Baptiste van, *A Ternary of Paradoxies: The Magnetick Cure of Wounds...*, W. Charleton (trans.), London, Printed by *James Flesher* for *William Lee...*, 1650.

Helmont, Jean Baptiste van, *Oriatrike or, Physick Refined...*, J. Chandler (trans.), London, Printed for Lodowick Loyd, 1662.

Henderson, John, "Epidemics in Renaissance Florence: Medical Theory and Government Response," *Maladies Et Société (XIIe-XVIIIe siècles)*, N. Bulst and R. Delort (eds.), Paris, Éditions Du Centre National De La Recherche Scientifique, 1989.

Henry, John, "Occult Qualities and the Experimental Philosophy: Active Principles in Pre-Newtonian Matter Theory," *History of Science*, 24 (1986), pp. 335-381.

Henry, John, "Newton, Matter, and Magic," *Let Newton Be!*, J. Fauvel, R. Flood, M. Shortland, and R. Wilson (eds.), Oxford, Oxford University Press, 1988.

Henry, John, "The Scientific Revolution in England," *The Scientific Revolution in National Context*, R. Porter and M. Teich (eds.), Cambridge, Cambridge University Press, 1992.

Henry, John, "'Pray Do Not Ascribe That Notion To Me': God and Newton's Gravity," *The Books of Nature and Scripture: Recent Essays on Natural Philosophy, Theology, and Biblical Criticism in the Netherlands of Spinoza's Time and the British Isles of Newton's Time*, J. E. Force and R. H. Popkin (eds.), Dordrecht, Kluwer Academic Publishers, 1994.

Hertodt, Johannis Ferdinandi, "Epistolâ *contra* Philalethæ Processus," *Bibliotheca Chemica Curiosa*, Volumes I-II, J. J. Manget (ed.), Genevæ, Sumpt...De Tournes, 1702.

Hillgarth, J. N., *Raymon Lull and Lullism in Fourteenth-Century France*, Oxford, Clarendon Press, 1971.

Hitchcock, Ethan Allen, *Remarks upon alchymists...*, Carlisle, Penn., Printed at the Herald Office, 1855.

Hitchcock, Ethan Allen, *Remarks upon Alchemy and the Alchemists, Indicating a Method of Discovering the True Nature of Hermetic Philosophy; and Showing that the Search after The Philosopher's Stone had not for its Object the Discovery of an Agent for the Transmutation of Metals...*, Boston, Crosby, Nichols, and Company, 1857.

Hodges, Nathaniel, *Vindiciæ Medicinæ & Medicorum: Or An Apology For the Profession and Professors of Physick*, London, Printed by *J. F.* for *Henry Broom*, 1666.

Hodges, Nathaniel, "Letter from Dr. Hodges to a Person of Quality," *A Collection of Very Valuable and Scarce Pieces Relating to the Last Plague in the Year 1665*, Second Edition, London, Printed for *J. Roberts...*, 1721.

Hodges, Nathaniel, *Loimologia: Or, An Historical Account of The Plague in London in 1665...*, J. Quincy (trans.), Third Edition, London, Printed for *E. Bell...*, 1721.

Hofmann, Albert, "Historical View on Ergot Alkaloids," *Pharmacology*, Supplement 1, 16 (1978), pp. 1-11.

Home, R. W., "'Newtonianism' and the Theory of the Magnet," *History of Science*, 15 (1977), pp. 252-266.

Home, R. W., "Force, electricity, and the powers of living matter in Newton's mature philosophy of nature," *Religion, Science and Worldview: Essays in Honor of Richard S. Westfall*, M. J. Osler and P. L. Farber (eds.), Cambridge, Cambridge University Press, 1985.

Home, R. W., "Newton's subtle matter: the *Opticks* queries and the mechanical philosophy," *Renaissance and Revolution: Humanists, scholars, craftsmen, and natural philosophers in early modern Europe*, J. V. Field and F. A. J. L. James (eds.), Cambridge, Cambridge University Press, 1993.

Hubicki, Wlodzimierz, "Maier, Michael," *Dictionary of Scientific Biography*, Volumes I-XIV, C. C. Gillispie (ed.), New York, Charles Scribner's Sons, 1974.

Hunter, Michael, "Alchemy, magic and moralism in the thought of Robert Boyle," *British Journal for the History of Science*, 23 (1990), pp. 387-410.

Hunter, Michael, "How Boyle Became a Scientist," *History of Science*, 33 (1995), pp. 59-103.

Hunter, Michael, "The Reluctant Philanthropist: Robert Boyle and the 'Communication of Secrets and Receits in Physick,'" *Religio Medici: Medicine and Religion in Seventeenth-Century England*, O. P. Grell and A. Cunningham (eds.), Aldershot, Eng., Scolar Press, 1996.

Hunter, Michael, "Boyle Versus the Galenists: A Suppressed Critique of Seventeenth-Century Medical Practice and its Significance," *Medical History*, 41 (1997), pp. 322-361.

Hunter, Michael, "Robert Boyle (1627-91): a suitable case for treatment?" *British Journal for the History of Science*, 32 (1999), pp. 261-275.

Hunter, Michael, and Davis, Edward B. (eds.), *The Works of Robert Boyle*, Volume XIII, London, Pickering & Chatto, 2000.

Hunter, Michael, and Clericuzio, Antonio, and Principe, Lawrence M. (eds.), *The Correspondence of Robert Boyle*, Volume I, London, Pickering & Chatto, 2001.

Huxham, John, *A Dissertation on the Malignant, Ulcerous Sore-Throat*, London, Printed for J. Hinton..., 1757.

Huygens, Christiaan, *Oeuvres Complètes*, Volume X, The Hague, Martinus Nijhoff, 1905.

Hyde, Edward, Earl of Clarendon, *Selections from Clarendon*, G. Huehns (ed.), London, Oxford University Press, 1955.

Iliffe, Rob, "Isaac Newton: Lucatello Professor of Mathematics," *Science Incarnate: Historical Embodiments of Natural Knowledge*, C. Lawrence and S. Shapin (eds.), Chicago, University of Chicago Press, 1998.

Iliffe, Rob, "Those 'Whose Business it is to Cavill': Newton's Anti-Catholicism," *Newton and Religion: Context, Nature, and Influence*, J. E. Force and R. H. Popkin (eds.), Dordrecht, Kluwer Academic Publishers, 1999.

Israel, Jonathan, *The Dutch Republic: Its Rise, Greatness, and Fall, 1477-1806*, Oxford, Clarendon Press, 1995.

Jantz, Harold, "America's First Cosmopolitan," *Proceedings of the Massachusetts Historical Society*, 84 (1972), pp. 3-25.

Jeake, Samuel, *An Astrological Diary of the Seventeenth Century: Samuel Jeake of Rye 1652-1699*, M. Hunter and A. Gregory (eds.), Oxford, Clarendon Press, 1988.

Jodziewicz, Thomas W., "A Stranger in the Land: Gershom Bulkeley of Connecticut," *Transactions of the American Philosophical Society*, 78 (1988), pp. 1-106.

Joffe, Abraham Z., "Toxicity of Overwintered Cereals," *Plant and Soil*, 18 (1963), pp. 31-44.

Joffe, Abraham Z., "Toxin Production by Cereal Fungi Causing Alimentary Toxic Aleukia in Man," *Mycotoxins in Foodstuffs*, G. N. Wogan (ed.) Cambridge, MIT Press, 1965.

Joffe, Abraham Z., "The Genus Fusarium," *Mycotoxic Fungi, Mycotoxins, Mycotoxicoses: An Encyclopedic Handbook*, Volume I, T. D. Wyllie and L. G. Morehouse (eds.), New York, Marcel Dekker, Inc., 1977.

Joffe, Abraham Z., "Environmental Conditions Conducive to *Fusarium* Toxin Formation Causing Serious Outbreaks in Animals and Man," *Veterinary Research Communications*, 7 (1983), pp. 187-193.

Joffe, Abraham Z., *Fusarium Species: Their Biology and Toxicology*, New York, John Wiley & Sons, Inc., 1986.

Johnson, L. W., and Wolbarsht, M. L., "Mercury Poisoning: A Probable Cause of Isaac Newton's Physical and Mental Ills," *Notes and Records of the Royal Society of London*, 34 (1979), pp. 1-9.

Johnstone, James, *An Historical Dissertation Concerning the Malignant Epidemical Fever of 1756*, London, Printed for W. Johnston..., 1758.

Johnstone, J., *A Treatise on the Malignant Angina: or, Putrid and Ulcerous Sore-Throat*, Worcester, Printed and Sold by E. Berrow..., 1779.

Jones, A. G. E., "Plagues in Suffolk in the Seventeenth Century," *Notes and Queries*, 198 (1953), pp. 384-386.

Jordan, Edward, *A Briefe Discourse of a Disease Called the Suffocation of the Mother*, London, Printed by *John Windet...*, 1603.

Josselin, Ralph, *The Diary of Ralph Josselin 1616-1683*, A. MacFarlane (ed.), London, Oxford University Press, 1976.

Jung, Carl Gustav, "Alchemical Studies," *The Collected Works of C. G. Jung*, Volume XIII, R. F. C. Hull (trans.), Princeton, Princeton University Press, 1967.

Kahr, Brett, "Robert Boyle: a Freudian perspective on an eminent scientist," *British Journal for the History of Science*, 32 (1999), pp. 277-284.

Katzung, Bertram G. (ed.), *Basic & Clinical Pharmacology*, Seventh Edition, Stamford, Ct., Appleton & Lange, 1998.

Kemp, William, *A Brief Treatise Of the Nature, Causes, Signes, Preservation From, and Cure Of The Pestilence*, London, Printed for...*D. Kemp...*, 1665.

Kendall, George, *An Appendix to the Unlearned Alchimist; Wherein is contained the true Receipt of that Excellent Diaphoretick and Diuretick Pill...*, London, Printed for *Joseph Leigh...*, 1664[?].

Kephale, Richard, *Medela Pestilentiæ: Wherein is contained several Theological Queries Concerning the Plague, With Approved Antidotes, Signes, and Symptoms...*, London, Printed by *J. C.* for *Samuel Speed...*, 1665.

Kerényi, Carl, *Eleusis: Archetypal Image of Mother and Daughter*, R. Manheim (trans.), New York, Bollingen Foundation, 1967.

Kerridge, Eric, *The Agricultural Revolution*, New York, Augustus M. Kelley, 1967.

Kerridge, Eric, *The Farmers of Old England*, Totowa, N.J., Rowman and Littlefield, 1973.

Keynes, John Maynard, "Newton, the Man," *Newton Tercentenary Celebrations 15-19 July 1946*, Cambridge, Cambridge University Press, 1947.

Keynes, Milo, "The Personality of Isaac Newton," *Notes and Records of the Royal Society of London*, 49 (1995), pp. 1-56.

Kim, Yung Sik, "Another Look at Robert Boyle's Acceptance of the Mechanical Philosophy: Its Limits and its Chemical and Social Contexts," *Ambix*, 38 (1991), pp. 1-10.

Kirby, David, *Northern Europe in the Early Modern Period: The Baltic World 1492-1772*, London, Longman Group UK Limited, 1990.

Kittredge, George Lyman, "George Stirk, Minister," *Transactions of the Colonial Society of Massachusetts*, 13 (1912), pp. 16-55.

Kittredge, George Lyman, "Dr. Robert Child the Remonstrant," *Transactions of the Colonial Society of Massachusetts*, 21 (1920), pp. 1-146.

Kittredge, George Lyman, *Witchcraft in Old and New England*, Cambridge, Harvard University Press, 1929.

Kochavi, Matania Z., "One Prophet Interprets Another: Sir Isaac Newton and Daniel," *The Books of Nature and Scripture: Recent Essays on Natural Philosophy, Theology, and Biblical Criticism in the Netherlands of Spinoza's Time and the British Isles of Newton's Time*, J. E. Force and R. H. Popkin (eds.), Dordrecht, Kluwer Academic Publishers, 1994.

Košir, B., and Smole, P., and Povšič, Z., ["Factors affecting the yield and quantity of sclerotia from *Claviceps purpurea*"], (in Slovenian, with English abstract), *Farmacevtski Vestnik*, 32 (1981), pp. 21-25.

Kozhamthadam, Job, *The Discovery of Kepler's Laws: The Interaction of Science, Philosophy, and Religion*, Notre Dame, University of Notre Dame Press, 1994.

Kritikos, P. G., and Papadaki, S. P., "The History of the Poppy and of Opium and their Expansion in Antiquity in the Eastern Mediterranean Area," G. Michalopoulos (trans.), *Bulletin on Narcotics*, 19 (1967), pp. 5-10, 17-38.

Kubrin, David, "Newton and the Cyclical Cosmos: Providence and the Mechanical Philosophy," *Journal of the History of Ideas*, 28 (1967), pp. 325-346.

Lamb, H. H., *Climate: Present, Past and Future*, Volume II, London, Methuen & Co. Ltd., 1977.

Langford, A. W., "The Plague in Herefordshire," *Transactions of the Woolhope Naturalists' Field Club*, 25 (1955-1957), pp. 146-153.

Leeuwen, Henry G. Van, *The Problem of Certainty in English Thought 1630-1690*, The Hague, Martinus Nijhoff, 1963.

Leigh, Charles, "Part of a Letter from *Dr. Charles Leigh* of *Lancashire* to the Publisher, giving an account of strange Epileptic Fits," *Philosophical Transactions*, 23 (1702), pp. 1174-1175.

Lemery, Nicholas, *A Course of Chymistry; containing An easie Method of Preparing those Chymical Medicines which are used in Physick...*, Third English Edition, J. Keill (trans.), London, W. Kettilby, 1698.

Lieber, Elinor, "Galen on Contaminated Cereals as a Cause of Epidemics," *Bulletin of the History of Medicine*, 44 (1970), pp. 332-345.

Linden, Stanton J., "Alchemy and Eschatology in Seventeenth-Century Poetry," *Ambix*, 31 (1984), pp. 102-124.

Lockyer, Lionel, *An Advertisement Concerning those most Excellent Pills Called Pillulæ Radiis Solis Extractæ. Being An Universal Medicine...*, London, s.n., 1664.

Lofft, Capel, "Food of the Poor of Ingleton," *Annals of Agriculture*, 26 (1796), p. 226.

Lorenz, Klaus, "Ergot in Cereal Grains," *CRC Critical Reviews in Food Science and Nutrition*, 11 (1979), pp. 311-354.

Luther, Martin, *Colloquia Mensalia*, H. Bell (trans.), London, Printed by *William Du Gard...*, 1652.

Macdonell, Arthur Anthony (trans.), *Hymns from the Rigveda*, London, Oxford University Press, 1922.

Maddison, R. E. W., "The Earliest Published Writing of Robert Boyle," *Annals of Science*, 17 (1961), pp. 165-173.

Maddison, R. E. W., "Studies in the Life of Robert Boyle, F.R.S.: Part VI. The Stalbridge Period, 1645-1655, and the Invisible College," *Notes and Records of the Royal Society of London*, 18 (1963), pp. 104-124.

Maddison, R. E. W., *The Life of the Honourable Robert Boyle*, London, Taylor and Francis Ltd., 1969.

Manley, Gordon, "Central England Temperatures: Monthly Means, 1659 to 1973," *Quarterly Journal of the Royal Meteorological Society*, 100 (1974), pp. 389-405.

Manuel, Frank E., *Isaac Newton: Historian*, Cambridge, Belknap Press of Harvard University Press, 1963.

Manuel, Frank E., *A Portrait of Isaac Newton*, Cambridge, Belknap Press of Harvard University Press, 1968.

Manuel, Frank E., *The Religion of Isaac Newton: The Fremantle Lectures 1973*, Oxford, Oxford University Press, 1974.

Matossian, Mary Kilbourne, "Mold poisoning: an unrecognized English health problem, 1550-1800," *Medical History*, 25 (1981), pp. 73-84.

Matossian, Mary Kilbourne, *Poisons of the Past: Molds, Epidemics, and History*, New Haven, Yale University Press, 1989.

Matossian, Mary Kilbourne, "Why the Quakers Quaked: The Influence of Climatic Change on Quaker Health, 1647-1659," *Quaker History*, 96 (2007), pp. 36-51.

McGuire, J. E., "Transmutation and Immutability: Newton's Doctrine on Physical Qualities," *Ambix*, 14 (1967), pp. 69-95.

McGuire, J. E., "Force, Active Principles, and Newton's Invisible Realm," *Ambix*, 15 (1968), pp. 154-208.

McKie, Douglas, "The Origins and Foundation of the Royal Society of London," *The Royal Society: Its Origins and Founders*, H. Hartley (ed.), London, Royal Society, 1960.

McMullin, Ernan, *Newton on Matter and Activity*, Notre Dame, University of Notre Dame Press, 1978.

Mellor, Joseph William, *A Comprehensive Treatise on Inorganic and Theoretical Chemistry*, Volume IX, London, Longmans, Green and Co., 1929.

Mendelsohn, J. Andrew, "Alchemy and Politics in England 1649-1665," *Past and Present*, 135 (1992), pp. 30-78.

Monfasani, John, "Marsilio Ficino and the Plato-Aristotle Controversy," *Marsilio Ficino: His Theology, His Philosophy, His Legacy*, M. J. B. Allen, V. Rees, and M. Davies (eds.), Leiden, Brill, 2002.

Mood, Fulmer, "John Winthrop, Jr., on Indian Corn," *New England Quarterly*, 10 (1937), pp. 131-133.

Moote, A. Lloyd, and Moote, Dorothy C., *The Great Plague: The Story of London's Most Deadly Year*, Baltimore, Johns Hopkins University Press, 2004.

Moran, Bruce T., "Privilege, Communication, and Chemiatry: The Hermetic-Alchemical Circle of Moritz of Hessen-Kassel," *Ambix*, 32 (1985), pp. 110-126.

Moran, Bruce T., *The Alchemical World of the German Court: Occult Philosophy and Chemical Medicine in the Circle of Moritz of Hessen (1572-1632)*, Stuttgart, Franz Steiner Verlag, 1991.

Moran, Bruce T., "The Herbarius of Paracelsus," *Pharmacy in History*, 35 (1993), pp. 99-127.

Moreau, Claude, *Moulds, Toxins and Food*, M. Moss (ed.), Chichester, Eng., John Wiley & Sons, 1979.

Morgan, John, *Godley Learning: Puritan Attitudes Towards Reason, Learning, and Education, 1560-1640*, Cambridge, Cambridge University Press, 1986.

Morison, Samuel Eliot, "The Library of George Alcock, Medical Student, 1676," *Transactions of the Colonial Society of Massachusetts*, 28 (1933), pp. 350-357.

Morison, Samuel Eliot, *The Founding of Harvard College*, Cambridge, Harvard University Press, 1935.

Morison, Samuel Eliot, *Harvard College in the Seventeenth Century*, Volumes I-II, Cambridge, Harvard University Press, 1936.

Morison, Samuel Eliot, *The Intellectual Life of Colonial New England*, New York, New York University Press, 1956.

Morris, Christopher, "The Plague in Britain" [Book Review], *The Historical Journal*, 14 (1971), pp. 205-215.

Mortimer, Cromwell, Dedication to John Winthrop, F. R. S., *Philosophical Transactions*, 40 (1741).

Morton, Charles, *Compendium Physicae*, T. Hornberger (ed.), *Publications of the Colonial Society of Massachusetts*, 33 (1940), pp. 1-210.

Moryson, Fynes, *An Itinerary*, Volume IV, Glasgow, James MacLehose and Sons, 1908.

Moss, William, *An Essay on the Management, Nursing and Diseases of Children...*, Second Edition, London, Printed by and for C. Boult..., 1794.

Mower, R. L., *et al.*, "Biological Control of Ergot by Fusarium," *Phytopatholgy*, 65 (1975), pp. 5-10.

Mullett, Charles F., *The Bubonic Plague and England*, Lexington, University of Kentucky Press, 1956.

Naphy, William, and Spicer, Andrew, *The Black Death and the history of plagues 1345-1730*, Stroud, Eng., Tempus Publishing Ltd., 2000.

"Necessary Directions For The Prevention and Cure Of The Plague in 1665. With divers Remedies of small Charge, by the College of Physicians," *A Collection of Very Valuable and Scarce Pieces Relating to the Last Plague in the Year 1665*, Second Edition, London, Printed for *J. Roberts...*, 1721.

Newman, William R., "Newton's *Clavis* as Starkey's *Key*," *Isis*, 78 (1987), pp. 564-574.

Newman, William R., "The Authorship of the *Introitus Apertus ad Occlusum Regis Palatium*," *Alchemy Revisited: Proceedings of the International Conference on the History of Alchemy at the University of Groningen 17-19 April 1989*, Z. R. W. M. von Martels (ed.), Leiden, E. J. Brill, 1990.

Newman, William R., "Prophecy and Alchemy: The Origin of Eirenaeus Philalethes," *Ambix*, 37 (1990), pp. 97-115.

Newman, William R., *Gehennical Fire: The Lives of George Starkey, an American Alchemist in the Scientific Revolution*, Cambridge, Harvard University Press, 1994.

Newman, William R., "George Starkey and the selling of secrets," *Samuel Hartlib and Universal Reformation: Studies in intellectual communication*, M. Greengrass, M. Leslie, and T. Raylor (eds.), Cambridge, Cambridge University Press, 1994.

Newman, William R., and Principe, Lawrence M., *Alchemy Tried in the Fire: Starkey, Boyle, and the Fate of Helmontian Chymistry*, Chicago, University of Chicago Press, 2002.

Newman, William R., "The background to Newton's chymistry," *The Cambridge Companion to Newton*, I. B. Cohen and G. E. Smith (eds.), Cambridge, Cambridge University Press, 2002.

Newman, William R., and Principe, Lawrence M., "The Chymical Laboratory Notebooks of George Starkey," *Reworking the Bench: Research Notebooks in the History of Science*, F. L. Holmes, J. Renn, and H. Rheinberger (eds.), Dordrecht, Kluwer Academic Publishers, 2003.

Newman, William R., and Principe, Lawrence M. (eds.), *George Starkey: Alchemical Laboratory Notebooks and Correspondence*, Chicago, University of Chicago Press, 2004.

Nicholson, Watson, *The Historical Sources of Defoe's Journal of the Plague Year*, Boston, Stratford Co., 1919.

Norton, Thomas, "The Ordinall of Alchimy," *Threatrum Chemicum Britannicum*, E. Ashmole (ed.), London, Printed by *J. Grismond* for Nath: Brooke, at the Angel in *Cornhill*, 1652.

Ogrinc, Will H. L., "Western society and alchemy from 1200 to 1500," *Journal of Medieval History*, 6 (1980), pp. 103-132.

Osler, Margaret J., "The New Newtonian Scholarship and the Fate of the Scientific Revolution," *Newton and Newtonianism: New Studies*, J. E. Force and S. Hutton (eds.), Dordrecht, Kluwer Academic Publishers, 2004.

Oster, Malcolm, "Biography, Culture, and Science: The Formative Years of Robert Boyle," *History of Science*, 31 (1993), pp. 177-226.

Oster, Malcolm, "Millenarianism and the new science: the case of Robert Boyle," *Samuel Hartlib and Universal Reformation: Studies in intellectual communication*, M. Greengrass, M. Leslie, and T. Raylor (eds.), Cambridge, Cambridge University Press, 1994.

Ottosson, Per-Gunnar, "Fear of the Plague and the Burial of Plague Victims in Sweden 1710-1711," *Maladies Et Société (XIIe-XVIIIe siècles)*, N. Bulst and R. Delort (eds.), Paris, Éditions Du Centre National De La Recherche Scientifique, 1989.

Pagel, Walter, "The Reaction to Aristotle in Seventeenth-Century Biological Thought," *Science Medicine and History: Essays on the Evolution of Scientific Thought and Medical Practice written in honour of Charles Singer*, Volumes I-II, E. A. Underwood (ed.), London, Oxford University Press, 1953.

Pagel, Walter, *Paracelsus: An Introduction to Philosophical Medicine in the Era of the Renaissance*, Basel, S. Karger, 1958.

Pagel, Walter, "Paracelsus and the Neoplatonic and Gnostic Tradition," *Ambix*, 8 (1960), pp. 125-166.

Pagel, Walter, "Johann Baptist van Helmont," *Aufgang der Artzney-Kunst*, Volume II, C. K. von Rosenroth (trans.), München, Kösel-Verlag, 1971.

Pagel, Water, "Paracelsus, Theophrastus Philippus Aereolus Bombastus Von Hohenheim," *Dictionary of Scientific Biography*, Volumes I-XIV, C. C. Gillispie (ed.), New York, Charles Scribner's Sons, 1970-1976.

Pagel, Walter, "Helmont, Johannes (Joan) Baptista Van," *Dictionary of Scientific Biography*, Volumes I-XIV, C. C. Gillispie (ed.), New York, Charles Scribner's Sons, 1970-1976.

Pagel, Walter, "Van Helmont's Concept of Disease—To Be or not to Be? The Influence of Paracelsus," *Bulletin of the History of Medicine*, 46 (1972), pp. 419-454.

Pagel, Walter, "The Paracelsian Elias Artista and the Alchemical Tradition," *Kreatur und Kosmos: Internationale Beiträge zur Paracelsusforschung*, Stuttgart, Gustav Fischer Verlag, 1981.

Pagel, Walter, *Joan Baptista Van Helmont: Reformer of science and medicine*, Cambridge, Cambridge University Press, 1982.

Pagel, Walter, *The Smiling Spleen: Paracelsianism in Storm and Stress*, Basel, S. Karger, 1984.

Paracelsus, *Sämtliche Werke. I. Abteilung: Medizinische naturwissenschaftliche und philosophische Schriften*, Volumes I-XIV, K. Sudhoff (ed.), München und Berlin, Druct und Verlag von R. Oldenbourg, 1922-1933.

Partington, James Riddick, *A History of Chemistry*, Volume II, New York, St. Martin's Press, 1961.

Patrick, Symon, *The Auto-Biography of Symon Patrick, Bishop of Ely*, Oxford, J. H. Parker, 1839.

Patterson, A. Temple, *A History of Southampton 1700-1914*, Volume I, Southampton, Southampton University Press, 1966.

Pepys, Samuel, *Memoirs of Samuel Pepys, Esq. F. R. S.*, Second Edition, R. L. Braybrooke (ed.), London, Frederick Warne and Co., 1870.

Pepys, Samuel, *The Diary of Samuel Pepys*, Volume VI, R. Latham and W. Matthews (eds.), London, G. Bell and Sons Ltd., 1972.

Pereira, Michela, *The Alchemical Corpus Attributed to Raymond Lull*, London, The Warburg Institute, University of London, 1989.

Pharmacopœia Londinensis, Londini, Typis G. Bowyer, Impensis R. Knaplock..., 1721.

Pharmacopœia Londinensis, Londini, Apud T. Longman, T. Shewell, et J. Nourse, 1746.

Philalethes, Eirenaeus (pseud.), *The Marrow Of Alchemy: Being an Experimental Treatise, Discovering the secret of the most hidden Mystery of the Philosophers Elixer*, London, Printed by *A. M.* for *Edw. Brewster* at the Signe of the Crane in *Pauls* Church-yard, 1654.

Philalethes, Eirenaeus (pseud.), *The Marrow Of Alchemy: Being an Experimental Treatise, Discovering the secret of the most hidden Mystery of the Philosophers Elixer. The Second Part*, London, Printed by *R. I.* for *Edw. Brewster* at the Sign of the Crane in *Pauls* Church-yard, 1655.

[Philalethes, Eirenaeus] (pseud.), "Sir George Riplye's Epistle to King Edward unfolded," *Chymical, Medicinal, and Chyrurgical Addresses: Made to Samuel Hartlib, Esquire*, London, Printed by *G. Dawson* for *Giles Calvert* at the *Black-spread Eagle* at the west end of *Pauls*, 1655.

Philalethes, Eirenaeus (pseud.), *Secrets Reveal'd: Or, An Open Entrance To The Shut-Palace of the King. Containing, The greatest Treasure in Chymistry, Never yet so plainly Discovered. Composed By a most famous English-man, Styling himself Anonymvs, or Eyrœneus Philaletha Cosmopolita: Who, by Inspiration and Reading, attained to the Philosophers Stone at his Age of Twenty three Years, Anno Domini, 1645...*, London, Printed by *W. Godbid* for *William Cooper* in Little St. *Bartholomews*, near *Little-Britain*, 1669.

Philalethes, Eirenaeus (pseud.), *Ripley Reviv'd: Or, An Exposition upon Sir George Ripley's Hermetico-Poetical Works*, London, Printed by *Tho. Ratcliff* and *Nat. Thompson*, for *William Cooper* at the *Pelican* in *Little-Britain*, 1678.

Philalethes, Eirenaeus (pseud.), "Ars Metallorum MetamorphoseΘs," *Three Tracts of the Great Medicine of Philosophers for Humane and Metalline Bodies*, London, Printed and sold by *T. Sowle*, at the *Crooked-Billet* in *Holy-well-Lane Shoreditch*, 1694.

Philalethes, Eirenaeus (pseud.), "Introitus Apertus, *ad* Occlusum Regis Palatium," *Bibliotheca Chemica Curiosa*, Volumes I-II, J. J. Manget (ed.), Genevæ, Sumpt...De Tournes, 1702.

Porter, Stephen, *Lord Have Mercy Upon Us: London's Plague Years*, Stroud, Eng., Tempus Publishing Ltd., 2005.

Prescott, Oliver, *A Disseration on the Natural History and Medicinal Effects of the Secale Cornutum, or Ergot*, Boston, Cummings & Hilliard, 1813.

Principe, Lawrence M., "Robert Boyle's Alchemical Secrecy: Codes, Ciphers and Concealments," *Ambix*, 39 (1992), pp. 63-74.

Principe, Lawrence M., "Boyle's alchemical pursuits," *Robert Boyle Reconsidered*, M. Hunter (ed.), Cambridge, Cambridge University Press, 1994.

Principe, Lawrence M., "Style and Thought of the Early Boyle: Discovery of the 1648 Manuscript of *Seraphic Love*," *Isis*, 85 (1994), pp. 247-260.

Principe, Lawrence M., "Newly Discovered Boyle Documents in the Royal Society Archive: Alchemical Tracts and his Student Notebook," *Notes and Records of the Royal Society of London*, 49 (1995), pp. 57-70.

Principe, Lawrence M., *The Aspiring Adept: Robert Boyle and his Alchemical Quest*, Princeton, Princeton University Press, 1998.

Principe, Lawrence M., "The Alchemies of Robert Boyle and Isaac Newton: Alternate Approaches and Divergent Deployments," *Rethinking the Scientific Revolution*, M. J. Osler (ed.), Cambridge, Cambridge University Press, 2000.

Quinn, Arthur, "On Reading Newton Apocalyptically," *Millenarianism and Messianism in English Literature and Thought 1650-1800; Clark Library Lectures 1981-1982*, R. H. Popkin (ed.), Leiden, E. J. Brill, 1988.

Race, William H. (trans.), *Pindar I*, Cambridge, Harvard University Press, 1997.

Rattansi, P. M., "Paracelsus and the Puritan Revolution," *Ambix*, 11 (1963), pp. 24-32.

Rattansi, P. M., "The Helmontian-Galenist Controversy in Restoration England," *Ambix*, 12 (1964), pp. 1-23.

Read, John, *Prelude to Chemistry...*, New York, Macmillan Company, 1937.

Řeháček, Zdeněk, and Sajdl, Přemysl, *Ergot Alkaloids: Chemistry, Biological Effects, Biotechnology*, Amsterdam, Elsevier, 1990.

Renaudot, [Theophraste], "A Conference Concerning the Philosophers' Stone," *Chymical, Medicinal, and Chyrurgical Addresses: Made to Samuel Hartlib, Esquire*, London, Printed by *G. Dawson* for *Giles Calvert* at the *Black-spread Eagle* at the west end of *Pauls*, 1655.

Reti, Ladislao, "Van Helmont, Boyle and the Alkahest," *Some Aspects of Seventeenth-Century Medicine & Science: Papers Read at a Clark Library Seminar, October 12, 1968*, Los Angeles, University of California, 1969.

Reusch, Franz Heinrich, *Der Index der Verbotenen Bücher*, Volumes I-II, Bonn, Verlag von Max Cohen & Sohn, 1883-1885.

Rice, Ch., "Historical Notes on Opium," *New Remedies*, 5 (1876), pp. 229-232, 6 (1877), pp. 144-145, 194-195.

Rist, Johann, *Die alleredelste Tohrheit der gantzen Welt...*, Hamburg, In Verlegung Ioh. Naumanns, Buchh., 1664.

Robertson, James, and Ashby, Hugh T., "Ergot Poisoning among Rye Bread Consumers," *British Medical Journal*, 1 (1928), pp. 302-303.

Romanell, Patrick, *John Locke and Medicine: A New Key to Locke*, Buffalo, N.Y., Prometheus Books, 1984.

Rowbottom, Margaret, "The Earliest Published Writing of Robert Boyle," *Annals of Science*, 6 (1950), pp. 376-387.

Ruck, Carl A. P., "The Wild and the Cultivated: Wine in Euripides' *Bacchae*," *Journal of Ethnopharmacology*, 5 (1982), pp. 231-270.

Salaman, Redcliffe N., *The History and Social Influence of the Potato*, Cambridge, Cambridge University Press, 1949.

Scarborough, John, "Hermetic and Related Texts in Classical Antiquity," *Hermeticism and the Renaissance*, I. Merkel and A. G. Debus (eds.), Washington, D.C., Folger Shakespeare Library, 1988.

Schott, Heinz, "Paracelsus and van Helmont on Imagination: Magnetism and Medicine before Mesmer," *Paracelsian Moments: Science, Medicine, & Astrology in Early Modern Europe*, G. S. Williams and C. D. Gunnoe, Jr. (eds.), Kirksville, Mo., Truman State University Press, 2002.

Schove, D. J., "Fire and Drought, 1600-1700," *Weather*, 21 (1966), pp. 311-314.

Scot, Reginald, *The Discovery of Witchcraft*, Third Edition, London, Printed for *Andrew Clark*..., 1665.

Scott, P. M., *et al.*, "Ergot Alkaloids in Grain Foods Sold in Canada," *Journal of AOAC International*, 75 (1992), pp. 773-779.

Scott, Susan, and Duncan, Christopher J., *Biology of Plagues: Evidence from Historical Populations*, Cambridge, Cambridge University Press, 2001.

Sennert, Daniel, *Of Agues and Fevers. Their Differences, Signes, and Cures. Divided into four Books: Made English by* N. D. B. M. *late of* Trinity *Colledge in* Cambridge, London, Printed by *J. M.* for *Lodowick Lloyd*, at the Castle in Cornhil, 1658.

Shapin, Steven, "The Philosopher and the Chicken: On the Dietetics of Disembodied Knowledge," *Science Incarnate: Historical Embodiments of Natural Knowledge*, C. Lawrence and S. Shapin (eds.), Chicago, University of Chicago Press, 1998.

Shapiro, Alan E., *Fits, Passions, and Paroxysms: Physics, method, and chemistry and Newton's theories of colored bodies and fits of easy reflection*, Cambridge, Cambridge University Press, 1993.

Shea, William R., "Galileo and the Church," *God and Nature: Historical Essays on the Encounter between Christianity and Science*, D. C. Lindberg and R. L. Numbers (eds.), Berkeley, University of California Press, 1986.

Shelley, William Scott, *The Elixir: An Alchemical Study of the Ergot Mushrooms*, Notre Dame, Ind., Cross Cultural Publications, 1995.

Shelley, William Scott, *The Origins of the Europeans: Classical Observations in Culture and Personality*, San Francisco, International Scholars Publications, 1998.

Shelley, William Scott, "Soma: Mead of the Gods," [unpublished].

Shelley, William Scott, "Chemistry and the *Kykeon* of the Eleusinian Mysteries," [unpublished].

Shelley, William Scott, "The 'Soma Vessels' of the Bactria-Margiana Archaeological Complex," [unpublished].

Sheppard, F. H. W. (ed.), *Survey of London: The Parish of St. James, Westminster*, Volume XXXI, London, Athlone Press, 1963.

Shipton, Clifford K., *Sibley's Harvard Graduates; Biographical Sketches of Those Who Attended Harvard College...*, Volumes IV-XIV, Cambridge, Harvard University Press [etc.], 1933-1968.

Short, Thomas, *A General Chronological History of the Air, Weather, Seasons, Meteors, &c. in Sundry Places and Different Times...*, Volume I, London, Printed for T. Longman..., 1749.

Shrewsbury, J. F. D., *A History of the Bubonic Plague in the British Isles*, Cambridge, Cambridge University Press, 1970.

Shutting up Infected Houses as it is practised in England Soberly Debated. By way of Address from the poor souls that are Visited, to their Brethren that are Free, The, London, s.n., 1665.

Sibley, John Langdon, *Biographical Sketches of Graduates of Harvard University*, Volumes I-III, Cambridge, Charles William Sever, 1873-1885.

Slack, Paul, *The Impact of Plague in Tudor and Stuart England*, London, Routledge & Kegan Paul, 1985.

Snobelen, Stephen D., "Isaac Newton, heretic: the strategies of a Nicodemite," *British Journal for the History of Science*, 32 (1999), pp. 381-419.

Snobelen, Stephen D., "'The Mystery of this Restitution of All Things': Isaac Newton and the Return of the Jews," *Millenarianism and Messianism in Early Modern European Culture*, Volume III, J. E. Force and R. H. Popkin (eds.), Dordrecht, Kluwer Academic Publishers, 2001.

Spargo, P. E., and Pounds, C. A., "Newton's 'Derangement of the Intellect': New Light on an Old Problem," *Notes and Records of the Royal Society of London*, 34 (1979), pp. 11-32.

Sprat, Thomas, *The History of the Royal-Society of London...*, London, Printed by *T. R.* for *J. Martyn...*, 1667.

Starkey, George, *Natures Explication and Helmont's Vindication. Or A short and sure way to a long and sound Life: Being, A necessary and full Apology for Chymical Medicaments, and a Vindication of their Excellency against those unworthy reproaches cast on the Art and its Profesors (such as were Paracelsus and Helmont) by Galenists...*, London, Printed by *E. Cotes* for *Thomas Alsop* at the two Sugar-loaves over against St. *Antholins* Church at the lower end of *Watling-street*, 1657.

Starkey, George, *Pyrotechny Asserted and Illustrated, To be the surest and safest means for Arts Triumph over Natures Infirmities. Being full and free Discovery of the Medicinal Mysteries studiously concealed by all Artists, and onely discoverable by*

Fire, London, Printed by *R. Daniel*, for *Samuel Thomson* at the *Whitehorse* in S. *Pauls* Church-yard, 1658.

[Starkey, George], *George Starkey's Pill Vindicated...*, London, s.n., 1660[?].

Starkey, George, *A Brief Examination and Censure Of Several Medicines, of late years Extol'd for Universal remedies, and Arcana's of the highest preparation...Namely, Lockyers pill, Hughes pouder, Constantines Spirit of Salt, with several other of their kind, by which the Art of Pyrotechny is in danger of being brought into Reproach and Contempt...*, London, s.n., 1664.

Starkey, George, *A Smart Scourge for a Silly, Sawcy Fool. Being An Answer to a Letter, at the End of a Pamphlet of* Lionell Lockyer..., London, s.n., 1664.

Starkey, George, *An Epistolar Discourse...*, London, Printed by *R. Wood*, for *Edward Thomas*, at the *Adam* and *Eve* in *Little Brittain*, 1665.

[Starkey, George], *Liquor Alchahest, Or A Discourse Of that Immortal Dissolvent Of Paracelsus & Helmont...*, London, Printed by *T. R.* & *N. T.* for *W. Cademan* at the *Popes-Head* in the Lower Walk of the *New Exchange*, 1675.

[Starkey, George], *The Dignity of Kingship Asserted, By G. S...*, W. R. Parker (ed.), New York, Columbia University Press, 1942.

Steele, Robert, "Alchemy in England," *Antiquary*, 24 (1891), pp. 99-105.

Smyth, Albert Henry (ed.), *The Writings of Benjamin Franklin*, Volume VI, New York, Macmillan Company, 1907.

Sydenham, Thomas, *Dr. Sydenham's Practice of Physick. The Signs, Symptoms, Causes and Cures of Diseases...*, W. Salmon (trans.), London, Printed for *Sam. Smith...*, 1695.

Sydenham, Thomas, *The Whole Works of that Excellent Practical Physician, Dr. Thomas Sydenham...*, J. Pechey (ed.), Seventh Edition, London, Printed for M. *Wellington...*, 1717.

Sydenham, Thomas, *The Entire Works of Dr. Thomas Sydenham*, J. Swan (trans.), Fifth Edition, London, Printed for F. Newbery..., 1769.

Sydenham, Thomas, *The Works of Thomas Sydenham, M.D. on Acute and Chronic Diseases; with their Histories and Modes of Cure*, B. Rush (ed.), Philadelphia, Published by B. & T. Kite..., 1815.

Sydenham, Thomas, *The Works of Thomas Sydenham, M.D.*, Volumes I-II, Dr. Greenhill (trans.), London, Printed for the Sydenham Society, 1848-1850.

Taswell, William, "Autobiography and Anecdotes," *Camden Miscellany*, Volume II, G. P. Elliott (ed.), London, Printed for the Camden Society, 1853.

Tenberge, Klaus B., "Biology and Life Strategy of the Ergot Fungi," *Ergot: The Genus Claviceps*, V. Křen and L. Cvak (eds.), Amsterdam, Harwood Academic Publishers, 1999.

Tétényi, Péter, "Opium Poppy (*Papaver somniferum*): Botany and Horticulture," *Horticultural Reviews*, 19 (1997), pp. 373-408.

Thackray, Arnold, *Atoms and Powers: An Essay on Newtonian Matter-Theory and the Development of Chemistry*, Cambridge, Harvard University Press, 1970.

Thatcher, James, *American Medical Biography...*, Volumes I-II, Boston, Richardson & Lord and Cottons & Barnard, 1828.

Theilmann, John, and Cate, Frances, "A Plague of Plagues: The Problem of Plague Diagnosis in Medieval England," *Journal of Interdisciplinary History*, 37 (2007), pp. 371-393.

Thirsk, Joan (ed.), *The Agrarian History of England and Wales 1500-1640*, Volume IV, Cambridge, Cambridge University Press, 1967.

Thirsk, Joan (ed.), *The Agrarian History of England and Wales 1640-1750*, Volume V.II, Cambridge, Cambridge University Press, 1985.

Thomas, Sir Henry, "The Society of Chymical Physitians: An Echo of the Great Plague of London, 1665," *Science Medicine and History: Essays on the Evolution of Scientific Thought and Medical Practice written in honour of Charles Singer*, Volumes I-II, E. A. Underwood (ed.), London, Oxford University Press, 1953.

Thomas, Sir Henry, "The Society of Chymical Physitians: An Echo of the Great Plague of London, 1665," *Science Medicine and Society in the Renaissance: Essays to Honor W. Pagel*, Volumes I-II, A. G. Debus (ed.), New York, Science History Publications, 1972.

Thomson, George, *Loimologia. Consolatory Advice, And some brief Observations Concerning the Present Pest*, London, Printed for *L. Chapman*, at his Shop in Exchange-ally, 1665.

Thomson, George, *Galeno-pale: Or, A Chymical Trial of the Galenists...*, London, Printed by *R. Wood*, for *Edward Thomas*, at the *Adam* and *Eve* in *Little Britain*, 1665.

Thomson, George, *Loimotomia: Or The Pest Anatomized*, London, Printed for *Nath: Crouch*, at the *Rose* and *Crown* in *Exchange*-Alley near *Lombard-street*, 1666.

Thorndike, Lynn, *A History of Magic and Experimental Science*, Volumes I-VIII, New York, Columbia University Press, 1923-1958.

Thumb, Thomas (pseud.), *The Monster of Monsters...*, Boston, Printed by Zechariah Fowle, 1754.

Thurneysser, Leonhard, *Historia sive Descriptio Plantarum omnium*, Berlini, Michael Hentzske, 1578.

Tilton, Hereward, *The Quest for the Phoenix: Spiritual Alchemy and Rosicrucianism in the Work of Count Michael Maier (1569-1622)*, Berlin, Walter de Gruyter, 2003.

Tissot, S. A. D., "An Account of the Disease, called *Ergot* in *French*, from its supposed Cause, *viz.* vitiated Rye," *Philosophical Transactions*, 55 (1765), pp. 106-126.

Trompf, Garry W., "On Newtonian History," *The Uses of Antiquity: The Scientific Revolution and the Classical Tradition*, S. Gaukroger (ed.), Dordrecht, Kluwer Academic Publishers, 1991.

Turnbull, George H., *Hartlib, Dury and Comenius: Gleanings from Hartlib's Papers*, London, University Press of Liverpool, 1947.

Turnbull, George H., "Robert Child," *Transactions of the Colonial Society of Massachusetts*, 38 (1947), pp. 21-53.

Turnbull, George H., "George Stirk, Philosopher by Fire," *Publications of the Colonial Society of Massachusetts*, 38 (1949), pp. 219-251.

Turnbull, H. W., and Scott, J. F., and Hall, A. R., and Tilling, L. (eds.), *The Correspondence of Isaac Newton*, Volumes I-VII, Cambridge, Published for the Royal Society at the University Press, 1959-1977.

Twigg, Graham I., *The Black Death: A Biological Reappraisal*, London, Batsford Academic and Educational, 1984.

Twigg, Graham I., "The Black Death in England: An Epidemiological Dilemma," *Maladies Et Société (XIIe-XVIIIe siècles)*, N. Bulst and R. Delort (eds.), Paris, Éditions Du Centre National De La Recherche Scientifique, 1989.

Twigg, Graham I., "Plague in London: spatial and temporal aspects of mortality," *Epidemic Disease in London*, J. A. I. Champion (ed.), London, Centre for Metropolitan History, University of London, 1993.

Tyre, Mary, *Medicatrix, or the Woman-Physician: Vindicating Thomas O'Dowde, a Chemical Physician...*, London, Printed by *T. R. & N. T...*, 1675.

Underwood, A. J. V., "The Historical Development of Distilling Plant," *Transactions of the Institution of Chemical Engineers*, 13 (1935), pp. 34-62.

Venner, Tobias, *Via Recta ad Vitam Longam...*, London, Printed by *R. Bishop...*, 1638.

Vincent, Thomas, *Gods Terrible Voice in the City of London...*, Cambridge, Printed by *Samuel Green*, 1667.

Vondung, Klaus, "Millenarianism, Hermeticism, and the Search for a Universal Science," *Science, Pseudo-Science, and Utopianism in Early Modern Thought*, S. A. McKnight (ed.), Columbia, University of Missouri Press, 1992.

Wallis, Patrick, "Plagues, Morality and the Place of Medicine in Early Modern England," *English Historical Review*, 121 (2006), pp. 1-24.

Wasson, R. Gordon, and Hofmann, Albert, and Ruck, Carl A. P., *The Road to Eleusis: Unveiling the Secret of the Mysteries*, New York, Harcourt Brace Jovanovich, Inc., 1978.

Watkins, Calvert, "Let Us Now Praise Famous Grains," *Proceedings of the American Philosophical Society*, 122 (1978), pp. 9-17.

Watson, Patricia A., *The Angelical Conjunction: The Preacher-Physicians of Colonial New England*, Knoxville, University of Tennessee Press, 1991.

Webster, Charles, "English Medical Reformers of the Puritan Revolution: A Background to the 'Society of Chymical Physitians,'" *Ambix*, 14 (1967), pp. 16-41.

Webster, Charles (ed.), *Samuel Hartlib and the Advancement of Learning*, Cambridge, Cambridge University Press, 1970.

Webster, Charles, "The Helmontian George Thomson and William Harvey: The Revival and Application of Splenectomy to Physiological Research," *Medical History*, 15 (1971), pp. 154-167.

Webster, Charles, "New Light on the Invisible College: The Social Relations of English Science in the Mid-Seventeenth Century," *Transactions of the Royal Historical Society*, Fifth Series, 24 (1974), pp. 19-42.

Webster, Charles, *The Great Instauration. Science, Medicine and Reform 1626-1660*, New York, Holmes & Meier Publishers, 1975.

Webster, Charles, "Alchemical and Paracelsian medicine," *Health, medicine and mortality in the sixteenth century*, C. Webster (ed.), Cambridge, Cambridge University Press, 1979.

Webster, Charles, *From Paracelsus to Newton: Magic and the Making of Modern Science*, Cambridge, Cambridge University Press, 1982.

Webster, Charles, "Paracelsus: medicine as popular protest," *Medicine and the Reformation*, O. P. Grell and A. Cunningham (eds.), London, Routledge, 1993.

Weeks, Andrew, *Paracelsus: Speculative Theory and the Crisis of the Early Reformation*, Albany, State University of New York Press, 1997.

[Weidenfeld, Johann Seger], *Four Books of Johannes Segerus Weidenfeld, Concerning the Secrets of the Adepts; Or, Of the Use of Lully's Spirit of Wine: A Practical Work*, London, Printed by *Will. Bonny*, for *Tho. Howkins* in *George-Yard* in *Lombard-Street*, 1685.

West, Richard, *The Life & Strange Surprising Adventures of Daniel Defoe*, London, HarperCollins, 1997.

Westfall, Richard S., *Force in Newton's Physics: The Science of Dynamics in the Seventeenth Century*, New York, American Elsevier, 1971.

Westfall, Richard S., "The Role of Alchemy in Newton's Career," *Reason, Experiment, and Mysticism in the Scientific Revolution*, M. L. R. Bonelli and W. R. Shea (eds.), New York, Science History Publications, 1975.

Westfall, Richard S., "Isaac Newton's Index Chemicus," *Ambix*, 22 (1975), pp. 174-185.

Westfall, Richard S., *Never at Rest: A Biography of Isaac Newton*, Cambridge, Cambridge University Press, 1980.

Westfall, Richard S., "The Influence of Alchemy on Newton," *Science, Pseudo-Science and Society*, M. P. Hanen, M. J. Osler, and R. G. Weyant (eds.), Waterloo, Ont., Wilfrid Laurier University Press, 1980.

Westfall, Richard S., "Newton's Marvelous Years of Discovery and Their Aftermath: Myth versus Manuscript," *Isis*, 71 (1980), pp. 109-121.

Westfall, Richard S., "The Rise of Science and the Decline of Orthodox Christianity: A Study of Kepler, Descartes, and Newton," *God and Nature: Historical Essays on the Encounter between Christianity and Science*, D. C. Lindberg and R. L. Numbers (eds.), Berkeley, University of California Press, 1986.

Westfall, Richard S., *Essays on the Trial of Galileo*, Vatican City State, Vatican Observatory Publications, 1989.

Westfall, Richard S., *The Life of Isaac Newton*, Cambridge, Cambridge University Press, 1993.

Westfall, Richard S., "The Scientific Revolution Reasserted," *Rethinking the Scientific Revolution*, M. J. Osler (ed.), Cambridge, Cambridge University Press, 2000.

Westman, Robert S., "The Copernicans and the Churches," *God and Nature: Historical Essays on the Encounter between Christianity and Science*, D. C. Lindberg and R. L. Numbers (eds.), Berkeley, University of California Press, 1986.

Wheelwright, Edith Grey, *The Physick Garden: Medicinal Plants and their History*, Boston, Houghton Mifflin Company, 1935.

Whitelocke, Bulstrode, *The Diary of Bulstrode Whitelocke 1605-1675*, R. Spalding (ed.), Oxford, Oxford University Press, 1990.

Wightman, W. P. D., *Science in a Renaissance Society*, London, Hutchinson University Library, 1972.

Wilkinson, Ronald Sterne, "New England's Last Alchemists," *Ambix*, 10 (1963), pp. 128-138.

Wilkinson, Ronald Sterne, "The Alchemical Library of John Winthrop, Jr. (1606-1676) and his Descendants in Colonial America," *Ambix*, 11 (1963), pp. 33-51.

Wilkinson, Ronald Sterne, "George Starkey, Physician and Alchemist," *Ambix*, 11 (1963), pp. 125-133.

Wilkinson, Ronald Sterne, "The Problem of the Identity of Eirenaeus Philalethes," *Ambix*, 12 (1964), p. 28.

Wilkinson, Ronald Sterne, "The Hartlib Papers and Seventeenth-Century Chemistry, Part I," *Ambix*, 15 (1968), pp. 54-69.

Wilkinson, Ronald Sterne, "The Hartlib Papers and Seventeenth-Century Chemistry, Part II: George Starkey," *Ambix*, 17 (1970), pp. 85-110.

Wilkinson, Ronald Sterne, "'Hermes Christianus:' John Winthrop, Jr. and Chemical Medicine in Seventeenth Century New England," *Science Medicine and Society in the Renaissance: Essays to Honor W. Pagel*, Volumes I-II, A. G. Debus (ed.), New York, Science History Publications, 1972.

Wilkinson, Ronald Sterne, "Further Thoughts on the Identity of 'Eirenaeus Philalethes'," *Ambix*, 19 (1972), pp. 204-208.

Wilkinson, Ronald Sterne, "Some Biographical Puzzles Concerning George Starkey," *Ambix*, 20 (1973), pp. 235-244.

Wilkinson, Ronald Sterne, "Starkey, George," *Dictionary of Scientific Biography*, Volumes I-XIV, C. C. Gillispie (ed.), New York, Charles Scribner's Sons, 1970-1976.

Willis, Thomas, *The London Practice Of Physick...*, London, Printed for *Thomas Basset...*, 1685.

Wilson, George, *A Compleat Course of Chymistry...*, London, Printed for *W. Turner...*, 1700.

Wilson, George, *A Compleat Course of Chymistry...*, London, Printed for John Bayley..., 1709.

"Winthrop Papers," *Collections of the Massachusetts Historical Society*, Fourth Series, 7 (1865), pp. 1-633.

Wojcik, Jan W., *Robert Boyle and the Limits of Reason*, Cambridge, Cambridge University Press, 1997.

Wood, G., and Coley-Smith, J. R., "Observations on the Prevalence and Incidence of Ergot Disease in Great Britain with Special Reference to Open-flowering and Male-sterile Cereals," *Annals of Applied Biology*, 95 (1980), pp. 41-46.

Y-worth, William, *Introitus Apertus ad Artem Distillationis...*, London, Printed for *J. Taylor* at the *Ship* in St. *Paul's Church-yard*, 1692.

Y-worth, William, *The Compleat Distiller: Or The Whole Art of Distillation Practically Stated, And Adorned with all the New Modes of Working now in Use. In which is Contained, The way of making Spirits, Aquavitæ, Artificial Brandy...*, London, Printed for J. Taylor, at the *Ship* in St. *Paul's* Church-Yard, 1705.

Yelling, J. A., "Changes in Crop Production in East Worcestershire 1540-1867," *Agricultural History Review*, 21 (1973), pp. 18-34.

Ziegler, Joseph, *Medicine and Religion c. 1300: The Case of Arnau de Vilanova*, Oxford, Clarendon Press, 1998.

Printed in the United States
By Bookmasters